# Frontiers in Probability and the Statistical Sciences

More information about this series at http://www.springer.com/series/11957

Mayer Alvo • Philip L.H. Yu

# Statistical Methods for Ranking Data

Springer

Mayer Alvo
Department of Mathematics and Statistics
University of Ottawa
Ottawa, ON, Canada

Philip L.H. Yu
Department of Statistics
and Actuarial Science
The University of Hong Kong
Hong Kong, China

ISBN 978-1-4939-4781-2 ISBN 978-1-4939-1471-5 (eBook)
DOI 10.1007/978-1-4939-1471-5
Springer New York Heidelberg Dordrecht London

Mathematics Subject Classification (2010): 62-07, 62F07, 62G10, 62G86, 62H11, 62H20, 62H30, 62K10, 62K15, 62P15



Printed on acid-free paper

Springer is part of Springer Science+Business Media (www.springer.com)

# Preface

This book grew out of a desire on the part of both authors to formally record in one volume some of their research on ranking methods. My own interest was sparked by a problem emanating from the School of Nursing at the University of Ottawa. It was desired to test if patients who were mobile differed from those who were not with respect to how they ranked certain sets of situations. The early research dealing with the one sample problem was conducted at the Technion where I spent a sabbatical and worked together with my friends and colleagues Paul Cabilio and Paul Feigin. It subsequently led to a series of papers written principally with Paul Cabilio which formed the basis of inference for ranking data. Other contributions were made by graduate students and various collaborators. It is our pleasure to thank Xin Gao from York University who generously permitted us to include the work on interaction and who shared his/her computer packages.

I am grateful to my teachers who have inspired me during both my undergraduate and graduate studies. Specifically I would like to thank Miklos Csorgo and Michael Stevens who taught me probability and statistics at McGill University and who instilled in me an interest to pursue the subjects. Herbert Robbins at Columbia University and Michael Woodroofe at the University of Michigan served as wonderful role models throughout my doctoral studies. They demonstrated convincingly that it is possible to combine depth of meaning and elegance in statistics.

It is a pleasure to thank my wife Helen and my family for their continued support and understanding during the preparation of this book. To my parents who have inspired me to pursue an academic career I owe a great deal.

Ottawa, ON, Canada Mayer Alvo

I am particularly grateful to Kin Lam and Richard Cowan for their teaching and guidance in my graduate years at The University of Hong Kong. Without their encouragement and support in analyzing contract bridge tournament data, my works on modeling ranking data and this book would not have gotten started. I would like to thank all my former graduate students in ranking data research, in particular Dorcas Lo, Wai-ming Wan, and Paul Lee. The second part of this book is based on our joint work on modeling of ranking data. A special thank you to Qinglong Li, Yuming Zhang, and Yiming Li for their outstanding assistance in writing R programs.

I thank God almighty for his/her support in all my endeavors; I especially owe thanks to my family, especially my wife Bonnie and our two lovely sons, for their continued support and understanding throughout the work. Sincere gratefulness goes to my beloved parents.

Hong Kong, China Philip L.H. Yu

Both authors are grateful to the publishers/ journals for allowing us to quote freely from our published papers. We are very happy to thank the staff at Springer for their patience and professionalism throughout the preparation of this book. In particular many thanks to Donna Chernyk for guiding us so well along the way

# Contents

# Chapter 1
# Introduction

This book was motivated by a desire to make available in a single volume many of the results on ranking methods developed by the authors and their collaborators that have appeared in the literature over a period of several years. In many instances, the presentations have a geometric flavor to them. As well there is a concerted effort to introduce real applications in order to exhibit the wide scope of ranking methods. Our hope is that the book will serve as a starting point and encourage students and researchers to make more use of nonparametric ranking methods. The statistical analysis of ranking data forms the main objective in this book.

Ranking data commonly arise from situations where it is desired to rank a set of individuals or objects in accordance with some criterion. Such data may be observed directly or it may come from a ranking of a set of assigned scores. Alternatively, ranking data may arise when transforming continuous or discrete data in a nonparametric analysis. Examples of ranking data may be found in politics (Inglehart 1977; Barnes and Kaase 1979; Croon 1989; Vermunt 2004; Moors and Vermunt 2007), voting and elections (Diaconis 1988; Koop and Poirier 1994; Kamishima and Akaho 2006; Stern 1993; Murphy and Martin 2003; Gormley and Murphy 2008; Skrondal and Rabe-Hesketh 2003), market research (Dittrich et al. 2000; Beggs et al. 1981; Chapman and Staelin (1982)), food preference (Kamishima and Akaho 2006; Nombekela et al. (1993); Vigneau et al. 1999), psychology (Regenwetter et al. 2007; Decarlo and Luthar 2000; Riketta and Vonjahr 1999; Maydeu-Olivares and Bockenholt 2005; Bockenholt 2001), health economics (Salomon 2003; Krabbe et al. 2007; McCabe et al. 2006; Craig et al. 2009; Ratcliffe et al. 2006, 2009), medical treatments (Plumb et al. 2009), types of sushi (Kamishima and Akaho 2006), place of living (Duncan and Brody 1982), choice of occupations (Goldberg 1975; Yu and Chan 2001), and even horse racing (Stern 1990b; Benter 1994; Henery 1981).

In some cases, incomplete ranking data are observed, particularly when assessing an object is time consuming or takes much effort. Instead of ranking all objects, each individual may be asked to rank the top $q$ objects only for $q \leq t$, called *top*



M. Alvo, P.L.H. Yu, *Statistical Methods for Ranking Data*, Frontiers in Probability and the Statistical Sciences, DOI 10.1007/978-1-4939-1471-5_1

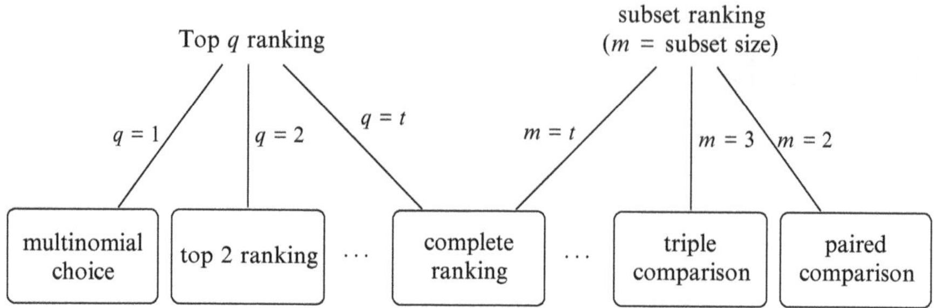


**Fig. 1.1** Classification of rankings of $t$ objects

*q partial rankings*. More generally, individuals are presented with a subset of the $t$ objects and rank the objects in the subset only, called *subset rankings*. Figure 1.1 shows the classification of rankings.

The analysis presented in this book follows two main themes. Beginning with an introduction to exploratory data analysis for ranking data in Chap. 2, we consider in the first part the inferential side of ranking methods. In Chap. 3, we define a distance-based notion of correlation between two complete rankings of the same set of objects. This notion plays an important role in developing tests of trend and of independence in the data. For example, we may test for monotone trends in river pH. Using the concept of compatibility introduced by Alvo and Cabilio (1992), we extend the notion of correlation to the case where some objects are unranked. As a consequence, this serves to widen the range of applicability and we may then test for trends in river pH when some monthly data are missing either randomly or by design. Correlation can also be defined for ranking data on a circle. Such data arise when one is interested in wind direction from an atmospheric site. In Chaps. 4 and 5, we make use of the average pairwise correlation among a set of rankings in order to test for randomness in complete and incomplete block designs. We exploit the notion of population diversity in order to develop tests of hypothesis that two or more groups come from the same population. Tests for interaction are discussed next. In Chap. 6, we develop a general theory of hypothesis testing and obtain generalizations of the Wilcoxon test of location for several populations. As well, we develop tests under umbrella alternatives applicable for dose response data that exhibit an increasing trend until it reaches a peak and a decreasing trend thereafter. The contents of Chaps. 2–7 are applicable to nonparametric analysis in general. We have not however attempted to provide a comprehensive treatment of nonparametric analysis. For more traditional texts which deal with such topics as analysis of variance in general, goodness-of-fit statistics, and regression and multivariate analysis, we refer the reader to Gibbons and Chakraborti (2011) and Higgins (2004) among others. Our goal instead has been to present a different approach for looking at a variety of inference problems using ranking methods.

The second main theme in the book deals with probability and statistical modeling for ranking data. Models can be categorized into four classes: (i) order statistics models (or random utility models), (ii) paired comparison models, (iii) distance-based models, and (iv) multistage models. Typical examples for (i) and (ii) include the Luce model (Luce 1959). They are reviewed in Chap. 8. Unlike the probability models which assume a homogeneous population of judges, predictive models assume that the judges' preferences are heterogeneous and then attempt to identify covariates that affect the judges' preferences or even to predict the ranking to be assigned by a new judge based on his/her socioeconomic variables. A popular example is the rank-ordered logit model (Chapman and Staelin 1982; Beggs et al. 1981; Hausman and Ruud 1987). In some situations, judges' rank-order preferences are derived by a number of common latent factors or form groups of different preferences. Multivariate normal order statistics models and factor analysis are considered in Chap. 9. We introduce decision tree models for ranking data in Chap. 10 in order to delve deeper into the judgment process. These models provide nonparametric methodology for prediction and classification problems. Based on the methodology of testing for agreement introduced in Chap. 4, a further refinement in building decision trees is introduced by considering the test for intergroup concordance at every split during the tree-growing stage. We come full circle in Chap. 11 where we consider extensions of distance-based models. Chapters 10–11 provide a substantial amount of detail and aim to present the researcher with an accurate picture of what is involved in attempting to apply the tools for analyzing ranking data.

The two themes in the book are complementary to one another. The work on inference can be used in a confirmatory analysis whereas the work on modeling would be appropriate in the non-null situation. We illustrate this difference using a small ranking data from C. Sutton's dissertation. In a survey conducted in Florida, Sutton asked a group of female elderly retired people aged 70–79 "with which sex do you prefer to spend your leisure?" Each elderly ranked the three choices: A: male(s), B: female(s), and C: both sexes, assigning rank 1 to the most desired choice, rank 2 to the next most desired choice, and rank 3 to the least desired choice. The ranking responses provided by 14 white females and 13 black females are listed in Table A.1 of Appendix A.2. The last row indicates that six white females and six black females preferred the response "C: both sexes" the most and the response "A: Males" the least and hence assigned rank 1 to C, rank 2 to B, and rank 3 to A.

To answer the question: is there a difference between the groups of females, we may use the tools of inference discussed in Chap. 4. However, if we wish to determine specifically where the differences lie, we would resort to the tools of modeling.

To cite another example, the English premier league (EPL) is a famous professional league for association soccer clubs in the UK. The so-called "Big Four" soccer clubs which are Arsenal, Chelsea, Liverpool, and Manchester United have dominated the top four spots since the 1996–1997 season. Wikipedia documented the results of the "Big Four" since the start of the Premier League in the 1992–1993 season. The rankings of these four EPL teams from the 1992–1993 season to the

2012–2013 season are listed in Table A.2 of Appendix A.3. The first row of the data means that there is one season that Arsenal ranked at the top of EPL, Chelsea the second, Manchester United the third, and Liverpool the fourth. We may test the hypothesis that the rankings observed are random. On the other hand, through the use of modeling, we may try to determine how the rankings cluster.

The notation in the book is as follows. For a set of $t$ objects, labeled $1, \ldots, t$, a ranking $\nu$ is a permutation of the integers $1, \ldots, t$, where $\nu(i)$ denotes the rank given to object $i$. Primes denote the transpose of either a vector or a matrix. In all cases, smaller ranks will be assigned to the more preferred objects. This is convenient for example when looking at the top $q$ objects. We write $\nu(2) = 3$ to mean that object 2 has rank equal to 3. The inverse of the ranking function (sometimes referred as ordering) $\nu^{-1}(i)$ is defined as the object whose rank is $i$. The anti-rank of the ranking $\nu$ is defined as $\tilde{\nu}(i) = (t+1) - \nu(i)$. For example suppose $t = 5$, $\nu^{-1}(3) = 4$ means that object 4 ranks third and $\tilde{\nu}(1) = 3$ means that object 1 ranks second $(= 5 - 3)$.

The book is written at the level of a research monograph aimed at a senior undergraduate or graduate student interested in using statistical methods to analyze ranking data. Such methods are by their nature nonparametric and consequently require no underlying assumptions on the distributions of the observed scores. It may also serve as a textbook for a course emphasizing statistical methods related to ranking data. In some cases we provide proofs of theorems while in others, we refer the reader to the original papers. The procedures are often illustrated by application to real data sets. At the end of each chapter we have a brief set of notes that provide further references. As a companion to the book, a web site is provided which will include some data sets and R programs to conduct some of the procedures described in the book.

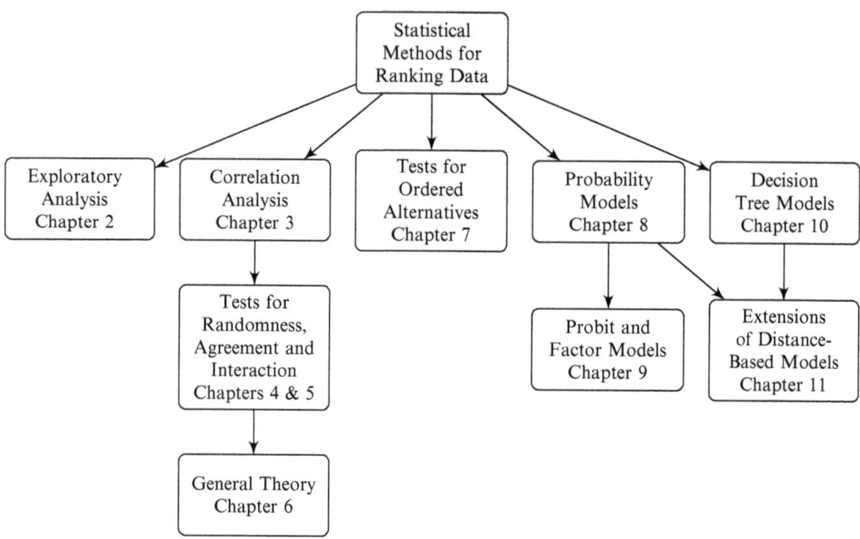


**Fig. 1.2** Schematic summary of the book

Chapters 2 and 3 are the starting points for all users. Readers interested in the foundations for inference could then proceed to Chaps. 4–7 where the emphasis is on a variety of tests of hypotheses including tests of randomness, trend, and those for ordered alternatives. As well, tests in connection with block designs provide researchers with methods for developing further tests involving more complex designs. On the other hand, readers more inclined to concentrate on modeling could proceed to Chaps. 8–11. Following a general introduction to ranking models, the reader is presented with applications involving the probit model as well as various decision tree models. A companion set of R programs located in

*http://web.hku.hk/~plhyu/StatMethRank/*

enables the user to perform the analyses described in the book (Fig. 1.2).

# Chapter 2
# Exploratory Analysis of Ranking Data

## 2.1 Descriptive Statistics

Descriptive statistics present an overall picture of ranking data. Not only do they provide a summary of the ranking data, but they are also often suggestive of the appropriate direction to analyze the data. Therefore, it is suggested that researchers consider descriptive analysis prior to any sophisticated data analysis.

We begin with a single measure of the popularity of an object. It is natural to use the mean rank attributed to an object to represent the central tendency of the ranks. The mean rank $\boldsymbol{m} = (m_1, \ldots, m_t)'$ is defined as the $t$-dimensional vector in which the $i$th entry equals

$$m_i = \sum_{j=1}^{t!} n_j \nu_j(i)/n,$$

where $\nu_j$, $j = 1, 2, \ldots, t!$ represents all possible rankings of the $t$ objects, $n_j$ is the observed frequency of ranking $j$, $n = \sum_{j=1}^{t} n_j$, and $\nu_j(i)$ is the rank score given to object $i$ in ranking $j$.

Apart from the mean ranks, the pairwise frequencies, that is, the frequency with which object $i$ is more preferred (i.e., ranked higher with a smaller rank score) than object $j$, for every possible $C_2^t$ object pairs $(i, j)$, are also often used. These pairwise frequencies can be summarized in a matrix called *a pair matrix* $P$ in which the $(a, b)$th entry equals

$$P_{ab} = \sum_{j=1}^{t!} n_j I(\nu_j(a) < \nu_j(b)),$$



M. Alvo, P.L.H. Yu, *Statistical Methods for Ranking Data*, Frontiers in Probability and the Statistical Sciences, DOI 10.1007/978-1-4939-1471-5_2

where $I(\cdot)$ is the indicator function. Note that $P_{ab}/n$ represents the empirical probability that object $a$ is more preferred than object. In addition to mean ranks and pairwise frequencies, one can look more deeply into ranking data by studying the so-called "marginal" distribution of the objects. A *marginal matrix*, specifically for this use, is the $t \times t$ matrix $M$ in which the $(a,b)$th entry equals

$$M_{ab} = \sum_{j=1}^{t!} n_j I[\nu_j(a) = b].$$

Note that $M_{ab}$ is the frequency of object $a$ being ranked $b$th. Marden (1995) called it a marginal matrix because the $a$th row gives the observed marginal distribution of the ranks assigned to object $a$ and the $b$th column gives the marginal distribution of objects given the rank $b$.

*Example 2.1.* The function `destat` in the R package `pmr` computes three types of descriptive statistics of a ranking data set, namely mean ranks, pairs, and marginals. Here, we will use Sutton's Leisure Time data (Table A.1) for illustration. The data set `leisure.black` in the `pmr` package contains the rank-order preference of spending leisure time with (1: male; 2: female; 3: both sexes) by 13 black women. By using the R code `destat(leisure.black)`, the function `destat` produces the following mean rank vector, pair matrix, and marginal matrix (Fig. 2.1):

From the above descriptive statistics, we can see that the object "3: both sexes" is clearly most preferred by the black females, and there is no strong preference between the other two objects.

```
> library("pmr")
> data(leisure.black)
> destat(leisure.black)

$mean.rank
[1] 2.308 2.462 1.231

$pair
     [,1] [,2] [,3]
[1,]    0    7    2
[2,]    6    0    1
[3,]   11   12    0

$mar
     [,1] [,2] [,3]
[1,]    2    5    6
[2,]    0    7    6
[3,]   11    1    1
```

**Fig. 2.1** Sutton's leisure time data: descriptive statistics

## 2.2 Visualizing Ranking Data

Visualization techniques for ranking data have drawn the attention of many researchers. Some of them are basically adopted from classical graphical methods for quantitative data while some are tailor-made for ranking data only. In this section, we will briefly review various graphical visualization methods and discuss the similarities and differences among them. Essentially, when a graphical method is developed for displaying ranking data, we would like this method to help answer the following questions:

1. What is the typical ranking of the $t$ objects (the general preference)?
2. To what extent is there an agreement among the judges (the dispersion)?
3. Are there any outliers among the judges and/or the objects?
4. What are the similarity and dissimilarity among the objects?

Note that when the size of the ranking data is large (e.g., $t \geq 8$ or $n \geq 100$), it is practically impossible to reveal the abovementioned pattern and characteristics by merely looking at the raw data or by using some simple descriptive statistics such as the means and standard deviations of the ranks. In this section, we will focus on several major visualization methods—permutation polytopes, multidimensional scaling (MDS) and unfolding (MDU), and multidimensional preference analysis (MDPREF). For other visualization methods, see the monograph by Marden (1995).

### 2.2.1 *Permutation Polytope*

To display a set of rankings, it is not advisable to use traditional graphical methods such as histograms and bar graphs because the elements of $\mathcal{P}$, the set of all possible permutations of the $t$ objects, do not have a natural linear ordering.

Geometrically, rankings of $t$ objects can be represented as points in $\mathbb{R}^{t-1}$. The set of all $t!$ rankings can then form a convex hull of $t!$ points in $\mathbb{R}^{t-1}$ known as a permutation polytope. The idea of using a permutation polytope to visualize ranking data was first proposed by Schulman (1979) and was considered later by McCullagh (1993a). Thompson (1993a,b) initiated the use of permutation polytopes to display the frequencies of a set of rankings in analogy with histograms for continuous data.

For complete ranking data, frequencies can be plotted on the vertices of a permutation polytope. Based on this polytope, Thompson found that the two most popular metrics for measuring distance between two rankings are the Kendall and Spearman distances which provide natural geometric interpretations of the rankings. More specifically, she showed that the minimum number of edges that must be traversed to get from one vertex of the permutation polytope to another reflects the Kendall distance between the two rankings labeled by the two vertices, whereas the Euclidean distance between any two vertices is proportional to the Spearman distance between the two rankings corresponding to the two vertices.

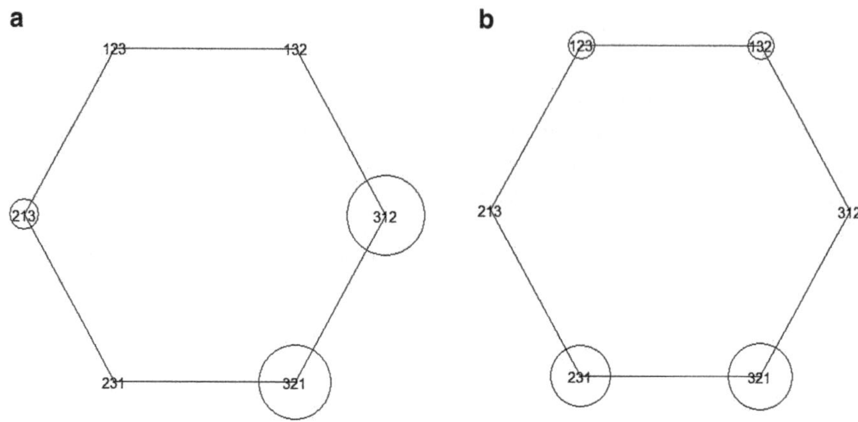


**Fig. 2.2** A hexagon displaying the leisure time data set (**a**) White females (**b**) Black females

*Example 2.2.* Note that a ranking of $t$ objects can be represented as a data point located in Euclidean space $\mathbb{R}^{t-1}$. Therefore, only rankings of three or four objects can be represented in a two-dimensional or three-dimensional graph without losing any information. For instance, ranking data with three objects can be displayed on a hexagon, in which each vertex represents a ranking and each edge connects two rankings a Kendall tau distance of 1 apart. To plot the leisure preferences given by black and white females in Sutton's Leisure Time data, we can make use of the `rankplot` function in the `pmr` package to produce the hexagons as shown in Fig. 2.2. Note that the area of the circle in each vertex is proportional to the frequency of the corresponding ranking. With a quick glance, we can see that the two rankings (2, 3, 1) and (3, 2, 1) have the largest frequencies for black females, indicating that many black females preferred to spend their leisure with "both sexes" the most while most of white females did not prefer to spend time with "male(s)."

*Example 2.3.* Consider the case of rankings of four objects. The 24(=4!) vertices form a permutation polytope in three dimensions called a truncated octahedron (Thompson 1993b). For illustration, a truncated octahedron of the four-object ranking data `big4`yr (the relative ranking of four teams in the English Premier League (EPL), namely Arsenal (1), Chelsea (2), Liverpool (3), and Manchester United (4), from 1992–1993 season to 2012–2013 season) is plotted in Fig. 2.3.

It has eight hexagonal faces and six square faces. Each face has its interpretation. Four of the hexagons refer to the rankings where a particular object is ranked first, the other four refer to the rankings where a particular object is ranked last, and the six square faces refer to the rankings where two particular objects are ranked among the top two. The hexagon face with Manchester United (4) ranked first implying that Manchester United was the best team in the EPL over many seasons.

In contrast to a complete ranking, a partial ranking is represented by a permutation of $t$ *nondistinct* numbers. For example, the top 2 partial ranking (2, –, –, 1)

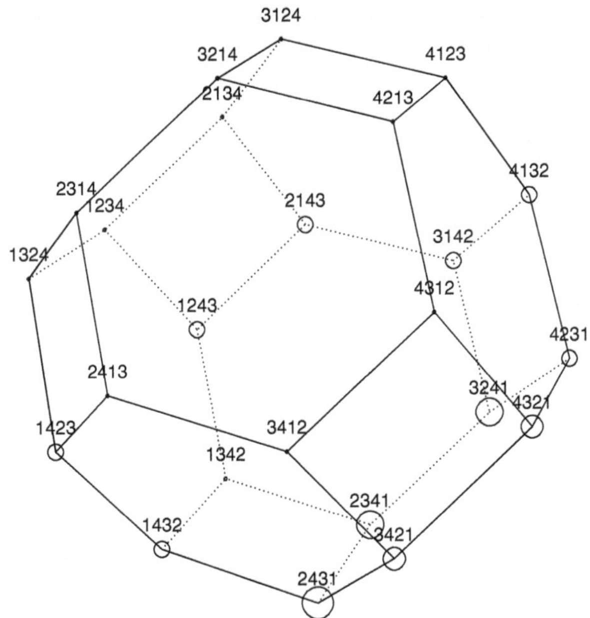


**Fig. 2.3** A truncated octahedron displaying the rankings of the Big Four EPL teams

can be represented by (2, 3.5, 3.5, 1). Therefore, the permutation polytope is not applicable to represent partial rankings. To tackle this problem, Thompson (1993a) defined a generalized permutation polytope as the convex hull of points in $\mathcal{P}$ whose coordinates are permutations of $t$ nondistinct numbers. The frequencies of a set of partial rankings can be plotted in a natural way on the vertices of a generalized permutation polytope. Similar to permutation polytopes, generalized permutation polytopes induce a new and very reasonable extension of Kendall and Spearman distances for top $q$ partially ranked data.

Owing to the fact that the generalized permutation polytope on which the frequencies are displayed is inscribed in a sphere in a $(t-1)$-dimensional subspace of $\mathcal{P}$, it is difficult to visualize all the points on a polytope in a high-dimensional space for $t \geq 5$. Thompson (1993a) proposed an approach to explore high-dimensional polytopes by examining the three-dimensional faces and portions of the four-dimensional faces. However, because drawing permutation polytopes is fairly difficult, this visualization method is not so commonly used.

### *2.2.2 Multidimensional Scaling and Unfolding*

Multidimensional scaling is a collection of graphical methods for representing data which are in the form of similarities, dissimilarities, or other measures of "closeness" between each pair of objects. Unless the data are already in that form a transformation on the raw data is required in order to obtain all pairs of objects in their (dis)similarity measurements. See Tables 1.1 and 1.2 of Cox and Cox (2001) for various transformations to calculate (dis)similarity measures for quantitative and binary data. The basic idea behind MDS is to search for a low-dimensional space, usually Euclidean, in which each object is represented by a point in the space, such that the distances between the points in the space "match" as well as possible with the original (dis)similarities. Applications of MDS can be found in fields such as behavioral science, marketing, and ecology.

The starting point consists of an $n \times n$ nonnegative symmetric matrix $\mathbf{\Delta} = (\delta_{ij})$ of dissimilarities among $n$ observations (e.g., products, people, or species) such that $\delta_{ij}$ indicates the perceived dissimilarity between observations $i$ and $j$. The goal of MDS is, given $\mathbf{\Delta}$, to find $n$ points $\{\mathbf{x}_1, \cdots, \mathbf{x}_n\}$ in a low-dimensional space such that the distances between the points approximate the given dissimilarities:

$$\delta_{ij} \approx \sqrt{(\mathbf{x}_i - \mathbf{x}_j)'(\mathbf{x}_i - \mathbf{x}_j)}.$$

Various approaches have been developed to determine the low-dimensional points $\mathbf{x}_i' s$. One typical approach is to formulate MDS as an optimization problem, where the $n$ points $\{\mathbf{x}_1, \cdots, \mathbf{x}_n\}$ are found by minimizing some loss function, commonly called *stress*, for instance,

$$\max_{x_1, \cdots, x_n} \sum_{i<j} (\delta_{ij} - d_{ij})^2,$$

where $d_{ij} = \sqrt{(\mathbf{x}_i - \mathbf{x}_j)'(\mathbf{x}_i - \mathbf{x}_j)}$.

In the context of ranking data, Kidwell et al. (2008) suggested computing the dissimilarity between any two complete or partial rankings on $t$ objects by Kendall distance proposed by Alvo and Cabilio (1995a) (see Chap. 3 for more details) and then applied MDS to find an embedding of a data set of $n$ rankings assigned by $n$ judges in a two- or three-dimensional Euclidean space.

*Example 2.4.* Consider a movie rating data set containing 72,979 possibly incomplete and tied rankings of 55 movies made by 5,625 raters who visited the web site MovieLens (Resnick et al. 1994) in 2000. We calculate the distance matrix by using a normalized version of the Kendall distance:

$$d^* = \frac{d^*_{orig} - m^*}{M^* - m^*} \tag{2.1}$$

where $d^*_{orig}$ is the Kendall distance, and $M^*$ and $m^*$ are the maximum and minimum values of the Kendall distance as defined later in Lemma 3.5.

Applying two-dimensional MDS to the distance matrix, we obtain a scatterplot of 5,625 points for the movie raters. However, the points are too densely clustered that the scatterplot is ineffective to visualize the patterns of the ranking data. Kernel smoothing is therefore used to produce a heat map for better identification of different clusters of movie raters (Fig. 2.4).

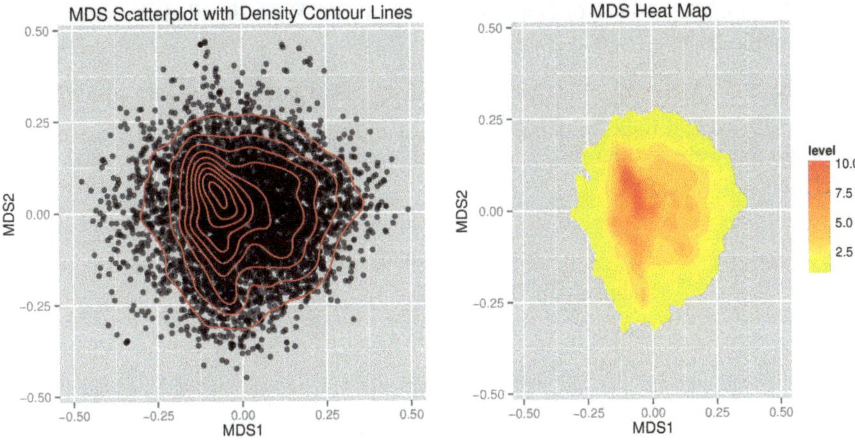


**Fig. 2.4** MDS scatterplot and heat map for the movie rating data

The movie-viewing population initially appears to be a single large cluster. A closer look identifies a dichotomy within the cluster where preferences vary from the left to the right. The right half of the cluster appears to appreciate action films with sci-fi elements—the *Star Wars* series and *The Matrix*. In the left half, the upper part mainly consists of people interested in romance, *Casablanca* and *The Graduate*, while the people in the lower part enjoy serious drama movies more—*Seven Samurai* and *To Kill a Mockingbird*.

Among the MDS techniques, the unfolding technique, formulated by Coombs (1950), is a typical variant designed for representing ranking data. Unlike the abovementioned MDS that only visualizes a set of judge points in a low-dimensional Euclidean space, the unfolding technique attempts to visualize a set of points in a low Euclidean space with both judges and objects being represented by the points in the same space. The points are sought so that the ranked order of the distances from a *judge* point to the object points matches as "close" as possible with the ranking assigned by the judge.

Mathematically, suppose that the $j$th judge is represented by the point $\mathbf{x}_j = (x_{j1}, \cdots, x_{jd})'$ in $\mathbb{R}^d$ $(j = 1, \cdots, n)$ and the $i$th object is represented by the point $\mathbf{y}_i = (y_{i1}, \cdots, y_{id})'$ in the same space $(i = 1, \cdots, t)$. Assume that the degree of preference of the $i$th object given by the $j$th judge is measured by the Euclidean distance, $d_{ij}$, between $\mathbf{x}_j$ and $\mathbf{y}_i$, where

$$d_{ij} = \sqrt{(\mathbf{y}_i - \mathbf{x}_j)'(\mathbf{y}_i - \mathbf{x}_j)}. \tag{2.2}$$

The smaller the value of $d_{ij}$, the more preferable for the $j$th judge is the $i$th object.

The problem of multidimensional unfolding (MDU) is to find $\mathbf{x}_j$'s and $\mathbf{y}_i$'s such that the distances $d_{ij}$'s match as much as possible with the ranks of objects given by the judges. In other words, this can be viewed as MDS for a rectangular dissimilarity matrix $\mathbf{\Delta}$ whose $(i, j)$ entry represents the rank of object $i$ assigned by judge $j$. Various methods are available to tackle this problem. See Chap. 8 of Cox and Cox

(2001) and Chaps. 14–16 of Borg and Groenen (2005) for detailed explanations on these methods.

*Example 2.5.* When $d = 1$, we have the so-called *unidimensional unfolding* for which objects and judges are represented by points on a straight line. For example, suppose there are two judges $\mathtt{J}_1$ and $\mathtt{J}_2$ who ranked the four objects A, B, C, and D and their rankings are

| | First | Second | Third | Fourth |
|---|---|---|---|---|
| Judge $\mathtt{J}_1$ | A | B | C | D |
| Judge $\mathtt{J}_2$ | C | D | B | A |

Then the unfolding result for this example is represented by the top line in Fig. 2.5. From the figure, we can see that for both judges, the distances from judge point ($\mathtt{J}_1$ or $\mathtt{J}_2$) to the four object points have the same ranking as his/her original ranking of the objects. It is interesting to note that when the line is folded from one side to the other side at any judge point, the judge's rankings can be observed and hence the name *unfolding* is termed. For instance, the folded line in Fig. 2.5 reveals that judge $\mathtt{J}_1$ prefers A the most, B the second, C the third, and D the least.

However, the ranking of the distances from a judge's point to all the objects and the judge's ranking cannot guarantee to be perfectly matched for every judge. For example, it is impossible to place a point for judge $\mathtt{J}_3$ who ranked the objects as DABC in Fig. 2.5.

A widely used approach of solving MDS problems is called SMACOF (Scaling by MAjorizing a COmplicated Function) which minimizes stress by means of majorization. de Leeuw and Mair (2009) extended the basic SMACOF theory to cover more types of data structures including MDU and they developed the `smacof` package in R.

*Example 2.6.* Consider the Big Four data (Table A.2) in which the seasonal rankings of four EPL teams, Arsenal, Chelsea, Liverpool, and Manchester United, from the 1992–1993 season to 2012–2013 season. Applying SMACOF to this $4 \times 21$ matrix, we obtain the unfolding solution in Fig. 2.6.

In Fig. 2.6, the configuration plot shows the coordinates of the Big Four teams and the seasons jointly. We can see that the Big Four teams are specially located at four distinct configurations with Manchester United located at the center and

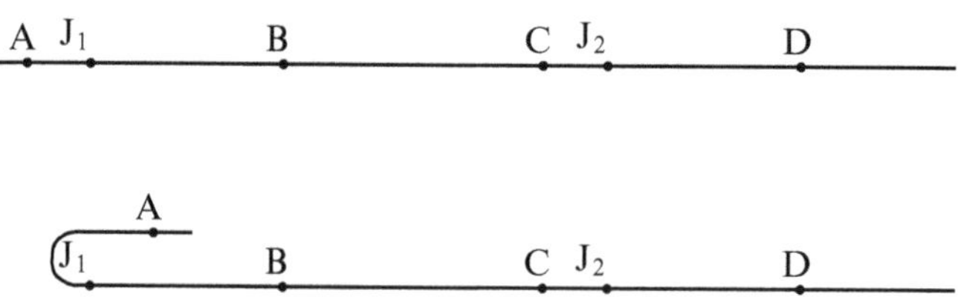


**Fig. 2.5** Unidimensional unfolding

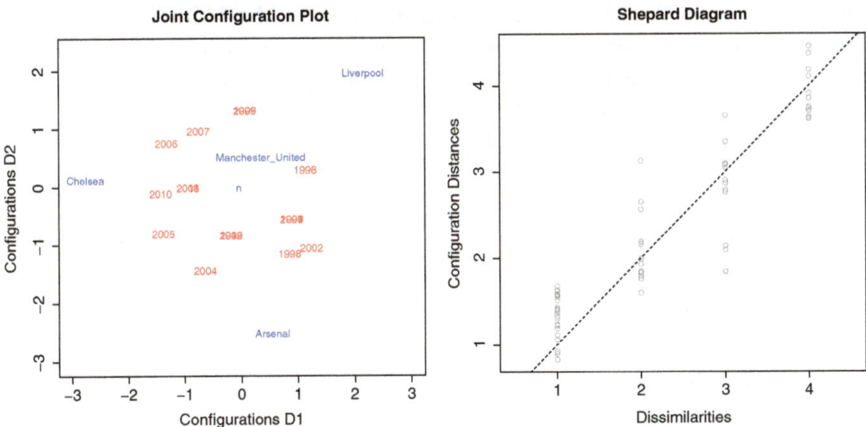


**Fig. 2.6** Joint configuration plot and Shepard diagram for the Big Four data

surrounded by the seasons. This is reasonable since out of 21 seasons, Manchester United ranked first in 15 seasons. Examining the configurations of teams and seasons, most of the seasons in the 1990s are on the right of the graph where Liverpool and Arsenal are located, and the seasons in the recent years are on the left and Chelsea and Manchester United are the nearest teams. This indicates that Chelsea had its increasing ability over Liverpool and Arsenal in the recent decade.

The Shepard diagram in Fig. 2.6 shows a scatterplot of the reconstructed distances obtained from the SMACOF solution against the original dissimilarities (i.e., ranks) and an isotonic regression fitted to the points in the graph. If all reconstructed distances lie on the fitted line, then the dissimilarities would be perfectly reproduced by the MDS solution. From Fig. 2.6, the Shepard diagram shows some lack of fit, particularly for ranks 2 and 3 in a few seasons.

### 2.2.3 *Multidimensional Preference Analysis*

Similar to MDU, MDPREF (Carroll 1972) displays the relationship between judges and their respective perceptions on objects by reducing the dimensionality of the data, while retaining their main features as much as possible. Instead of using points as in the unfolding technique, each judge is now represented by a vector in a low-dimensional space while each object is represented by a point in the same space. The vectors and points are so chosen that the projections of the object points onto the judge vector indicate the rankings of the judge as closely as possible.

More specifically, let $\boldsymbol{y}^{(j)} = (y_{j1}, \cdots, y_{jt})'$, $j = 1, \cdots, n$ be a $t \times 1$ vector of anti-ranks for $t$ objects given by judge $j$. Also, let $\boldsymbol{Y} = [\boldsymbol{y}^{(1)} \cdots \boldsymbol{y}^{(n)}]'$ be a collection of the rankings given by $n$ judges. Notice that

$$y_{ji} = t + 1 - \delta_{ij}, i = 1, \cdots, t, j = 1, \cdots, n,$$

where $\delta_{ij}$ is the dissimilarity (rank) measure used in the MDU in the previous section. Since each row of $\boldsymbol{Y}$ is just a particular permutation of the integers $1, 2, \cdots, t$, the mean of the ranks in any row equals $\frac{t+1}{2}$. For the sake of convenience, each row of $\boldsymbol{Y}$ is centered by $\frac{t+1}{2}$, resulting in a new matrix, $\boldsymbol{Y}_c$:

$$\boldsymbol{Y}_c = \boldsymbol{Y} - \frac{t+1}{2}\mathbf{1}_n\mathbf{1}_t'$$

where $\mathbf{1}_m$ is a $m \times 1$ vector with all entries being 1.

The problem of MDPREF is to factorize $\boldsymbol{Y}_c$ as

$$\boldsymbol{Y}_c = \boldsymbol{G}\boldsymbol{H}', \tag{2.3}$$

where $\boldsymbol{G}$ and $\boldsymbol{H}$ are $n \times d$ and $t \times d$ matrices, respectively. By comparing entries on both sides of (2.3), we obtain

$$y_{ij} - \frac{t+1}{2} = \boldsymbol{g}_i'\boldsymbol{h}_j,$$

where $\boldsymbol{g}_i$ and $\boldsymbol{h}_j$ are the rows of $\boldsymbol{G}$ and $\boldsymbol{H}$, respectively, and we have a geometric representation of $\boldsymbol{Y}_c$ in terms of $d$-dimensional vectors.

Suppose the rectangular matrix $\boldsymbol{Y}_c$ has rank $r(\leq \min(n, t-1))$. A natural approach to determine $\boldsymbol{G}$ and $\boldsymbol{H}$ is to use the singular value decomposition:

$$\boldsymbol{Y}_c = \boldsymbol{P}\boldsymbol{\Lambda}\boldsymbol{Q}',$$

where $\boldsymbol{P}$ is an $n \times r$ orthogonal matrix of rank $r$, $\boldsymbol{Q}$ is an $t \times r$ orthogonal matrix of rank $r$, $\boldsymbol{\Lambda} = diag(\lambda_1, \lambda_2, \cdots, \lambda_r)$, and $\lambda_1 \geq \lambda_2 \geq \cdots \geq \lambda_r > 0$ are the positive eigenvalues of $\boldsymbol{Y}_c$. Here, $\boldsymbol{P}, \boldsymbol{Q}, \boldsymbol{\Lambda}$ can also be obtained using the spectral decomposition, since $\boldsymbol{Y}_c\boldsymbol{Y}_c' = \boldsymbol{P}\boldsymbol{\Lambda}^2\boldsymbol{P}'$ and $\boldsymbol{Y}_c'\boldsymbol{Y}_c = \boldsymbol{Q}\boldsymbol{\Lambda}^2\boldsymbol{Q}'$.

Using only the $d$ $(\leq r)$ largest eigenvalues, $\boldsymbol{Y}_c$ can be approximated by

$$\hat{\boldsymbol{Y}}_c = \boldsymbol{P}_d\boldsymbol{\Lambda}_d\boldsymbol{Q}_d', \tag{2.4}$$

where $\boldsymbol{P}_d, \boldsymbol{Q}_d$ denote the matrices consisting of the first $d$ columns of $\boldsymbol{P}$ and $\boldsymbol{Q}$, respectively, and $\boldsymbol{\Lambda}_d = diag(\lambda_1, \lambda_2, \cdots, \lambda_d)$. In fact, it was shown by many researchers that the approximation is the least squares solution of the problem of minimizing

$$trace[(\boldsymbol{Y}_c - \boldsymbol{X})(\boldsymbol{Y}_c - \boldsymbol{X})'] = \sum_{i=1}^{n}\sum_{j=1}^{t}(y_{ij} - \frac{t+1}{2} - x_{ij})^2$$

among all $n \times t$ matrices $\boldsymbol{X}$ of rank $d$ or less.

A $d$-dimensional MDPREF solution includes the following steps: (a) label object $i$ in $\mathbb{R}^d$ by a point represented by the $i$th row of $\boldsymbol{Q}_d$, and (b) label judge $j$ on the same graph represented by the $j$th row of $\boldsymbol{P}_d\boldsymbol{\Lambda}_d$, by an arrow drawn from the origin on the graph. To give a better graphical display, the length of the judge vectors can be scaled to fit the position of the objects. It is not difficult to see that the perpendicular projection of all $t$ object points onto a judge vector will closely approximate the ranking of the $t$ objects by that judge if the current MDPREF solution fits the data well. Otherwise, we may look for a higher-dimension solution.

*Example 2.7.* Let us revisit the data on ranking the Big Four EPL teams (Table A.2). Applying MDPREF to this Big Four data gives the two-dimensional MDPREF solution as shown in Fig. 2.7.

```
> mdpref(big4yr,rank.vector=T)

$item
        [,1]    [,2]
[1,] -0.1208  1.0514
[2,] -1.3187 -1.9222
[3,] -1.1504  1.2467
[4,]  2.5898 -0.3759

$ranking
      [,1] [,2] [,3] [,4] [,5]     [,6]    [,7]
 [1,]    3    4    2    1    1  0.83970  0.5716
 [2,]    2    4    3    1    1  1.00136  0.5254
 [3,]    4    3    2    1    1  0.65162 -0.1316
 [4,]    3    4    2    1    1  0.83970  0.5716
 [5,]    2    4    3    1    1  1.00136  0.5254
 [6,]    1    4    3    2    1  0.57576  0.8630
 [7,]    2    3    4    1    1  0.97494 -0.2240
 [8,]    2    4    3    1    1  1.00136  0.5254
 [9,]    2    4    3    1    1  1.00136  0.5254
[10,]    1    4    2    3    1 -0.01150  1.2467
[11,]    2    3    4    1    1  0.97494 -0.2240
[12,]    1    2    4    3    1 -0.06435 -0.2521
[13,]    2    1    4    3    1 -0.25244 -0.9554
[14,]    4    1    3    2    1  0.01150 -1.2467
[15,]    4    2    3    1    1  0.62519 -0.8810
[16,]    3    2    4    1    1  0.78685 -0.9272
[17,]    4    3    2    1    1  0.65162 -0.1316
[18,]    3    1    4    2    1  0.17317 -1.2929
[19,]    3    2    4    1    1  0.78685 -0.9272
[20,]    2    3    4    1    1  0.97494 -0.2240
[21,]    3    2    4    1    1  0.78685 -0.9272

$explain
[1] 0.7232
```

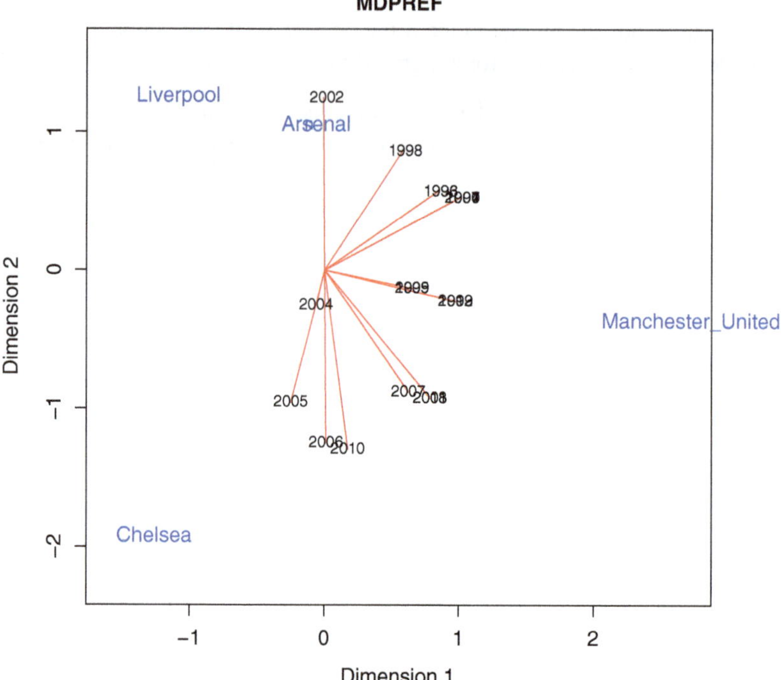


**Fig. 2.7** The two-dimensional MDPREF solution for the Big Four data

The coordinates of the objects and season rankings and the proportion of variance explained by the first two dimensions are stored in the values `$item`, `$ranking`, and `$explain`, respectively. It can be seen that two dimensions can explain about 72.32 % of the total variance. Figure 2.7 shows the two-dimensional MDPREF solution. The first dimension can be interpreted as the overall preference of the four teams. Manchester United performed best, whereas Liverpool and Chelsea performed poorly. The second dimension represents the contrast between Chelsea and the two teams, Liverpool and Arsenal. Examining their rankings in Table A.2, we see that Chelsea had its increasing performance over Liverpool and Arsenal in the recent decade.

Note that the season vectors point to the direction of the best performing team. For example, the ranking (3, 2, 4, 1) was observed in three seasons in 2008, 2011, and 2013, meaning that Manchester United ranked first, then Chelsea second, Arsenal third, and finally Liverpool the last. They all have the same season vector pointing to (0.79, −0.93). By projecting the four team points to the season vector in Fig. 2.7, we obtain correctly the ordering Manchester United>Chelsea>Arsenal>Liverpool.

### 2.2.4 *Comparison of the Methods*

So far, we have described three methods of visualizing ranking data. Each method has its distinct features. Permutation polytopes provide the frequency distribution of the observed rankings. Unfolding and MDPREF attempt to reveal relationships among the judges and the objects such as clustering of objects and unusual judges.

Note that permutation polytopes do not make any assumptions while unfolding and MDPREF assume that the ranking assigned by each judge can be represented in terms of the ordering of distances and projections, respectively.

Despite their features and assumptions, each of them has their own weaknesses. Permutation polytopes become difficult to interpret for large numbers of objects, say more than 5. Unfolding and MDPREF rely on the validity of assumptions made in each method which may not be true in general. Nevertheless, unfolding and MDPREF can often provide a very informative graph which gives a spatial representation of objects and judges and from which the between-object and between-judge relationships could then be identified (see Cohen and Mallows 1980).

## 2.3 Tests for Randomness on Ranking of a Set of Objects

When we say that selecting a ranking of $t$ objects is completely random, we mean all possible rankings of $t$ objects have the same probability (i.e., a uniform distribution) of being selected. In that case, the expected frequencies of each ranking of $t$ objects in a data set of size $n$ should be $n/t!$, and the standard chi-square goodness-of-fit test can be applied to test for uniformity. However, it is not always applicable when $t!$ is too large as compared to $n$, because we may encounter the problem of observed frequencies on some rankings being less than 5. If such a case occurs, Marden (1995) suggested using mean ranks, pairs, or marginals (see Sect. 2.1), instead of ranking proportions, to test for randomness.

Under the null hypothesis $H_0$ of randomness, the expected mean ranks, pairs, and marginals should be $(t+1)/2$, $0.5n$, and $n/t$, respectively. Under $H_0$, the test statistics based on the sample mean ranks, pairs, and marginals are

$$\text{Mean rank:}\quad \frac{12n}{t(t+1)}\sum_{j=1}^{t}(m_j-\frac{t+1}{2})^2$$

$$\text{Pairs:}\quad 12n\left\{\sum_{a>b}^{t}(P_{ab}-0.5)^2-\frac{1}{t+1}\sum_{j=1}^{t}(m_j-\frac{t+1}{2})^2\right\}$$

$$\text{Marginals:}\quad n(t+1)\sum_{a>b}^{t}(M_{ab}-\frac{1}{t})^2$$

which follow a $\chi^2$ distribution with degrees of freedom $t-1$, $\binom{t}{2}$, and $(t-1)^2$, respectively.

*Example 2.8.* To assess for randomness for Sutton's Leisure Time data based on the mean ranks, the following R code could be used:

```
> data(leisure.black)
> de1<-destat(leisure.black)
> mean<-rep(2,3)
> chi<-12*13*sum((de1$mean.rank - mean)^2)/3/4
> chi
```

```
[1] 11.69
```

```
> dchisq(chi,2)
```

```
[1] 0.001445
```

The value of the test statistic is 11.69 which has a p-value of 0.0014. This indicates that black females had uneven preference on the three choices.

*Example 2.9.* The above $\chi^2$ tests can also be extended to evaluate the agreement or diversity between two populations of judges who ranked the same set of objects. We may test for the agreement of leisure time preferences between black females and white females based on the marginals. The R code used is given as follows:

```
> de1 <- destat(leisure.black)
> data(leisure.white)
> de2 <- destat(leisure.white)
> chisq.test(cbind(as.vector(de1$mar),
       as.vector(de2$mar)))
```

```
Pearson's Chi-squared test

data:  cbind(as.vector(de1$mar), as.vector(de2$mar))
X-squared = 27.22, df = 8, p-value = 0.0006479
```

The value of the test statistic is 27.22 which has a p-value of 0.00065. It is evident that black and white females did not agree on their preference with which sex they prefer to spend their leisure time. Note that the marginal matrix has only $(t-1)^2 = 4$ df and hence the proper degrees of freedom of the chi-square test here should be 4. However, this will make p-value even smaller and hence this does not affect our conclusion.

## Chapter Notes

Alvo and Ertas (1992) extended MDPREF to visualize rankings obtained from more than one population. Yu and Chan (2001) and Leung (2003) developed a probabilistic extension of MDPREF and MDU, respectively, so that statistical inference on model parameters can be made. Other graphical representation techniques include Ye and McCullagh (1993), Han and Huh (1995), Baba (1986), and Hirst and Naes (1994). For examining agreement or diversity among three or more populations of judges, see Chap. 4 and Marden (1995) for some distance-based methods and MANOVA-like methods.

# Chapter 3
# Correlation Analysis of Paired Ranking Data

## 3.1 Notion of Distance Between Two Rankings

A ranking represents the order of preference one has with respect to a set of $t$ objects. If we label the objects by the integers 1 to $t$, a ranking can then be thought of as a permutation of the integers $(1, 2, \ldots, t)$. We may denote such a permutation by $\mu = (\mu(1), \mu(2), \ldots, \mu(t))'$ which may also be conceptualized as a point in t-dimensional space. It is natural to measure the spread between two individual permutations $\mu, \nu$ by means of a distance function. There are several examples of distance functions that have been proposed in the literature. Here are a few:

Spearman

$$d_S(\mu, \nu) = \frac{1}{2} \sum_{i=1}^{t} (\mu(i) - \nu(i))^2 . \tag{3.1}$$

Kendall

$$d_K(\mu, \nu) = \sum_{i<j} \{1 - sgn\,(\mu(j) - \mu(i))\, sgn\,(\nu(j) - \nu(i))\}, \tag{3.2}$$

where $sgn(x)$ is either 1 or $-1$ depending on whether $x > 0$ or $x < 0$.

Hamming

$$d_H(\mu, \nu) = t - \sum_{i=1}^{t} \sum_{j=1}^{t} I\,(\mu(i) = j)\, I\,(\nu(i) = j) \tag{3.3}$$

where $I(.)$ is the indicator function taking values 1 or 0 depending on whether the statement in brackets holds or not.



M. Alvo, P.L.H. Yu, *Statistical Methods for Ranking Data*, Frontiers in Probability and the Statistical Sciences, DOI 10.1007/978-1-4939-1471-5_3

Spearman Footrule

$$d_F(\mu, \nu) = \sum_{i=1}^{t} |\mu(i) - \nu(i)|. \tag{3.4}$$

The Spearman measure is not a proper "distance" in that it does not obey the triangular inequality property. We shall nonetheless refer to it as a distance function in this book. It is based upon squared Euclidean distance whereas the Footrule is based on the absolute deviations. The Kendall distance counts the number of "discordant" pairs whereas the Hamming distance counts the number of "mismatches." The Hamming distance has found uses in coding theory. These distances have the property of being invariant under any permutation relabeling of the objects. That is, for any permutations $\sigma, \mu, \nu$,

$$d\,(\mu, \nu) = d\,(\mu \circ \sigma, \nu \circ \sigma)$$

where $\mu \circ \sigma\,(i) = \mu\,(\sigma\,(i))$. This property is known as right invariance. Let $\Delta = \left(d\left(\mu_i, \mu_j\right)\right)$ denote the matrix of all pairwise distances. If $d$ is right invariant, then it follows that there exists a constant $c > 0$ for which

$$\Delta \mathbf{1} = (ct!)\mathbf{1}$$

where $\mathbf{1} = (1, 1, \ldots, 1)'$ is of dimension $t!$. Hence, $c$ is equal to the average distance. It is straightforward to show that for the Spearman and Kendall distances

$$c_S = \frac{t(t^2-1)}{12}, c_K = \frac{t(t-1)}{2}.$$

Turning attention to the Hamming distance, we note that if $e = (1, 2, \ldots, t)'$, then

$$\begin{aligned} \Sigma_\mu d_H\,(\mu, e) &= \Sigma_\mu t - \Sigma_\mu \Sigma_i \Sigma_j I\,(\mu\,(i) = j)\, I\,(e\,(i) = j.) \\ &= t\,(t!) - t! \end{aligned}$$

and hence $c_H = (t-1)$.

*Example 3.1.* Suppose that $t = 3$ and that the complete rankings are denoted by

$$\mu_1 = (1, 2, 3)', \mu_2 = (1, 3, 2)', \mu_3 = (2, 1, 3)', \mu_4 = (2, 3, 1)', \mu_5 = (3, 1, 2)',$$
$$\mu_6 = (3, 2, 1)'.$$

Using the above order of the permutations, we may write the matrix $\Delta$ of pairwise Spearman, Kendall, Hamming, and Footrule distances respectively as

$$\Delta_S = \begin{pmatrix} 0 & 1 & 1 & 3 & 3 & 4 \\ 1 & 0 & 3 & 1 & 4 & 3 \\ 1 & 3 & 0 & 4 & 1 & 3 \\ 3 & 1 & 4 & 0 & 3 & 1 \\ 3 & 4 & 1 & 3 & 0 & 1 \\ 4 & 3 & 3 & 1 & 1 & 0 \end{pmatrix}$$

$$\Delta_K = \begin{pmatrix} 0 & 2 & 2 & 4 & 4 & 6 \\ 2 & 0 & 4 & 2 & 6 & 4 \\ 2 & 4 & 0 & 6 & 2 & 4 \\ 4 & 2 & 6 & 0 & 4 & 2 \\ 4 & 6 & 2 & 4 & 0 & 2 \\ 6 & 4 & 4 & 2 & 2 & 0 \end{pmatrix}$$

$$\Delta_H = \begin{pmatrix} 0 & 2 & 2 & 3 & 3 & 2 \\ 2 & 0 & 3 & 2 & 2 & 3 \\ 2 & 3 & 0 & 2 & 2 & 3 \\ 3 & 2 & 2 & 0 & 3 & 2 \\ 3 & 2 & 2 & 3 & 0 & 2 \\ 2 & 3 & 3 & 2 & 2 & 0 \end{pmatrix}$$

$$\Delta_F = \begin{pmatrix} 0 & 2 & 2 & 4 & 4 & 4 \\ 2 & 0 & 4 & 2 & 4 & 4 \\ 2 & 4 & 0 & 4 & 2 & 4 \\ 4 & 2 & 4 & 0 & 4 & 2 \\ 4 & 4 & 2 & 4 & 0 & 2 \\ 4 & 4 & 4 & 2 & 2 & 0 \end{pmatrix}$$

These distances may alternatively be written in terms of a similarity function in the form

$$d(\mu, \nu) = c - \mathcal{A}(\mu, \nu), \tag{3.5}$$

Spearman:

$$\mathcal{A}_S = \mathcal{A}_S(\mu, \nu) = \sum_{i=1}^{t} \left( \mu(i) - \frac{t+1}{2} \right) \left( \nu(i) - \frac{t+1}{2} \right). \tag{3.6}$$

Kendall:

$$\mathcal{A}_K = \mathcal{A}_K(\mu, \nu) = \sum_{i<j} sgn\left(\mu(j) - \mu(i)\right) sgn\left(\nu(j) - \nu(i)\right). \tag{3.7}$$

Hamming:

$$\mathcal{A}_H(\mu, \nu) = \sum_{i=1}^{t}\sum_{j=1}^{t} I\left([\mu(i) = j] - \frac{1}{t}\right) I\left([\nu(i) = j] - \frac{1}{t}\right). \tag{3.8}$$

Footrule:

$$\mathcal{A}_F(\mu, \nu) = \sum_{i=1}^{t}\sum_{j=1}^{t} I\left([\mu(i) \leq j] - \frac{j}{t}\right) I\left([\nu(i) \leq j] - \frac{j}{t}\right).$$

The similarity measures may be also interpreted geometrically as inner products which sets the groundwork for defining correlation in the next section.

## 3.2 Correlation Between Two Rankings

The notion of correlation occurs frequently in statistics. For example, in regression analysis, one is interested in the correlation between two variables such as height and weight. Similarly, in nonparametric statistics, we shall be interested in the correlation between two rankings. Let $\mathcal{P}$ be the space of all possible permutations of the integers $1, 2, \ldots, t$. We may define the correlation between two rankings $\mu, \nu$ as

$$\alpha(\mu, \nu) = 1 - \frac{2d(\mu, \nu)}{M} \tag{3.9}$$

where $M$ is the maximum value of the distance $d(\mu, \nu)$ taken over all possible pairs $\mu, \nu$ in $\mathcal{P}$ (Diaconis and Graham 1977). In the case of the Spearman and Kendall distance, the maximum values occur when

$$\left(\mu(i) - \frac{t+1}{2}\right) = -\left(\nu(i) - \frac{t+1}{2}\right) \textit{ for all } i,$$

whereas the minimum occurs when

$$\left(\mu(i) - \frac{t+1}{2}\right) = \left(\nu(i) - \frac{t+1}{2}\right)$$

This is a consequence of the rearrangement inequality given as a lemma below.

**Lemma 3.1.** *Let* $a_1, \ldots, a_t$ *and* $b_1, \ldots, b_t$ *be real numbers, not necessarily positive with*

$$a_1 \leq \ldots \leq a_t, b_1 \leq \ldots \leq b_t$$

*and let $\sigma$ be a permutation of the integers* $1, \ldots, t$. *Then*

$$a_1 b_t + \ldots + a_t b_1 \leq a_1 b_{\sigma(1)} + \ldots + a_t b_{\sigma(t)} \leq a_1 b_1 + \ldots + a_t b_t.$$

*Proof.* The proof follows by induction on $t$. □

It can be shown that for the Spearman and Kendall distances, the maximum is equal to twice the mean,

$$M_S = 2c_S, M_K = 2c_K. \tag{3.10}$$

In view of (3.10) we have

$$\alpha_S(\mu, \nu) = \frac{\mathcal{A}_S}{c_S}, \alpha_K(\mu, \nu) = \frac{\mathcal{A}_K}{c_K}. \tag{3.11}$$

*Example 3.2 (Lehmann 1975, p. 298).* Consider the test scores in Language and Arithmetic for a group of 9 students as shown in Table 3.1. The right-invariance property shared by the Spearman and Kendall distances enables us to rewrite the table in a more convenient fashion with one of the rankings in natural order as in Table 3.2. The Spearman and Kendall correlations are respectively 0.683 and 0.500. Here $c_S = 60, c_K = 36$.

The correlation coefficients based on these distances are of the multiplicative type (Kendall and Gibbons 1990); that is, there exists a function $g$ such that

$$\alpha(\mu, \nu) = k_\mu k_\nu \sum_i \sum_j g(\mu(i), \mu(j))\, g(\nu(i), \nu(j)) \tag{3.12}$$

**Table 3.1** Language and Arithmetic scores

| Student | 1 | 2 | 3 | 4 | 5 | 6 | 7 | 8 | 9 |
|---|---|---|---|---|---|---|---|---|---|
| Language | 50 | 23 | 28 | 34 | 14 | 54 | 46 | 52 | 53 |
| Arithmetic | 38 | 28 | 14 | 26 | 18 | 40 | 23 | 30 | 27 |
| Language ranks | 6 | 2 | 3 | 4 | 1 | 9 | 5 | 7 | 8 |
| Arithmetic ranks | 8 | 6 | 1 | 4 | 2 | 9 | 3 | 7 | 5 |

**Table 3.2** Language and Arithmetic scores rearranged

| Student | 5 | 2 | 3 | 4 | 7 | 1 | 8 | 9 | 6 |
|---|---|---|---|---|---|---|---|---|---|
| Language | 14 | 23 | 28 | 34 | 46 | 50 | 52 | 53 | 54 |
| Arithmetic | 18 | 28 | 14 | 26 | 23 | 38 | 30 | 27 | 40 |
| Language ranks | 1 | 2 | 3 | 4 | 5 | 6 | 7 | 8 | 9 |
| Arithmetic ranks | 2 | 6 | 1 | 4 | 3 | 8 | 7 | 5 | 9 |

where $k_\mu, k_\nu$ are normalizing constants. The constants may be different depending on whether the coefficient is of type a or b. A type a correlation is used above. For Spearman and Kendall, the functions are, respectively,

$$g_S(\mu(i), \mu(j)) = (\mu(i) - \mu(j))$$

$$g_K(\mu(i), \mu(j)) = sgn[\mu(i) - \mu(j)].$$

For a type b correlation, the constants are given by

$$k_\mu = \sqrt{\Sigma_i \Sigma_j [g(\mu(i), \mu(j))]^2}.$$

We shall make use of a type b correlation when defining angular correlations in Sect. 3.6.

For a multiplicative index, it can be shown that the correlation matrix is necessarily positive semidefinite (Quade 1972). Setting

$$Q = \left(J - \frac{2}{M}\Delta\right) \tag{3.13}$$

where $J = \mathbf{1}\mathbf{1}'$ and $\frac{M}{2} = c$, this implies that there exists a matrix $\mathbf{T}$ for which

$$Q = \frac{1}{c}(\mathbf{T}'\mathbf{T}). \tag{3.14}$$

It follows that the distance matrix for both Spearman and Kendall can be expressed as

$$\Delta = cJ - \mathbf{T}'\mathbf{T}. \tag{3.15}$$

From the form of the Spearman and Kendall similarity measures (3.12), it can be seen that the matrices $\mathbf{T}$ are respectively

$$\mathbf{T}_S = (t_S(\mu_1), \ldots, t_S(\mu_{t!}))' \tag{3.16}$$

where

$$t_S(\mu) = \left(\mu(1) - \frac{t+1}{2}, \ldots, \mu(t) - \frac{t+1}{2}\right)'$$

is the centered rank vector and

$$\mathbf{T}_K = (t_K(\mu_1), \ldots, t_K(\mu_{t!}))' \tag{3.17}$$

is of dimension $\binom{t}{2} \times t!$ where the $q$th element for $q = (i-1)\left(t - \frac{i}{2}\right) + (j-i)$, $1 \leq i < j \leq t$,

$$(t_K(\mu))_q = sgn\left[\mu(j) - \mu(i)\right].$$

For Hamming, we may write the $t^2$-dimensional vector where the $(i,j)$th element is

$$(t_H(\mu))_{ij} = \left(I\left[\mu(i) = j\right] - \frac{1}{t}\right)$$

for $1 \leq i, j \leq t$.

For the Footrule we have the $t^2$-dimensional vector where the $q$th element for $q = (i-1)t + j, 1 \leq i < j \leq t$

$$(t_F(\mu))_q = \left(I\left[\mu(i) \leq j\right] - \frac{j}{t}\right).$$

*Example 3.3.* Suppose that $t = 3$. Then, placing the rankings in the natural order of Example 3.1, we have that

$$\mathbf{T}_S = \begin{pmatrix} -1 & -1 & 0 & 0 & 1 & 1 \\ 0 & 1 & -1 & 1 & -1 & 0 \\ 1 & 0 & 1 & -1 & 0 & -1 \end{pmatrix}$$

and

$$\mathbf{T}_K = \begin{pmatrix} 1 & 1 & -1 & 1 & -1 & -1 \\ 1 & 1 & 1 & -1 & -1 & -1 \\ 1 & -1 & 1 & -1 & 1 & -1 \end{pmatrix}.$$

The notion of correlation is particularly useful in problems wherein one wishes to test for the independence of two variables as in Example 3.2 or for the existence of long-term monotone trend in the pH of a river. We will postpone a discussion of these important topics later in this chapter where it will be addressed in the general context of incomplete rankings.

## 3.3 Incomplete Rankings and the Notion of Compatibility

A judge may rank a complete set of candidates in accordance with some criterion. On occasion, however, data may be missing either at random or by design. For example, one or more candidates may not be ranked. In another example, the pH data on a lake may not be available for certain months in a year, thereby making it impossible to test for a long-term trend using traditional nonparametric rank-based statistics. The option to ignore the missing data is unsatisfactory because it distorts the time scale. As we shall see later on, this option is always suboptimal when testing for trend. We address the topic in this section by first introducing the notion of compatibility.

**Notation.** Incomplete ranks will be denoted by "–" and corresponding incomplete rankings will be written with an upper script "*".

For example, the ranking $\mu^* = (2, -, 3, 4, 1)'$ indicates that object 2 is unranked among the five objects presented.

**Definition 3.1.** The complete ranking $\mu$ of $t$ objects is said to be compatible with an incomplete ranking $\mu^*$ of a subset of $k$ of these objects, $2 \leq k \leq t$, if the relative ranking of every pair of objects ranked in $\mu^*$ coincides with their relative ranking in $\mu$.

An incomplete ranking gives rise to a class of order preserving complete rankings. Denoting by $C(\mu^*)$ the set of complete permutations compatible with $\mu^* = (2, -, 3, 4, 1)'$, we have that

$$C(\mu^*) = \left\{(2,5,3,4,1)', (2,4,3,5,1)', (2,3,4,5,1)', (3,2,4,5,1)', (3,1,4,5,2)'\right\}.$$

The total number of complete rankings of $t$ objects compatible with an incomplete ranking of a subset of $k$ objects is given by $t!/k!$. This follows from the fact that there are $\binom{t}{k}$ ways of choosing $k$ integers for the ranked objects, one way in placing them to preserve the order and then $(t-k)!$ ways of rearranging the remaining integers. The product is thus

$$a = \binom{t}{k}(t-k)! = t!/k! \tag{3.18}$$

The notion of compatibility establishes a connection between an incomplete ranking and the class of complete rankings from which the incomplete ranking could have arisen. It seems natural as a result to extend the notion of distance to incomplete rankings by referring to the corresponding compatibility classes.

**Definition 3.2.** The distance $d^*(\mu^*, \nu^*)$ between two incomplete rankings $\mu^*$ and $\nu^*$ is defined to be the average of all values of the distances $d(\mu_{i}, _{j})$ taken over all pairs of complete rankings $\mu_{i}, _{j}$ compatible with $\mu^*$ and $\nu^*$, respectively.

*Example 3.4.* Suppose that $t = 3, k = 2$. In that case, the possible incomplete rankings are denoted by

$$\nu_{11}^* = (1,2,-)', \nu_{12}^* = (2,1,-)', \nu_{21}^* = (1,-,2)', \nu_{22}^* = (2,-,1)',$$
$$\nu_{31}^* = (-,1,2)', \nu_{32}^* = (-,2,1)'$$

We may associate with every incomplete ranking a ($t!$ x 1) compatibility vector, also denoted by $C(\nu^*)$, whose $i$th component is 1 or 0 according to whether $\mu_i$ is compatible with $\nu^*$. A summary can be provided by a compatibility matrix as follows.

$$C = \begin{array}{c|cccccc} & \nu_{11}^* & \nu_{12}^* & \nu_{21}^* & \nu_{22}^* & \nu_{31}^* & \nu_{32}^* \\ \mu_1 & 1 & 0 & 1 & 0 & 1 & 0 \\ \mu_2 & 1 & 0 & 1 & 0 & 0 & 1 \\ \mu_3 & 0 & 1 & 1 & 0 & 1 & 0 \\ \mu_4 & 1 & 0 & 0 & 1 & 0 & 1 \\ \mu_5 & 0 & 1 & 0 & 1 & 1 & 0 \\ \mu_6 & 0 & 1 & 0 & 1 & 0 & 1 \end{array}$$

Consequently, the matrix of average pairwise Spearman distances for the incomplete rankings is given by the product $C_S' \Delta C_S / a^2$ where $a = t!/\,k! = 3$ and

$$C_S' \Delta C_S = \begin{array}{c|cccccc} & \nu_{11}^* & \nu_{12}^* & \nu_{21}^* & \nu_{22}^* & \nu_{31}^* & \nu_{32}^* \\ \nu_{11}^* & 10 & 26 & 14 & 22 & 22 & 14 \\ \nu_{12}^* & 26 & 10 & 22 & 14 & 14 & 22 \\ \nu_{21}^* & 14 & 22 & 10 & 26 & 14 & 22 \\ \nu_{22}^* & 22 & 14 & 26 & 10 & 22 & 14 \\ \nu_{31}^* & 22 & 14 & 14 & 22 & 10 & 26 \\ \nu_{32}^* & 14 & 22 & 22 & 14 & 26 & 10 \end{array}$$

We note from this example that the distance of an incomplete ranking to itself is 10 and not 0. In extending the notion of correlation to incomplete rankings, it will be necessary to take this into account.

For the Spearman and Kendall distances, we may re-express the distance $d^*(\mu^*, \nu^*)$ as

$$d^*(\mu^*, \nu^*) = \frac{1}{a^2}\left[C(\mu^*)\right]' \Delta \left[C(\nu^*)\right] \tag{3.19}$$

$$= \frac{1}{a^2}\left[C(\mu^*)\right]' \left(cJ - \mathbf{T}'\mathbf{T}\right)\left[C(\nu^*)\right] \tag{3.20}$$

$$= c - \mathcal{A}^*(\mu^*, \nu^*)$$

where

$$\mathcal{A}^*(\mu^*, \nu^*) = \frac{1}{a^2}\left[C(\mu^*)\right]' \mathbf{T}'\mathbf{T}\left[C(\nu^*)\right].$$

The latter may be viewed as the average of the $\mathcal{A}(\mu_i, \nu_j)$ taken over all complete rankings $\mu_i, \nu_j$ compatible with $\mu^*$ and $\nu^*$, respectively.

## 3.4 Correlation for Incomplete Rankings

At this point it is useful to derive an expression for an incomplete ranking $\mu^*$ given knowledge of its compatibility class $C(\mu^*)$. We shall assume that each complete ranking has the same probability of being selected, i.e., they are uniformly distributed over the $t!$ permutations of $(1, 2, \ldots, t)$.

**Lemma 3.2.** *The conditional distribution of the rank $\mu(i)$ given the compatibility class $C(\mu^*)$ generated by $\mu^*$ is given by*

$$P\{\mu(i) = j|C(\mu^*)\} = \binom{j-1}{\mu^*(i)-1}\binom{t-j}{k-\mu^*(i)}\binom{t}{k}^{-1}\delta(i) + \frac{1}{t}(1-\delta(i))$$

*where $\delta(i)$ is either 1 or 0 depending on whether the object $i$ is or is not ranked in the incomplete ranking. Here $\mu^*(i) \leq j \leq (t-k) + \mu^*(i)$, if object $i$ is ranked whereas $1 \leq j \leq t$, if object $i$ is not ranked.*

*Proof.* If an object $i$ is ranked in an incomplete ranking $\mu^*$ of $k$ objects, then the number of complete rankings compatible with $\mu^*$ which assign rank $j$ to object $i$ is

$$\binom{j-1}{\mu^*(i)-1}\binom{t-j}{k-\mu^*(i)}(t-k)!$$

This consists of the number of ways of picking a set of $(\mu^*(i) - 1)$ from the first $(j-1)$ integers and a set of $(k - \mu^*(i))$ from the last $(t-j)$ integers while allowing all possible permutations of the $(t-k)$ integers not picked. On the other hand, if object $i$ is not ranked in $\mu^*$ then the number of such complete compatible rankings is given by

$$\binom{t-1}{k}(t-k-1)!$$

the number of ways of picking $k$ from the $t-1$ integers not equal to $j$ and allowing all possible permutations of the remaining $(t-k-1)$ integers. Dividing these by $\frac{t!}{k!}$ the number of complete rankings compatible with $\mu^*$ gives the result. □

In the next lemma, we show that it is possible to compute the value of a score function corresponding to an incomplete ranking from knowledge of the compatibility class. To this end, we make use of the conditional distribution of a complete ranking given its compatibility class and the fact that the conditional expectation of the score function corresponds to its projection onto that class. We apply this approach to compute the form of score functions for both the Spearman and Kendall distances.

**Lemma 3.3.** *Suppose that we select a complete ranking $\mu$ at random from the class of compatible rankings $C(\mu^*)$. Suppose that object s is ranked. Then (a)*

$$E\left[\left(\mu(s)-\frac{t+1}{2}\right)\mid \mathcal{C}(\mu^*)\right]=\frac{t+1}{k+1}\left(\mu^*(s)-\frac{k+1}{2}\right), \tag{3.21}$$

*and (b) for any pair of objects $i<j$,*

$$E\left[sgn\left(\mu(j)-\mu(i)\right)\mid \mathcal{C}(\mu^*)\right]=a(i,j), \tag{3.22}$$

*where*

$$a(i,j)=\begin{cases} sgn(\mu^*(j)-\mu^*(i)) & \textit{if both objects } i \textit{ and } j \textit{ are ranked}\\ 1-\frac{2\mu^*(i)}{(k+1)} & \textit{if only object } i \textit{ is ranked}\\ \frac{2\mu^*(j)}{(k+1)}-1 & \textit{if only object } j \textit{ is ranked}\\ 0 & \textit{otherwise}\end{cases} \tag{3.23}$$

*Proof.* To prove (a), recall the identity

$$\sum_{j=l}^{t-k+l}\binom{j-1}{l-1}\binom{t-j}{k-l}=\binom{t}{l}. \tag{3.24}$$

Consequently, we have that

$$\begin{aligned}E\left[\left(\mu(s)-\frac{t+1}{2}\right)\mid \mathcal{C}(\mu^*)\right]&=\sum_{j=\mu^*(s)}^{t-k+\mu^*(s)}\left(j-\frac{t+1}{2}\right)\binom{j-1}{\mu^*(s)-1}\binom{t-j}{k-\mu^*(s)}\Big/\binom{t}{l}\\&=\frac{t+1}{k+1}\left(\mu^*(s)-\frac{k+1}{2}\right).\end{aligned}$$

For the proof of (b), let

$$\delta(s,j)=\begin{cases}1 & \text{if judge } j \text{ ranks object } s\\ 0 & \text{otherwise}\end{cases}$$

and define

$$\varpi_j(s)=\mu_j^*(s)\,\delta(s,j)+\left(\frac{k+1}{2}\right)(1-\delta(s,j)) \tag{3.25}$$

so that the incomplete ranking takes value $\frac{k+1}{2}$ when an object is unranked. Note that for any complete ranking,

$$\mu(j)=\frac{t+1}{2}+\frac{1}{2}\sum_{i=1}^{t}sgn\left(\mu(j)-\mu(i)\right). \tag{3.26}$$

It is clear that if objects $i$ and $j$ are both ranked, then $a(i, j)$ is as stated. Suppose now only object $j$ is ranked. The adjusted score becomes on using (3.21)

$$E\left[\mu(j) \mid C(\mu^*)\right] = \frac{t+1}{2} + \frac{1}{2}E\left[\sum_{i=1}^{t} sgn\left(\mu(j) - \mu(i)\right) \mid C(\mu^*)\right]$$

$$\frac{t+1}{k+1}\mu^*(j) = \frac{t+1}{2} + \frac{1}{2}\sum_{i=1}^{k} sgn\left(\mu^*(j) - \mu^*(i)\right) + \frac{(t-k)}{2}a(i, j)$$

$$= \frac{t+1}{2} + \left(\mu^*(j) - \frac{k+1}{2}\right) + \frac{(t-k)}{2}a(i, j).$$

Hence, $a(i, j) = \left(\frac{2\mu^*(j)}{k+1} - 1\right)$. The case where only object $i$ is ranked is dealt with similarly. □

In describing visualization techniques for incomplete ranking data, Kidwell et al. (2008) have noted the efficiency for computing the Kendall scores in (3.23). Next, we proceed to find the maximum and minimum distances when only $k$ objects are ranked among the incomplete rankings.

**Lemma 3.4.** *(a) For the Spearman distance,*

$$m_S^* = c_S - \frac{(t+1)^2}{12}\frac{k(k-1)}{(k+1)}, M_S^* = c_S + \frac{(t+1)^2}{12}\frac{k(k-1)}{(k+1)}$$

*where* $c_S = \frac{t(t^2-1)}{12}$.

*(b) For the Kendall distance,*

$$m_K^* = c_K - \frac{(2t+k+3)}{6}\frac{k(k-1)}{(k+1)}, M_K^* = c_K + \frac{(2t+k+3)}{6}\frac{k(k-1)}{(k+1)}$$

*where* $c_K = \frac{t(t-1)}{2}$. *It follows that the correlation between the incomplete rankings* $\mu_1^*, \mu_2^*$ *can be defined to be*

$$\alpha\left(\mu_1^*, \mu_2^*\right) = 1 - \frac{2\left[d_K^*\left(\mu_i^*, \mu_j^*\right) - m^*\right]}{M^* - m^*}. \tag{3.27}$$

*Proof.* The right-hand side of (3.21) provides a general expression for an incomplete ranking. It follows that the Spearman distance between two incomplete rankings with the same number of ranked objects is

$$d_S^*\left(\mu_i^*, \mu_j^*\right) = \frac{t(t+1)(2t+1)}{6} - \left(\frac{t+1}{k+1}\right)^2 \sum_{s=1}^{t} \varpi_i(s)\,\varpi_j(s)$$

and in the Kendall case, the distance may be written as

$$d_K^*\left(\mu_i^*, \mu_j^*\right) = \frac{t(t-1)}{2} - \sum_{q_1<q_2} a_i\left(q_1, q_2\right) a_j\left(q_1, q_2\right)$$

where $a_i\left(q_1, q_2\right)$is defined as in (3.23) and $\varpi_i\left(s\right)$ is given in 3.25. An application of the Cauchy–Schwarz inequality indicates that the upper bound of the Spearman distance occurs when $\mathbf{T}_S C\left(\mu_i^*\right) = -\mathbf{T}_S C\left(\mu_j^*\right)$ whereas the lower bound is achieved when $\mathbf{T}_S C\left(\mu_i^*\right) = \mathbf{T}_S C\left(\mu_j^*\right)$. If we let $\mu_j^*$ be the inverted ranking, that is, $\mu_j^*\left(s\right) = k+1-\mu_i^*\left(s\right)$ when object $s$ is ranked by $i$, then $\varpi_j\left(s\right) = k+1-\varpi_i\left(s\right)$ and $\mathbf{T}_S C\left(\mu_i^*\right) = -\mathbf{T}_S C\left(\mu_j^*\right)$. Furthermore, for the Kendall scores, $a_j\left(q_1, q_2\right) = -a_i\left(q_1, q_2\right)$ and thus $\mathbf{T}_K C\left(\mu_i^*\right) = -\mathbf{T}_K C\left(\mu_j^*\right)$. A straightforward calculation of these distances using the incomplete ranking $(1, 2, \ldots, k, -, -, \ldots, -)'$ and its inversion yields the minimum and maximum for each distance. □

We quote without proof a result in Alvo and Cabilio (1995a) which allows for different numbers of observations missing at random.

**Lemma 3.5.** *For fixed $k_1 \leq k_2$ suppose the pattern of missing observations is randomly selected from the set of all possible patterns. Then, for the Spearman and Kendall cases, the minimum and maximum values of the distance are of the form*

$$m^* = c - \gamma\left(i\right), \; M^* = c + \gamma\left(i\right)$$

*where the $\gamma\left(i\right)$ are given as*

$$\gamma_S\left(1\right) = \frac{(t+1)^2\left(k_1-1\right)\left(3k_2-k_1\right)}{24\left(k_2+1\right)}, \; k_1\, odd$$

$$\gamma_S\left(2\right) = \frac{(t+1)^2 k_1\left(k_1\left(3k_2-k_1\right)-2\right)}{24\left(k_1+1\right)\left(k_2+1\right)}, \; k_1\, even$$

$$\gamma_K\left(1\right) = \frac{\left(k_1-1\right)\left(t\left(3k_2-k_1\right)+k_2\left(k_1+3\right)\right)}{6\left(k_2+1\right)}, \; k_1\, odd$$

$$\gamma_K\left(2\right) = \frac{k_1\left(3k_1k_2\left(t+1\right)-\left(k_1^2+2\right)\left(t-k_2\right)-3\left(k_2+1\right)\right)}{6\left(k_1+1\right)\left(k_2+1\right)}, \; k_1\, even$$

Consider now two independent rankings of length $k_1, k_2$, respectively, with $2 \leq k_1 \leq k_2 \leq t$. It follows from (3.6) and Lemma 3.3 that

$$\mathcal{A}_S^*(\mu^*, \nu^*) = E\left[\mathcal{A}_S(\mu, \nu) \mid \mathcal{C}(\mu^*), \mathcal{C}(\nu^*)\right] \tag{3.28}$$

$$= \frac{(t+1)^2}{\left(k_1+1\right)\left(k_2+1\right)} \sum_{s=1}^{t} \left(\mu^*\left(s\right) - \frac{k_2+1}{2}\right)\left(\nu^*\left(s\right) - \frac{k_1+1}{2}\right)\delta\left(s, \mu^*\right)\delta\left(s, \nu^*\right)$$

$$= \frac{(t+1)^2}{\left(k_1+1\right)\left(k_2+1\right)} \sum_{i=1}^{k^*} \left(o_i - \frac{k_2+1}{2}\right)\left(\mu^*\left(o_i\right) - \frac{k_1+1}{2}\right) \tag{3.29}$$

**Table 3.3** Language and arithmetic scores revisited

| Student | 1 | 2 | 3 | 4 | 5 | 6 | 7 | 8 | 9 |
|---|---|---|---|---|---|---|---|---|---|
| Arithmetic (2) | 14 | 18 | 23 | 26 | 27 | 30 | 40 | – | – |
| Language (1) | 28 | 14 | 46 | – | 53 | – | 54 | 50 | – |
| Ranking (2) | 1 | 2 | 3 | 4 | 5 | 6 | 7 | – | – |
| Ranking (1) | 2 | 1 | 3 | – | 5 | – | 6 | 4 | – |

where $k^*$ is the number of objects ranked in ranking 1 among the $k_2$ objects ranked in ranking 2 and $o_i$ is the label of the $i$th object ranked in ranking 1. Here, $\delta(s, \mu^*)$ takes value 1 if object s is ranked by $\mu^*$ and value 0 otherwise. Note that

$$o_i = i + l_i,$$

where $l_i$ = number of objects unranked in ranking 1 which are to the left of the object being ranked. Similarly from (3.7) we have that

$$\mathcal{A}_K^*(\mu^*, \nu^*) = E\left[\mathcal{A}_K(\mu, \nu) \mid \mathcal{C}(\mu^*), \mathcal{C}(\nu^*)\right] \tag{3.30}$$

$$= \sum_{i<j} a_1(i, j)\, a_2(i, j)\,. \tag{3.31}$$

*Example 3.5.* Consider the test scores in Language (ranking 1) and Arithmetic (ranking 2) of a group of nine students in Table 3.3. The original data was altered by removing certain values, with the remaining observations reordered and ranked as follows.

Here $t = 9, k_1 = 6, k_2 = 7, k^* = 5, o_1 = 1, o_2 = 2, o_3 = 3, o_4 = 5, o_5 = 7, o_6 = 8$, and $l_1 = l_2 = l_3 = 0, l_4 = 1, l_5 = l_6 = 2$. Further,

$$\mu^*(o_1) = 2, \mu^*(o_2) = 1, \mu^*(o_3) = 3, \mu^*(o_4) = 5, \mu^*(o_5) = 6, \mu^*(o_6) = 4.$$

Hence $A_S^* = 33.9286$ and $A_K^* = 4$.

### 3.4.1 Asymptotic Normality of the Spearman and Kendall Test Statistics

The main objective of this section is to demonstrate the asymptotic normality of the similarity measures due to Spearman and Kendall in the case of incomplete rankings. Specifically, we shall be concerned with the asymptotic distributions of both $\mathcal{A}_S^*$,$\mathcal{A}_K^*$ under each of two possible null hypotheses $H_1$ and $H_2$. For both hypotheses we assume that $k_1, k_2$, the number of ranked observations, are fixed and the rankings for which we have (possibly) incomplete data are uniformly distributed over the $t!$ permutations of $(1, 2, \ldots, t)$.

- Under hypothesis $H_1$, we assume that the pattern of missing observations is fixed, so that all inference in this case is conditional on such a pattern.
- Under $H_2$, we assume that the patterns of missing observations are randomly selected from the set of all possible patterns. The latter situation would arise in practice if unranked objects occur by chance. An example would be testing for trend in water quality data when the historical data is incomplete.

We begin with the definition of a linear rank statistic.

**Definition 3.3.** Let $\{a(i)\}$ and $\{c(i)\}$ be two sets of constants. A statistic of the form

$$S = \Sigma_{i=1}^{N} c(i)\, a(R_i)$$

where $R = (R_1, \ldots, R_N)$ is a vector of ranks is called a linear rank statistic. The constants $a(i)$ are called scores whereas the $c(i)$ are called regression coefficients.

Many test statistics are of this form. For example, suppose that we have a random sample of n observations from a population and N-n from another. We are interested in testing the null hypothesis that the two populations are the same against the alternative that they differ only in location. Rank all N observations together. The Wilcoxon statistic then considers only the ranks of one of the populations by choosing

$$c(i) - \begin{cases} 0 & i = 1, \ldots, n \\ 1 & i = n+1, \ldots, N. \end{cases}$$

**Lemma 3.6.** *Suppose that R is uniformly distributed over the set of permutations in $\mathcal{P}$. Then*

*(i) for* $i = 1, \ldots, N$, $E(R_i) = \frac{N+1}{2}$, $Var(R_i) = \frac{(N^2-1)}{12}$ *and for* $i \neq j$, $Cov(R_i, R_j) = -\frac{N+1}{12}$ *and*

*(ii)*

$$ES = N\bar{c}\bar{a}$$

*and*

$$Var\, S = \frac{1}{N-1} \Sigma (c(i) - \bar{c})^2\, \Sigma (a(i) - \bar{a})^2$$

*where $\bar{a}$ and $\bar{c}$ represent the corresponding means.*

*Proof.* The proof of this lemma is given in (Hájek and Sidak 1967). □

The following theorem states that under certain conditions, linear rank statistics are asymptotically normally distributed. We shall consider square integrable functions $\phi$ defined on $(0, 1)$ which have the property that they can be written as the difference of two nondecreasing functions and satisfy

$$0 < \int_0^1 \left[\phi(u) - \bar{\phi}\right]^2 du < \infty$$

where $\bar{\phi} = \int_0^1 \phi(u)\, du$.

**Theorem 3.1.** *Suppose that $R$ is uniformly distributed over the set of permutations in $\mathcal{P}$. Let the score function be given by $a(i) = \phi\left(\frac{i}{N}\right)$ where $\phi()$ is a square integrable score function. Then $S$ is asymptotically normally distributed as $N \to \infty$ with mean $N\bar{c}\bar{a}$ and variance*

$$Var\ S = \frac{1}{N-1} \Sigma_{i=1}^N (c(i) - \bar{c})^2\, \Sigma_{i=1}^N (a(i) - \bar{a})^2$$

*provided*

$$\frac{\sum_{i=1}^N (c(i) - \bar{c})^2}{\max_{1 \le i \le N} (c(i) - \bar{c})^2} \to \infty.$$

*Proof.* The proof of this important result is given in (Hájek and Sidak 1967). □

We may now apply Theorem 3.1 to obtain the asymptotic normality of the Spearman test statistic in the case of incomplete rankings under Hypothesis 1 wherein the pattern of missing data is fixed. Set

$$\sigma_S^2 = \frac{1}{12}\left[\frac{(t+1)^2}{(k_2+1)}\right]^2 \sum_{i=1}^{k_1} \left(o_i^* - \bar{o}_1\right)^2, \tag{3.32}$$

where

$$o_i^* = \begin{cases} o_i & \text{if } 1 \le i \le k^* \\ \frac{k_2+1}{2} & \text{if } k^* + 1 \le i \le k_1 \end{cases} \tag{3.33}$$

and $\bar{o}_1 = \left(\sum_{i=1}^{k_1} o_i^*\right)/k_1$. Also set $\bar{o}^* = \left(\sum_{i=1}^{k^*} o_i\right)/k^*$.

**Theorem 3.2.** *Assume that $k^* \to \infty$ (and hence $k_1 \to \infty, k_2 \to \infty, t \to \infty$) with $k^*/t \to \lambda > 0$, where $\lambda$ is a finite constant. Then, under $H_1$, whereby the pattern of missing data is fixed, $\mathcal{A}_S^*$ given in (3.28) is asymptotically normal with mean 0 and variance $\sigma_S^2$.*

*Proof.* The proof hinges on the fact that $\mathcal{A}_S^*$ is a linear rank statistic. In fact

$$\mathcal{A}_S^* = \frac{(t+1)^2}{(k_1+1)(k_2+1)} \sum_{i=1}^{k_1} \left(o_i^* - \frac{k_2+1}{2}\right)\left(\mu^*(o_i) - \frac{k_1+1}{2}\right)$$

$$= \frac{(t+1)^2}{(k_1+1)(k_2+1)} \sum_{i=1}^{k_1} \left(o_i^* - \bar{o}_1\right)\left(\mu^*(o_i)\right).$$

The normality follows provided

$$\frac{\sum_{i=1}^{k_1} \left(o_i^* - \bar{o}_1\right)^2}{\max\left(o_i^* - \bar{o}_1\right)^2} \to \infty.$$

Now

$$\Sigma_{i=1}^{k_1} \left(o_i^* - \bar{o}_1\right)^2 = \Sigma_{i=1}^{k^*} \left(o_i^* - \bar{o}_1\right)^2 + k^*\left(\bar{o}^* - \bar{o}_1\right)^2 + \left(k_1 - k^*\right)\left(\frac{k_2+1}{2} - \bar{o}_1\right)^2$$

$$\geq k^*\left(k^{*2} - 1\right)/12.$$

Further, $\left(o_i^* - \bar{o}_1\right)^2 \leq (t-1)^2$, so that the result follows on letting $k^* \to \infty$ with $k^*/t \to \lambda$. □

The exact variance of $\mathcal{A}_S^*$ under $H_1$, which is recommended in applications of Theorem 3.2, is related to $\sigma_S^2$ by

$$Var(\mathcal{A}_S^*) = \frac{k_1}{k_1+1}\sigma_S^2$$

(Lehmann 1975 (A. 49) p. 334). That is, the asymptotic variance given in the theorem is essentially the actual variance of $\mathcal{A}_S^*$. In any application, the calculation of the variance of $\mathcal{A}_S^*$ is a straightforward computation. Next, we consider the asymptotic distribution of $\mathcal{A}_S^*$ and $\mathcal{A}_K^*$ when the pattern of missing observations is random.

**Theorem 3.3.** *Let $k_1 \to \infty$ (and hence $k_2 \to \infty, t \to \infty$) with $k_1/t \to \lambda > 0$, where $\lambda$ is a finite constant. Then, under $H_2$, whereby the pattern of missing data is random, $\mathcal{A}_S^*$ is asymptotically normal with mean 0 and variance*

$$Var\left(\mathcal{A}_S^*\right) = \frac{(t+1)^4}{144(t-1)}\kappa_1\kappa_2, \tag{3.34}$$

*with*

$$\kappa_i = \frac{k_i(k_i-1)}{(k_i+1)}, i = 1, 2.$$

*Proof.* Define $\mathbf{U} = (U_1, U_2, \ldots, U_t)$ as the random vector uniformly distributed over the permutations of $(1, 2, \ldots, k_1, \frac{k_1+1}{2}, \ldots, \frac{k_1+1}{2})$. In this case, the extended Spearman distance may be written as

$$d_S^* = \frac{t(t+1)(2t+1)}{6} - \mathcal{A}_S^* \tag{3.35}$$

$$= \frac{(t+1)^2}{(k_1+1)(k_2+1)} \left[ \sum_{i=1}^{k_2} i U_i + \frac{k_2+1}{2} \sum_{i=k_2+1}^{t} U_i \right]. \tag{3.36}$$

The result follows from the combinatorial central limit theorem of Hoeffding (see Appendix B.1) applied to the quantity within square brackets above. □

**Theorem 3.4.** *$\mathcal{A}_K^*$ is asymptotically equivalent to $\mathcal{A}_S^*$ under both hypotheses $H_1$ and $H_2$. Hence, $\mathcal{A}_K^*$ is asymptotically normal with mean 0 and variance $\left(\frac{16}{t^2}\right) Var\left(\mathcal{A}_S^*\right)$.*

*Proof.* We know from (Hájek and Sidak 1967) that for the complete case

$$E\left(\mathcal{A}_K - \frac{4}{t}\mathcal{A}_S\right)^2 = \frac{(t-1)(t-2)}{18}$$

and that, moreover,

$$\frac{12\mathcal{A}_S}{t(t+1)\sqrt{t-1}} \Rightarrow N(0,1) \text{ as } t \to \infty.$$

Consequently, we have

$$\frac{6\mathcal{A}_K}{\sqrt{2t(t-1)(2t+5)}} \Rightarrow N(0,1).$$

From Jensen's inequality

$$E\left(\mathcal{A}_K^* - \frac{4}{t}\mathcal{A}_S^*\right)^2 = E\left(E^2\left(\left(\mathcal{A}_K - \frac{4}{t}\mathcal{A}_S\right) | \mathcal{C}(\mu^*), \mathcal{C}(\nu^*)\right)\right)$$

$$\leq E\left(E\left(\mathcal{A}_K - \frac{4}{t}\mathcal{A}_S\right)^2 | \mathcal{C}(\mu^*), \mathcal{C}(\nu^*)\right) = o\left(t^2\right)$$

and consequently the asymptotic normality of $\mathcal{A}_S^*$ will imply the asymptotic normality of $\mathcal{A}_K^*$. □

*Example 3.6.* We return to Example 3.2 wherein we wish to test the hypothesis of independence against the alternative of a positive correlation. For the complete data, the value of $\mathcal{A}_S$ is 41, and from the tables, under the randomness hypothesis,

$P(\mathcal{A}_S \geq 41) = 0.0252$, whereas the use of the asymptotic result gives a p-value of $1 - \Phi(1.9328) = 0.0266$, where $\Phi$ is the cumulative distribution function of a standard normal. For the data in Example 3.5, the value of $\mathcal{A}_S^*$ for the reduced data is calculated to be 33.9286. An application of the theorem yields that under $H_1$, the p-value is $P\left(\mathcal{A}_S^* \geq 33.9286\right) = 0.0178$. On the other hand, if all observations with missing values are deleted, we obtain a reduced value of $\mathcal{A}_S = 9$ with $t = 5$, and from the tables $P(\mathcal{A}_S \geq 9) = 0.0417$.

### 3.4.2 Asymptotic Efficiency

We now turn to the question of the efficiency which is further discussed in Appendix B.4. Let $X_1, X_2, \ldots, X_t$ be independent random variables whose joint density under the alternative is described by

$$q_d = \prod_{i=1}^{t} f_0\left(x_i - d_i\right)$$

where $f_0$ is a known density having finite Fisher information $I\left(f_0\right)$ and $\mathbf{d} = (d_1, d_2, \ldots, d_t)$ is an arbitrary vector. In the notation of our tests, $k_2 = t$, and write $k_1 = k$, the actual number of $X_i$'s observed. Recalling that $o_i$ is the label of the $i$th object ranked, the Spearman test which deletes all missing observations is based on the Spearman correlation of the reduced sample of $k$ pairs, and the test statistic may be written as

$$A_{RS} = (t+1)\sum_{i=1}^{k}\left(i - \frac{k+1}{2}\right)\left(\frac{\mu^*\left(o_i\right)}{t+1}\right).$$

Since $k = k_1 = k^*$ and consequently $o_i = o_i^*$, the statistic $\mathcal{A}_S^*$ may be written as

$$\mathcal{A}_S^* = \frac{(t+1)}{(k+1)}\sum_{i=1}^{k}\left(o_i - \frac{t+1}{2}\right)\left(\mu^*\left(o_i\right) - \frac{k+1}{2}\right).$$

Hence,

$$\mathcal{A}_S^* = \frac{(t+1)}{(k+1)}\left\{A_{RS} + \sum_{i=1}^{k}\left(\mu^*\left(o_i\right) - \frac{k+1}{2}\right)\left(o_i - i\right)\right\}.$$

The weight $(o_i - i)$ represents the number of time points to the left of $o_i$ for which there are no observations. Similarly,

$$\mathcal{A}_K^* = A_{RK} + \frac{4}{k+1}\sum_{i=1}^{k}\left(\mu^*\left(o_i\right) - \frac{k+1}{2}\right)\left(o_i - i\right)$$

where

$$A_{RK} = \sum_{i<j}^{k} sgn\left(\mu^*\left(o_j\right) - \mu^*\left(o_i\right)\right).$$

Set $d_i^* = d_{o_i}$ and $\overline{d} = \sum_{i=1}^{t} d_i/t$. Under the alternative $q_d$, provided

$$\max_{1\le i\le t}\left(d_i - \overline{d}\right)^2 \to 0 \text{ and } I\left(f_0\right)\sum_{i=1}^{t}\left(d_i - \overline{d}\right)^2 \to b^2, 0 < b^2 < \infty,$$

both $A_{RS}$ and $\mathcal{A}_S^*$ are asymptotically normal with means and variances given respectively by $\left(\mu_R, \sigma_{RS}^2\right)$ and $\left(\mu_S, \sigma_S^2\right)$, where

$$\mu_{RS} = (t+1)\sum_{i=1}^{k}\left(i - \frac{k+1}{2}\right)\left(d_i^* - \overline{d}\right)\int_0^1 u\,\phi\left(u, f_0\right)du$$

$$\mu_S = \frac{(t+1)^2}{(k+1)}\sum_{i=1}^{k}\left(o_i - \overline{o}\right)\left(d_i^* - \overline{d}\right)\int_0^1 u\,\phi\left(u, f_0\right)du.$$

$$\sigma_{RS}^2 = \frac{(t+1)^2}{12}\sum_{i=1}^{k}\left(i - \frac{k+1}{2}\right)^2, \quad \sigma_S^2 = \frac{(t+1)^4}{12\,(k+1)^2}\sum_{i=1}^{k}\left(o_i - \overline{o}\right)^2.$$

Here $\phi(u, f) = \left[f'\left(F^{-1}(u)\right)\right]/\left[f\left(F^{-1}(u)\right)\right], 0 < u < 1$, and $F$ is the cumulative distribution of $f$.

Shifting now to the efficiencies, it is seen that the asymptotic efficiencies as $k \to \infty$, for $A_{RS}$ and $\mathcal{A}_S^*$ are respectively given by

$$e_{RS} = \lim \frac{\left[\sum_{i=1}^{k}\left(i - \frac{k+1}{2}\right)\left(d_i^* - \overline{d}\right)\right]^2}{\sum_{i=1}^{k}\left(i - \frac{k+1}{2}\right)^2 \sum_{i=1}^{t}\left(d_i - \overline{d}\right)^2} Q_1$$

$$e_S = \lim \frac{\left[\sum_{i=1}^{k}\left(o_i - \overline{o}\right)\left(d_i^* - \overline{d}\right)\right]^2}{\sum_{i=1}^{k}\left(o_i - \overline{o}\right)^2 \sum_{i=1}^{t}\left(d_i - \overline{d}\right)^2} Q_1,$$

where $Q_1$ is a positive function of $f_0$ and the limit is taken as $t \to \infty, k \to \infty$, with $k/t \to \lambda > 0$. The asymptotic relative efficiency of $\mathcal{A}_S^*$ relative to $A_{RS}$ is then given by the ratio $e_S/e_{RS}$ (Appendix B.4).

Now consider the case where $d_i^* = o_i, \bar{d} = \bar{o}, i = 1, \ldots, k$ and the remaining $d_i$ are arbitrary, a situation which includes alternatives of the form $EX_i = \beta_0 + \beta i, \beta > 0$. It can be shown that irrespective of the density $f_0$, the asymptotic relative efficiency of $\mathcal{A}_S^*$ relative to $A_{RS}$ is given by

$$ARE\left(\mathcal{A}_S^*, A_{RS}\right) = \lim_{k\to\infty} R\left(k, \mathbf{o_k}\right),$$

where $\mathbf{o_k} = (o_1, \ldots, o_k)$ and

$$R(k, \mathbf{o_k}) = \frac{\sum_{i=1}^{k} \left(i - \frac{k+1}{2}\right)^2 \sum_{i=1}^{k} (o_i - \overline{o})^2}{\left[\sum_{i=1}^{k} \left(i - \frac{k+1}{2}\right)(o_i - \overline{o})\right]^2} \geq 1.$$

Note that $R(k, \mathbf{o_k}) > 1$ unless the $o_i$'s are equally spaced.

In order to illustrate the magnitude of this efficiency, suppose for example that $t = 19, k = 7, o_1 = 1, o_2 = 2, o_3 = 3, o_4 = 10, o_5 = 17, o_6 = 18, o_7 = 19,$ then the ratio of the efficacies of $A_S^*$ to $A_{RS}$ is 1.086. On the other hand, if $o_1 = 1,$ $o_2 = 8, o_3 = 9, o_4 = 10, o_5 = 11, o_6 = 12, o_7 = 19$, then that ratio is 1.176.

## 3.5 Tied Rankings and the Notion of Compatibility

The notion of compatibility may also be extended to deal with tied rankings. As an example, suppose that objects 1 and 2 are equally preferred whereas object 3 is least preferred. Such a ranking would be compatible with the rankings $(1, 2, 3)$ and $(2, 1, 3)$ in that both are plausible. The average of the rankings in the compatibility class, which as we shall see results from the use of the Spearman distance, will then be the ranking

$$\frac{1}{2}[(1, 2, 3) + (2, 1, 3)] = (1.5, 1.5, 3)$$

to be presented in this case. It is seen that the notion of compatibility serves to justify the use of the midrank when ties exist. Formally we can define tied orderings as follows.

**Definition 3.4.** A tied ordering of t objects is a partition into $e$ sets, $1 \leq e \leq t$, each containing $d_i$ objects, $d_1 + d_2 + \ldots + d_e = t$, so that the $d_i$ objects in each set share the rank $i$,$1 \leq i \leq e$. Such a tie pattern is denoted by $\delta = (d_1, d_2, \ldots, d_e)$. The ranking denoted by $\mu_\delta = (\mu_\delta(1), \ldots, \mu_\delta(t))$ resulting from such an ordering is a tied ranking and is one of $\frac{t!}{d_1!d_2!\ldots d_e!}$ possible permutations.

Associated with every tied ranking we may define a $t! \times (\frac{t!}{d_1!d_2!\ldots d_e!})$ matrix of compatibility $D_\delta$. Yu et al. (2002) considered the problem of testing for independence between two random variables when the tie patterns and the pattern of missing observations are fixed. Specifically, let $\mu^*$ be an incomplete ranking of $k_1$ out of $t$ objects with tie pattern $\delta_1 = (d_{11}, \ldots, d_{1e_1})$. Similarly, let $\nu^*$ be an incomplete ranking of $k_2$ out of $t$ objects with tie pattern $\delta_2 = (d_{21}, \ldots, d_{2e_2})$. The Spearman similarity measure between two incomplete rankings $\mu^*, \nu^*$ is defined to be

$$A_S^* = \frac{(t+1)^2}{(k_1+1)(k_2+1)} \sum_{j=1}^{t} \delta(j) \left[\mu^*(j) - \frac{k_1+1}{2}\right]\left[\nu^*(j) - \frac{k_2+1}{2}\right]$$

where $\delta(j) = 1$ if both rankings of object j are not missing and 0 otherwise.

**Table 3.4** Data from the public opinion survey

| | Response | | | | | | |
|---|---|---|---|---|---|---|---|
| Education level | 1 | 2 | 3 | 4 | 5 | Missing | Subtotal |
| Primary or below | 2 | 35 | 23 | 7 | 3 | 33 | 103 |
| Secondary | 2 | 72 | 129 | 37 | 6 | 53 | 299 |
| Matriculated | 0 | 9 | 9 | 7 | 0 | 3 | 28 |
| Tertiary, nondegree | 1 | 9 | 6 | 6 | 0 | 5 | 27 |
| Tertiary, degree | 0 | 22 | 28 | 7 | 6 | 6 | 69 |
| Missing | 0 | 2 | 3 | 0 | 0 | 1 | 6 |
| Subtotal | 5 | 149 | 198 | 64 | 15 | 101 | 532 |

**Theorem 3.5.** *Let $k^*$ be the number of objects ranked in ranking 1 among the $k_2$ objects ranked in ranking 2. Let $2 \leq k_1 \leq k_2 \leq t$. Assume that*

*(i)* $k^* \to \infty$, *(and hence* $k_1 \to \infty, k_2 \to \infty, t \to \infty$*) with* $k^*/t \to \lambda > 0$.
*(ii)* $\max_{j=1,\cdots,e_1} \frac{g_{1j}}{k^*}$ *is bounded away from 1.*
*(iii)* $\max_{j=1,\cdots,e_2} \frac{g_{2j}}{k^*}$ *is bounded away from 1.*

*Then, under the null hypothesis of independence whereby the pattern of ties and missing data is fixed, $A_S^*$ is asymptotically normal with mean 0 and exact variance*

$$Var\left(A_S^*\right) = \left[\frac{(t+1)^2 k_1}{(k_1+1)(k_2+1)}\right]^2 \frac{\sum_{j=1}^{k_1}\left(o_j^* - \bar{o}\right)^2}{12}\left\{1 - \frac{\sum_{j=1}^{e_1}\left(g_{1j}^3 - g_{1j}\right)}{k_1^3 - k_1}\right\}.$$

*Proof.* See Yu et al. (2002). □

*Example 3.7.* In a public opinion survey held in 1999 in Hong Kong, it was of interest to determine whether the education level of the respondents is related to the level of dissatisfaction of the Policy Address of the Chief Executive of the Hong Kong Special Administrative Region. The response is an ordinal variable having seven options as follows: (1), very satisfied; (2), satisfied; (3), neutral; (4), unsatisfied; (5), very unsatisfied; (6), not sure; and (7), refuse to answer. Options (6) and (7) were combined and listed as "missing." Table 3.4 displays the frequencies of the respondents listed by option and by education level.

It is noted that about 19.9 % of the respondents did not respond either to one or to both questions. Moreover, since the education levels are grouped into a few categories, the problem of ties cannot be ignored. One alternative approach for analyzing this data is as a contingency table. In that case, however, the ordering among the education levels and separately among the responses would not be taken into account. The results of the analysis shown in Table 3.5 reveal that at the 5 % significance level, the test based on the reduced sample (which discards all observations with at least one missing variable) cannot reject the hypothesis of

**Table 3.5** Results of the analyses

| Test | Statistic | Standardized statistic | p-value |
|---|---|---|---|
| Reduced sample | 494,132.0 | 1.9075 | 0.0564 |
| Complete sample | 786,633.2 | 1.9690 | 0.0490 |

**Table 3.6** Wind direction in degrees

| | | | | | | | | | | | |
|---|---|---|---|---|---|---|---|---|---|---|---|
| 6 a.m. | 356 | 97 | 211 | 262 | 343 | 292 | 157 | 302 | 324 | 85 | 324 |
| Noon | 119 | 162 | 221 | 259 | 270 | 29 | 97 | 292 | 40 | 313 | 94 |
| 6 a.m. | 85 | 324 | 340 | 157 | 238 | 254 | 146 | 232 | 122 | 329 | |
| Noon | 45 | 47 | 108 | 221 | 248 | 270 | 45 | 23 | 270 | 119 | |
| Data replaced by their ranks | | | | | | | | | | | |
| 6 a.m. | 21 | 3 | 8 | 12 | 20 | 13 | 6.5 | 14 | 16 | 1.5 | 16 |
| Noon | 10.5 | 12 | 13.5 | 16 | 18 | 2 | 8 | 20 | 3 | 21 | 7 |
| 6 a.m. | 1.5 | 16 | 19 | 6.5 | 10 | 11 | 5 | 9 | 4 | 18 | |
| Noon | 4.5 | 6 | 9 | 13.5 | 15 | 18 | 4.5 | 1 | 18 | 10.5 | |

independence whereas the one based on the complete sample can. Since the test statistic is positive, this implies that there is a positive association between education level and level of dissatisfaction. More highly educated respondents tend to be less satisfied with the Policy Address. The analysis by means of a contingency table whereby the missing categories for education and response were dropped leads to a chi-square statistic with a value of 35.2161 on 16 degrees of freedom and a p-value of 0.0037.

## 3.6 Angular Correlations

There has been a great deal of interest in directional statistics in the literature. Consider the following example on wind directions whereby we are interested in testing for independence between the 6 a.m. and the noon readings. The data shown in Table 3.6 can be viewed as points on the unit circle and cannot be dealt with by simply computing the usual rank correlation. The reason is that the larger ranks are close to the smaller ranks. Hence, for example, for the noon readings, angle 23 is closer to angle 313 than to angle 248. Yet, the ranks imply an opposite interpretation. In the table, tied ranks were replaced by their midranks.

*Example 3.8 (Johnson and Wehrly 1977).* Wind directions were recorded at 6 a.m. and at 12 noon on each day at a weather station for 21 consecutive days. It is desired to test for independence. Tied rankings were replaced by their midranks (Table 3.6).

Excellent review articles along with additional references are given by Mardia (1975, 1976) and Jupp and Mardia (1989). Typically, data is provided in the form of directions either in two- or three-dimensional space or as rotations in such a space. The data may take on a variety of forms. It may consist of a unit vector of directions, pairs of such vectors, or a vector of directions along with a corresponding random variable on the line. Examples of applications are to be found in the fields of astronomy, biology, geology, medicine, and meteorology (Downs 1973; Johnson and Wehrly 1977; Breckling 1989). A large number of the works presented deal with the study of inference from parametric models. In this section, we define a corresponding notion of angular correlation using the ranks of the data.

Let $X$ and $Y$ be random vectors with covariance matrix $\Sigma$ partitioned as

$$\Sigma = \begin{pmatrix} \Sigma_{11} & \Sigma_{12} \\ \Sigma_{21} & \Sigma_{22} \end{pmatrix}$$

and suppose $\Sigma_{11}$ and $\Sigma_{22}$ are non-singular of ranks p and q, respectively.

**Definition 3.5 (Jupp and Mardia 1989).** The correlation coefficient $\gamma_{XY}$ between $X$ and $Y$ is defined to be the trace $\gamma$ of the matrix

$$\gamma_{XY} = Tr[\Sigma_{11}^{-1}\Sigma_{12}\Sigma_{22}^{-1}\Sigma_{21}].$$

It follows that, $\gamma_{XY} = \sum_{i=1}^{s} \lambda_i^2$ where the $\lambda_i$ are the canonical correlations and $s = min(p,q)$. This coefficient satisfies the property of invariance under rotation and reflection in addition to the usual properties of a correlation.

Suppose now that $\theta$ and $\varphi$ are circular variables with $0 \le \theta, \varphi \le 2\pi$. Define the directional vectors $t_1^{'}(\theta) = (\cos\theta, \sin\theta)$, $t_2^{'}(\varphi) = (\cos\varphi, \sin\varphi)$, and let $\Sigma$ be the covariance matrix of $t_1$ and $t_2$. It is seen that

$$\begin{aligned} \gamma_{\theta\varphi} = [&\rho_{cc}^2 + \rho_{cs}^2 + \rho_{sc}^2 + \rho_{ss}^2 + 2(\rho_{cc}\rho_{ss} - \rho_{cs}\rho_{sc})\rho_1\rho_2 - 2(\rho_{cc}\rho_{cs} \\ &+\rho_{sc}\rho_{ss})\rho_1 - 2(\rho_{cc}\rho_{sc} + \rho_{cs}\rho_{ss})\rho_2]/[(1-\rho_1^2)(1-\rho_2^2)]. \end{aligned} \tag{3.37}$$

where $\rho_{cc} = corr(\cos\theta, \cos\varphi)$, $\rho_{cs} = corr(\cos\theta, \sin\varphi)$, etc., and $\rho_1 = corr(\cos\theta, \sin\theta)$, $\rho_2 = corr(\cos\varphi, \sin\varphi)$.

Let $(\theta_i, \varphi_i)$ for $i = 1, \ldots, n$ be a random sample of $n$ pairs of angles which define points on the unit circle. Without loss in generality assume that the ranks of the $\theta$'s are the natural integers 1,…,n whereas the corresponding ranks of the $\varphi$'s are denoted by $R_1, \ldots, R_n$. Let

$$\eta^{(1)} = (\cos\frac{2\pi}{n}, \cos\frac{4\pi}{n}, \ldots, \cos 2\pi)', \eta^{(2)} = (\sin\frac{2\pi}{n}, \sin\frac{4\pi}{n}, \ldots, \sin 2\pi)'$$

$$\nu^{(1)} = (\cos\frac{2\pi R_1}{n}, \cos\frac{2\pi R_2}{n}, \ldots, \cos\frac{2\pi R_n}{n})', \nu^{(2)} = (\sin\frac{2\pi R_1}{n}, \sin\frac{2\pi R_2}{n}, \ldots, \sin\frac{2\pi R_n}{n})'.$$

We may formally construct on the basis of the sample the matrix of pairwise correlations

$$\Upsilon_{12} = \begin{pmatrix} \rho(\eta^{(1)}, \nu^{(1)}) & \rho(\eta^{(1)}, \nu^{(2)}) \\ \rho(\eta^{(2)}, \nu^{(1)}) & \rho(\eta^{(2)}, \nu^{(2)}) \end{pmatrix}$$

where $\rho(\eta, \nu)$ is a measure of correlation between $\eta$ and $\nu$. We shall consider correlations based on the Spearman and Kendall distance functions in subsequent sections and we will determine the corresponding asymptotic distributions of the correlation coefficients as $n \to \infty$.

### 3.6.1 Spearman Distance

We shall consider the Kendall notion of a type b correlation (Kendall and Gibbons 1990) given by

$$\begin{aligned} \rho_S(\eta, \nu) &= \frac{\sum_{i \neq j} (\eta_i - \eta_j)(\nu_i - \nu_j)}{\sqrt{\sum_{i \neq j} (\eta_i - \eta_j)^2 \sum_{i \neq j} (\nu_i - \nu_j)^2}} \\ &= \frac{2}{n} \eta' \nu. \end{aligned}$$

It is straightforward to show

$$\sum_{i=1}^{n} \cos \frac{2\pi i}{n} = \sum_{i=1}^{n} \sin \frac{2\pi i}{n} = \sum_{i=1}^{n} \cos \frac{2\pi i}{n} \sin \frac{2\pi i}{n} = 0$$

and

$$\sum_{i=1}^{n} \cos^2 \frac{2\pi i}{n} = \sum_{i=1}^{n} \sin^2 \frac{2\pi i}{n} = \frac{n}{2}.$$

It follows that $\Sigma_{11} = \Sigma_{22} = \frac{n}{2} I$. The sample estimate of $\Sigma_{12}$ is given by

$$\Upsilon_{12}^S = \frac{2}{n} \begin{pmatrix} T_{cc} & T_{cs} \\ T_{sc} & T_{ss} \end{pmatrix}$$

where $T_{cc} = \eta^{(1)'} \nu^{(1)}, T_{cs} = \eta^{(1)'} \nu^{(2)}, T_{sc} = \eta^{(2)'} \nu^{(1)}, T_{ss} = \eta^{(2)'} \nu^{(2)}$.

We recognize that the $\mathbf{T}'s$ are measures of correlation in the Spearman sense. Consequently, the sample correlation using Spearman distance becomes

$$\gamma_S = \frac{4}{n^2} \left( \mathbf{T}_{cc}^2 + \mathbf{T}_{ss}^2 + \mathbf{T}_{cs}^2 + \mathbf{T}_{sc}^2 \right).$$

### 3.6.2 *Kendall Distance*

Recalling the Kendall measure of distance defined by

$$d_K(\eta, \nu) = \sum_{i<j} \{1 - sgn(\eta_i - \eta_j) sgn(\nu_i - \nu_j)\}$$

where sgn indicates the sign function, we may define a corresponding type b correlation as

$$\rho_K(\eta, \nu) = \frac{\sum_{i \neq j} sgn(\eta_i - \eta_j) sgn(\nu_i - \nu_j)}{\sqrt{\sum_{i \neq j} (sgn(\eta_i - \eta_j))^2} \sqrt{\sum_{i \neq j} (sgn(\nu_i - \nu_j))^2}}$$
$$= \frac{\sum_{i \neq j} sgn(\eta_i - \eta_j) sgn(\nu_i - \nu_j)}{\sqrt{A(\eta) A(\nu)}},$$

where $A(\eta) = \# \left(\text{pairs } (i, j), i \neq j | \eta_i \neq \eta_j\right)$. It is easy to see that $\Sigma_{11}$ and $\Sigma_{22}$ are diagonal matrices. In fact, the off-diagonal terms are equal to

$$\sum_{i \neq j} sgn\left(\cos \frac{2\pi i}{n} - \cos \frac{2\pi j}{n}\right) sgn\left(\sin \frac{2\pi i}{n} - \sin \frac{2\pi j}{n}\right)$$
$$= -4 \sum_{i \neq j} sgn\left(\sin \frac{\pi(i + j)}{n} \sin \frac{\pi(i - j)}{n}\right) sgn\left(\cos \frac{\pi(i + j)}{n} \sin \frac{\pi(i - j)}{n}\right)$$
$$= -2 \sum_{i \neq j} sgn\left(\sin \frac{2\pi(i + j)}{n}\right) = 0.$$

The normalization in the Kendall case is somewhat delicate and depends in part on the parity of n. For example, for $n = 10$, there are five pairs of equal values in the set $\{\sin \frac{2\pi i}{n}\}$ whereas for $n = 11$, all the values are distinct. In general, the number of equal pairs is at most $O(n)$. The sample estimate of $\Upsilon_{12}$ is given by

$$\Upsilon_{12}^K = \begin{pmatrix} K_{cc} & K_{cs} \\ K_{sc} & K_{ss} \end{pmatrix}$$

where $K_{cc} = \rho_K(\eta^{(1)}, \nu^{(1)})$, $K_{cs} = \rho_K(\eta^{(1)}, \nu^{(2)})$, $K_{sc} = \rho_K(\eta^{(2)}, \nu^{(1)})$, $K_{ss} = \rho_K(\eta^{(2)}, \nu^{(2)})$.

It follows that the sample correlation coefficient in the Kendall case is given by

$$\gamma_K = \left(K_{cc}^2 + K_{ss}^2 + K_{cs}^2 + K_{sc}^2\right).$$

In the following sections, we shall derive the asymptotic null distributions of the test statistics induced by the Spearman and Kendall distances.

### 3.6.3 Asymptotic Distributions

We are interested in testing the null hypothesis that the circular variables $\theta, \varphi$ are independent. In terms of the ranks, assuming no ties, this translates into the hypothesis $H_0$ that all permutations of the integers $1, \ldots, n$ are equally likely.

**Theorem 3.6.** *The asymptotic null distribution of $n\gamma_S$ as $n \to \infty$ is $\chi^2_4$.*

*Proof.* The joint distribution of $T_{cc}, T_{ss}, T_{cs}, T_{sc}$ is asymptotically normal. In fact, for arbitrary $\{a_i\}$, consider the linear combination

$$a_1 T_{cc} + a_2 T_{ss} + a_3 T_{cs} + a_4 T_{sc} =$$

$$\sum_{i=1}^{n} [\cos \frac{2\pi R_i}{n} (a_1 \cos \frac{2\pi i}{n} + a_2 \sin \frac{2\pi i}{n}) + \sin \frac{2\pi R_i}{n} (a_3 \cos \frac{2\pi i}{n} + a_4 \sin \frac{2\pi i}{n})].$$

Let

$$d(i,j) = \cos \frac{2\pi i}{n} (a_1 \cos \frac{2\pi j}{n} + a_2 \sin \frac{2\pi j}{n}) + \sin \frac{2\pi i}{n} (a_3 \cos \frac{2\pi j}{n} + a_4 \sin \frac{2\pi j}{n}).$$

Since

$$max d_n^2(i,j) \leq 4(a_1^2 + a_2^2 + a_3^2 + a_4^2)$$

and the variance

$$\frac{1}{n} \sum_{i=1}^{n} \sum_{j=1}^{n} d_n^2(i,j) = \frac{n}{4}(a_1^2 + a_2^2 + a_3^2 + a_4^2)$$

we have that

$$\frac{max\ d_n^2(i,j)}{\frac{1}{n} \sum_{i=1}^{n} \sum_{j=1}^{n} d_n^2(i,j)} \to 0\ as\ n \to \infty.$$

The result follows on using Hoeffding's combinatorial central limit theorem (see Appendix B.1). Hence $\Upsilon_{12}^S$ is multivariate normal and the theorem follows. □

A similar result holds for the Kendall tau statistic.

**Theorem 3.7.** *The asymptotic null distribution of $\frac{9}{4} n\gamma_K$ as $n \to \infty$ is $\chi^2_4$.*

*Proof.* See Alvo (1998) for the proof. A different proof can make use of the asymptotic equivalence between the Kendall and Spearman coefficients in general. □

*Example 3.9.* We revisit the wind direction data. We calculate

$$\Upsilon_{12}^S = \begin{pmatrix} -0.246 & 0.306 \\ -0.376 & -0.452 \end{pmatrix}$$

and hence $n\gamma_S = 21(0.50047) = 10.51$ with a p-value of 0.0327. Consequently, we conclude that there is evidence that the 6 a.m. and noon wind directions are significantly correlated.

It is interesting to compare this result with the usual product moment correlation between the two angular measurements. The latter yields a value equal to –0.04, thereby implying that the variables are independent. On the other hand, restricting attention only to the pairs of measurements for which the 6 a.m. readings are below 180° the value of the product moment correlation is 0.512 while for pairs for which the 6 a.m. readings are above 180° it is –0.475. These results taken separately imply a fair degree of dependence. The test statistic $\gamma_S$ takes into account the fact that very small and very large angles (mod $2\pi$) are close to one another.

For the Kendall statistic, we may also calculate

$$\Upsilon_{12}^{K} = \begin{pmatrix} -0.1822 & 0.2097 \\ -0.3106 & -0.3637 \end{pmatrix}$$

and hence $\frac{9n}{4}\gamma_K = \frac{9(21)}{4}(0.3056) = 14.44$ with a p-value of 0.006. It is clear that with either the Spearman or the Kendall statistic, the hypothesis of independence is in doubt.

## 3.7 Angle-Linear Correlation

Suppose that we are now interested in defining the correlation between an angle $\theta$ and a real valued random variable $X$. It can be shown that the correlation coefficient in that case is given by

$$\gamma_L = [\rho_{xc}^2 + \rho_{xs}^2 - 2\rho_{xc}\rho_{xs}\rho_{cs}]/(1 - \rho_{cs}^2)$$

where

$$\rho_{xc} = corr(X, \cos\theta), \rho_{xs} = corr(X, \sin\theta), \rho_{cs} = corr(\cos\theta, \sin\theta).$$

In the nonparametric context, let $(X_i, \theta_i)$ for $i = 1, \ldots, n$ be a random sample of linear-angular measurements. Let $\{R_i\}$ be the ranks of the $\{X_i\}$ and let $\{S_i\}$ be the ranks of the $\{\theta_i\}$. We may assume without loss in generality that the $S_i$ are in natural order $1, 2, \ldots, n$. Based on the Spearman measure of distance, the sample angular-linear correlation is defined by

$$\gamma_{LS} = \frac{[T_{xc}^2 + T_{xs}^2]}{\frac{n}{2}\left(\frac{n(n^2-1)}{12}\right)}$$

where $T_{xc} = \sum R_i \cos\left(\frac{2\pi i}{n}\right)$, $T_{xs} = \sum R_i \sin\left(\frac{2\pi i}{n}\right)$. Similarly, for the Kendall measure, the angular-linear correlation is then given by

$$\gamma_{LK} = [K_{xc}^2 + K_{xs}^2]$$

where

$$K_{xc} = \frac{\sum_{i \neq j}[sgn(R_i - R_j)sgn(\cos\left(\frac{2\pi i}{n}\right) - \cos\left(\frac{2\pi j}{n}\right))]}{\sqrt{[n(n-1)]}\sqrt{\sum_{i \neq j}(sgn(\eta_i^{(1)} - \eta_j^{(1)}))^2}}$$

$$K_{xs} = \frac{\sum_{i \neq j}[sgn(R_i - R_j)sgn(\sin\left(\frac{2\pi i}{n}\right) - \sin\left(\frac{2\pi j}{n}\right))]}{\sqrt{[n(n-1)]}\sqrt{\sum_{i \neq j}(sgn(\eta_i^{(2)} - \eta_j^{(2)}))^2}}.$$

We may now prove a theorem giving the asymptotic distributions of $\gamma_{LS}$ and $\gamma_{LK}$ under the null hypothesis that all vectors of ranks $(R_1, \ldots, R_n)$ are equally likely.

**Theorem 3.8.** *The asymptotic null distribution of $n\gamma_{LS}$ as $n \to \infty$ is $\chi_2^2$.*

*Proof.* The joint distribution of $T_{xc}, T_{xs}$ is asymptotically normal. In fact, for arbitrary constants $a_1, a_2$, consider the linear combination

$$a_1 T_{xc} + a_2 T_{xs} = \sum_{i=1}^{n} [R_i (a_1 \cos \frac{2\pi i}{n} + a_2 \sin \frac{2\pi i}{n})$$

This is a linear rank statistic for which the conditions in Hoeffding (1951) are satisfied. In fact, let

$$d_n(i, j) = (i - \frac{n+1}{2})(a_1 \cos \frac{2\pi i}{n} + a_2 \sin \frac{2\pi i}{n}).$$

The variance is then equal to

$$\frac{1}{n} \sum_{i=1}^{n} \sum_{j=1}^{n} d_n^2(i, j) = \frac{1}{4}(a_1^2 + a_2^2)\frac{n(n^2 - 1)}{12}$$

and we have that

$$\frac{max\ d_n^2(i, j)}{\frac{1}{n}\sum_{i=1}^{n}\sum_{j=1}^{n} d_n^2(i, j)} \to 0$$

as $n \to \infty$. The result follows. □

**Theorem 3.9.** *The asymptotic null distribution of $\frac{9n}{4}\gamma_{LK}$ as $n \to \infty$ is $\chi_2^2$.*

**Table 3.7** Wind direction and ozone concentration

| | | | | | | | | | | |
|---|---|---|---|---|---|---|---|---|---|---|
| Wind direction | 327 | 91 | 88 | 305 | 344 | 270 | 67 | 21 | 281 | |
| Ozone concentration | 28.0 | 85.2 | 80.5 | 4.7 | 45.9 | 12.7 | 72.5 | 56.6 | 31.5 | |
| Wind direction | 8 | 204 | 86 | 333 | 18 | 57 | 6 | 11 | 27 | 84 |
| Ozone concentration | 112.0 | 20.0 | 72.5 | 16.0 | 45.9 | 32.6 | 56.6 | 52.6 | 91.8 | 55.2 |

*Proof.* For arbitrary constants $a_1, a_2$, consider the linear combination

$$\sum_{i \neq j}^{n} sgn(R_i - R_j) b_{ij}$$

where

$$b_{ij} = [(a_1 sgn(\cos \frac{2\pi i}{n} - \cos \frac{2\pi j}{n}) + a_2 sgn(\sin \frac{2\pi i}{n} - \sin \frac{2\pi j}{n})].$$

Using a result of Daniels (1950), the asymptotic normality of $K_{xc}$ and $K_{xs}$ follows. □

*Example 3.10 (Johnson and Wehrly 1977).* We consider data on wind direction and ozone concentration collected at a weather station for 19 days at 4-day intervals. The readings are given in Table 3.7.

The Spearman test statistic to be $n\gamma_{LS} = 19\,(0.3751) = 7.13$ which has a p-value equal to 0.0283. On the other hand the Kendall statistic is given by $\frac{9n}{4}\gamma_{LK} = \frac{9(19)}{4}(0.1595) = 6.82$ for a p-value of 0.033. Both statistics imply that there is a fair degree of dependence between wind direction and ozone concentration.

## Chapter Notes

In this chapter, the traditional rank correlation has been extended to include incomplete rankings. This was made possible using the notion of compatibility which was developed by Alvo and Cabilio in a series of papers. Cabilio and Tilley (1999) report the results of a simulation study where they considered linear, quadratic, and square root trends. They observed that when there were no missing observations, the Spearman statistic was more powerful than Kendall's. In the incomplete case, however, the new Kendall statistic has superior power for more patterns.

The calculation of the exact variance of $\mathcal{A}_K^*$ under $H_2$, in Theorem 3.4, is more involved, and the reader is referred to Alvo and Cabilio (1992), where it is shown that

$$Var\left(\mathcal{A}_K^*\right) = \frac{\kappa_1\kappa_2}{9t\,(t-1)}\left[\frac{(2t+k_1+3)\,(2t+k_2+3)}{2} + \frac{\left(t^2-k_1-2\right)\left(t^2-k_2-2\right)}{(t-2)}\right].$$

An important application of the results presented above which do not discard missing data is in tests of trend where $k_2 = t$ and $k_1 < t$. It is seen that in this context, the superiority of the extended Spearman statistic is established through the calculation of its asymptotic relative efficiency relative to the "naive" statistic. (Alvo and Cabilio 1994) applied these methods to test for trend in precipitation data for St John and Fredericton (NB) and showed that the extended statistic based on Spearman distance is more sensitive in detecting trends than the statistic which ignores the missing observations. Tables of selected critical values of $\mathcal{A}_S^*$ and $\mathcal{A}_K^*$ for the trend case when $k \geq t/2$ have been developed for both hypotheses (Alvo and Cabilio 1993). The results of this section have been extended to the case of ties (Yu et al. 2002) and applied to deal with tests of independence in opinion surveys. A further extension to assess trend in proportions appears in Chap. 7.

Alvo and Smrz (2005) proposed an arc model which serves as a good approximation to Kendall distance.

Although not considered in this book, Alvo and Park (2002) were concerned with multivariate tests of trend when the data are partially incomplete. Such is the case in environmental studies when pH data for one or more lakes are often recorded over regular time intervals and examined for monotone increasing or decreasing trends in order to test for trend in acidification. In monitoring recovering patients, one looks for trends in their vital signs which are often multivariate data in nature. There may be as many as 20–30 blood constituents measured weekly over a period of several months or years. In those case, the use of separate tests on each constituent is inefficient.

# Chapter 4
# Testing for Randomness, Agreement, and Interaction

Suppose that $n$ judges are asked to rank $t$ contestants in accordance with some predetermined criterion. One immediate question that comes to mind is: are the judges ranking the contestants by selecting a ranking at random or is there some specific pattern for their choices? Placing this problem in a geometric setting, we may represent each ranking as a point in a $t$-dimensional space. If indeed the judges act in accordance with some specific nonrandom manner, the points would tend to cluster close together in one or more groups. Intuitively then, a test of randomness could be based on the average pairwise distance between points with large values of that statistic displaying evidence of the random pattern of the points.

In the literature, the Kendall $W$ has been a widely used statistic whose asymptotic distribution was derived by Friedman (1937). Treating each judge as a block, it consists of calculating for each object the average of the ranks assigned by the judges and computing the variance of the averages. Small values of the test statistic are considered consistent with the null hypothesis of randomness. This test statistic is not always sensitive to patterns that may exist in the data. For example, if half the judges assign rankings in the natural order, $1, 2, \ldots, t$ and the other half assign rankings in the reverse order, $t, t-1, \ldots, 1$, then the value of the Kendall $W$ statistic will be small and the null hypothesis will not be rejected. Such considerations lead one to inquire as to whether or not there are other test statistics with better performance.

## 4.1 Tests for Randomness

We begin with some notation. Let $\mathcal{P} = \{\nu_j\}$ be the set of $t!$ possible rankings of $t$ objects and denote the rankings by

$$\nu_j = \left(\nu_j(1), \ldots, \nu_j(t)\right)', j = 1, \ldots, t!$$



M. Alvo, P.L.H. Yu, *Statistical Methods for Ranking Data*, Frontiers in Probability and the Statistical Sciences, DOI 10.1007/978-1-4939-1471-5_4

Suppose that we have a random sample of $n$ rankings denoted by $R, \ldots, R_n$ observed from some population of rankers and suppose that each judge chooses a ranking in accordance with some distribution $\boldsymbol{p}$,

$$\boldsymbol{p} = (p_1, \ldots, p_{t!})'$$

where

$$p_j = P\left(R = \nu_j\right).$$

Our interest is in developing a test of the null hypothesis of randomness, namely

$$H_0 : \boldsymbol{p} = \boldsymbol{p}_0 = \mathbf{1}/t!$$

against the alternative

$$H_1 : \boldsymbol{p} \neq \boldsymbol{p}_0.$$

The null hypothesis indicates that each judge chooses a ranking at random from the population of possible rankings. Select a distance function, $d\left(R_k, R_l\right)$, between two rankings, $R_k, R_l$. A possible test statistic for testing the null hypothesis consists of computing the average pairwise distance between all the observed rankings

$$\bar{d}_n = \frac{1}{n\left(n-1\right)} \Sigma \Sigma_{k,l} d\left(R_k, R_l\right). \tag{4.1}$$

Under the null hypothesis, one would expect the average pairwise distance to be large or, equivalently from (4.1), the average pairwise correlation

$$\bar{\alpha}_n = 1 - \frac{2\bar{d}_n}{M} \tag{4.2}$$

to be small. Equivalently, one should reject the null hypothesis whenever $\bar{\alpha}_n$ is large. Note that we may write

$$d\left(R_k, R_l\right) = \Sigma_i \Sigma_j d\left(\nu_i, \nu_j\right) I\left[R_k = \nu_i\right] I\left[R_l = \nu_i\right]$$

where $I[B]$ is the indicator function taking value 1 if the event $B$ is true and 0 otherwise. It follows that

$$\begin{aligned}
n\left(n-1\right)\bar{d}_n &= \Sigma_k \Sigma_l d\left(R_k, R_l\right) && (4.3)\\
&= \Sigma_k \Sigma_l \Sigma_i \Sigma_j d\left(\nu_i, \nu_j\right) I\left[R_k = \nu_i\right] I\left[R_l = \nu_j\right] && (4.4)\\
&= \Sigma_i \Sigma_j \left(\Sigma_k I\left[R_k = \nu_i\right]\right)\left(\Sigma_l I\left[R_l = \nu_j\right]\right) d\left(\nu_i, \nu_j\right) && (4.5)\\
&= \Sigma_i \Sigma_j N_i N_j d\left(\nu_i, \nu_j\right) && (4.6)\\
&= N' \Delta N && (4.7)
\end{aligned}$$

where $\Delta = \left(d\left(\nu_i, \nu_j\right)\right)$ is the matrix of pairwise distances and $N' = (N_1, \ldots, N_{t!})$ is the vector of frequencies with

$$N_i = \Sigma_k I\left[R_k = \nu_i\right].$$

We recognize that (4.7) is a quadratic form and that $N$ is a multinomial random variable with mean and covariance respectively given by

$$EN = n\boldsymbol{p}, Cov\left(N\right) = n\sum,$$

where $\sum = (diag\left(\boldsymbol{p}\right) - \boldsymbol{pp}')$ and $diag\left(\boldsymbol{p}\right)$ is a $t! \times t!$ diagonal matrix having entries $p_i$ along the diagonal. Let $\hat{\boldsymbol{p}}_n = N/n$ and recall from (3.13)

$$Q = J - \left(\frac{2}{M}\right)\Delta.$$

**Theorem 4.1.** *(a) Under $H_0$, for $n \to \infty$, and $Q\,\boldsymbol{p}_0 = c^*\mathbf{1}$, we have that*

$$(n-1)\left(\bar{\alpha}_n - c^*\right) \Rightarrow_{\mathcal{L}} Z_0' Q Z_0 - 1 + c^*$$

*where $Z$ has a $t!$-variate normal distribution with mean 0 and covariance matrix $\Sigma_0 = (t!)^{-2}\left((t!I - J\right)$. Here $I$ is the identity matrix and $J$ is a $t! \times t!$ matrix of ones.*

*(b) Under $H_1$, for $n \to \infty$,*

$$\sqrt{n}\left(\overline{\alpha}_n - \boldsymbol{p}'Q\,\boldsymbol{p}\right) \Rightarrow_{\mathcal{L}} 2Z'Q\,\boldsymbol{p}$$

*where $Z$ has a $t!$-variate normal distribution with mean 0 and covariance matrix*

$$\sum = \left(diag\left(\boldsymbol{p}\right) - \boldsymbol{pp}'\right).$$

*Proof.* (a) Define

$$Z_n = n^{-1/2}\left(N - n\boldsymbol{p}\right).$$

A Taylor series expansion around $\hat{\boldsymbol{p}}_n = \boldsymbol{p}$ reveals the identity

$$\begin{aligned} \hat{\boldsymbol{p}}_n' Q\,\hat{\boldsymbol{p}}_n - \boldsymbol{p}'Q\,\boldsymbol{p} &= 2\left(\hat{\boldsymbol{p}}_n - \boldsymbol{p}\right)' Q\,\boldsymbol{p} + \left(\hat{\boldsymbol{p}}_n - \boldsymbol{p}\right)' Q\left(\hat{\boldsymbol{p}}_n - \boldsymbol{p}\right) \qquad (4.8) \\ &= \frac{2}{\sqrt{n}} Z_n' Q\,\boldsymbol{p} + \frac{1}{n} Z_n' Q Z_n. \end{aligned}$$

Under the null hypothesis $\boldsymbol{p} = \boldsymbol{p}_0$ and $Q\boldsymbol{p}_0 = c^*\mathbf{1}$, so that $\boldsymbol{p}_0' Q \boldsymbol{p}_0 = c^*$. Now the relationships

$$Q = J - \frac{2}{M}\Delta$$

$$\overline{\alpha}_n = 1 - \binom{n}{2}^{-1} \left(N'\Delta N\right)/M$$

imply

$$(n-1)\bar{\alpha}_n + 1 = \left(n\hat{\boldsymbol{p}}_n' Q \hat{\boldsymbol{p}}_n\right)$$

On using the multivariate central limit theorem for multinomial random variables (Timm 1975), it follows from (4.8)

$$\begin{aligned}(n-1)\left(\overline{\alpha}_n - c^*\right) + 1 - c^* &= n\left(\hat{\boldsymbol{p}}_n' Q \hat{\boldsymbol{p}}_n - \boldsymbol{p}_0' Q \boldsymbol{p}_0\right) \\ &= Z_n' Q Z_n \\ &\Rightarrow {}_{\mathcal{L}} Z' Q Z.\end{aligned}$$

(b) On the other hand if $\boldsymbol{p} \neq \boldsymbol{p}_0$ we have from (4.8)

$$\sqrt{n}\left(\hat{\boldsymbol{p}}_n' Q \hat{\boldsymbol{p}}_n - \boldsymbol{p}' Q \boldsymbol{p}\right) \Rightarrow_{\mathcal{L}} 2Z' Q \boldsymbol{p}$$

$$(n-1)\overline{\alpha}_n = -1 + n\hat{\boldsymbol{p}}_n' Q \hat{\boldsymbol{p}}_n$$

and it follows that

$$\sqrt{n}\left(\overline{\alpha}_n - \boldsymbol{p}' Q \boldsymbol{p}\right) \Rightarrow_{\mathcal{L}} 2Z' Q \boldsymbol{p}.$$

□

The distribution of $Z'QZ$ under the null hypothesis is that of a weighted chi square where the weights are given by the eigenvalues of the matrix

$$Q\Sigma_0 = (t!)^{-1}\left(Q - c^* J\right).$$

In what follows, we shall obtain properties of that matrix for both the Spearman and Kendall cases. For these cases, the constant $c^* = 0$ and hence

$$Q\Sigma_0 = (t!)^{-1} Q.$$

Before dealing with the specific distributions of the Spearman and Kendall statistics, we will need the following lemmas which are useful in their own right.

**Lemma 4.1.** *Let* $A(s,s',t,t') = \Sigma_\nu sgn(\nu(s)-\nu(t))\, sgn(\nu(s')-\nu(t'))$. *Then, under* $H_0$

$$A\left(s,s',t,t'\right) = \begin{cases} 0 & s\neq s', t\neq t', \\ t! & s=s', t=t', \\ \frac{t!}{3} & s=s', t\neq t', \\ -\frac{t!}{3} & s=t', s'\neq t. \end{cases} \tag{4.9}$$

*Proof.* Let $R$ be a random ranking of t objects. Note that in distribution

$$sgn\left[R(s)-R(t)\right] =_d sgn\left[U-V\right]$$

where $U, V$ are independent uniform random variables on $(0,1)$.

Let $Z = sgn[U_1-V_1]\, sgn[U_2-V_2]$ where $U_1, U_2, V_1, V_2$ are independent uniform random variables on $(0,1)$. It follows that

$$\begin{aligned} P(Z>0) &= P(U_1-V_1>0, U_2-V_2>0) + P(U_1-V_1<0, U_2-V_2<0) \\ &= P(U_1-V_1>0)\,P(U_2-V_2>0) + P(U_1-V_1<0)\,P(U_2-V_2<0) \\ &= \left(\frac{1}{2}\right)^2 + \left(\frac{1}{2}\right)^2 = \frac{1}{2}. \end{aligned}$$

Similarly, $P(Z<0) = \frac{1}{2}$. Also, if $Z_1 = sgn[U_1-V_1]\, sgn[U_1-V_2]$, then

$$\begin{aligned} P(Z_1>0) &= P(U_1-V_1>0, U_1-V_2>0) + P(U_1-V_1<0, U_1-V_2<0) \\ &= \int_0^1 P(x-V_1>0, x-V_2>0)\,dx + \int_0^1 P(x-V_1<0, x-V_2<0)\,dx \\ &= \int_0^1 P(x-V_1>0)\,P(x-V_2>0)\,dx + \int_0^1 P(x-V_1<0)\,P(x-V_2<0)\,dx \\ &= \int_0^1 x^2 dx + \int_0^1 (1-x)^2\,dx = \frac{2}{3}. \end{aligned}$$

It now follows that

$$\begin{aligned} \frac{1}{t!}\Sigma_\nu sgn(\nu(s)-\nu(t))\, sgn(\nu(s)-\nu(t')) &= E\left[sgn(R(s)-R(t))\, sgn(R(s)-R(t'))\right] \\ &= P(Z_1>0) - P(Z_1<0) = \frac{1}{3}. \end{aligned}$$

The other cases follow in a similar way. □

**Lemma 4.2.** *The matrices* $Q_S, Q_K$ *satisfy*

(i) $Q_S^2 = \frac{t!}{t-1} Q_S$ *and hence,* $\frac{t-1}{t!} Q_S$ *is idempotent.*
(ii) $Q_K Q_S = \frac{2t!(t+1)}{3t(t-1)} Q_S$.
(iii) $Q_K = \frac{2(t+1)}{3t} Q_S + A, Q_S A = 0$.
(iv) $Q_K^2 = \frac{4t!(t+1)^2}{9t^2(t-1)} Q_S + \frac{2t!}{3t(t-1)} A$.

*Proof.* Setting $c_S = \frac{t(t^2-1)}{12}$ the matrix

$$c_S Q_S = \mathbf{T}_S' \mathbf{T}_S = (t-2)! c_S [tI - J].$$

In fact, the diagonal elements are equal to

$$(t-1)! \sum \left(i - \frac{t+1}{2}\right)^2 = (t-1)! c_S,$$

whereas the off-diagonal elements are equal to

$$(t-2)! \sum_{i \neq j} \left(i - \frac{t+1}{2}\right)\left(j - \frac{t+1}{2}\right) = -(t-2)! c_S.$$

The matrix $Q_S$ is singular since the rows sum to 0. A generalized inverse of $c_S Q_S$ is given by $\frac{1}{(t-2)! t c_S} [I + J]$.

Now to show idempotency in (i), we see that

$$\begin{aligned} (Q_S)^2 &= \tfrac{1}{c_S^2} \mathbf{T}_S' \mathbf{T}_S \mathbf{T}_S' \mathbf{T}_S \\ &= \frac{1}{c_S} (t-2)! \mathbf{T}_S' [tI_t - J] \mathbf{T}_S \\ &= \frac{1}{c_S} (t-2)! \left[t \mathbf{T}_S' \mathbf{T}_S\right] \\ &= \frac{t!}{t-1} (Q_S). \end{aligned}$$

Next we prove (ii). We first note that the $i$th rank can be represented in terms of the remaining $(t-1)$ ranks as

$$\left[\nu(i) - \frac{t+1}{2}\right] = \frac{1}{2} \Sigma_l sgn [\nu(i) - \nu(l)]. \tag{4.10}$$

Part (ii) is equivalent to showing

$$(\mathbf{T}_K' \mathbf{T}_K)(\mathbf{T}_S' \mathbf{T}_S) = \frac{1}{c_K} \frac{t!(t+1)}{3} (\mathbf{T}_S' \mathbf{T}_S)$$

where $c_K = \frac{t(t-1)}{2}$.

For simplicity, note that the first row and column entry in the matrix $\left(\mathbf{T}_K'\mathbf{T}_K\right)\mathbf{T}_S'$ is given by

$$\Sigma_{h=1}^{t!}\Sigma_{i<j}sgn\left[\nu_1(j)-\nu_1(i)\right]sgn\left[\nu_h(j)-\nu_h(i)\right]\left[\nu_h(1)-\frac{t+1}{2}\right]=$$

$$\frac{1}{2}\Sigma_{l=2}^{t}\Sigma_{i<j}sgn\left[\nu_1(j)-\nu_1(i)\right]\Sigma_{h=1}^{t!}sgn\left[\nu_h(j)-\nu_h(i)\right]sgn\left[\nu_h(1)-\nu_h(l)\right].$$

There are two cases to consider, namely $(i=1, j=l)$ and $(i\neq 1, j=l)$. It follows that

$$-\frac{1}{2}\left\{t!\Sigma_{l=2}^{t}sgn\left[\nu_1(l)-\nu_1(1)\right]+\frac{t!}{3}\Sigma_{i\neq 1,j}\Sigma_{l\neq j}sgn\left[\nu_h(j)-\nu_h(l)\right]\right\}=$$

$$\left\{\left[t!\left(\nu_1(1)-\frac{t+1}{2}\right)\right]-\frac{t!}{3}(t-2)\Sigma_{j\neq 1}\left(\nu_1(l)-\frac{t+1}{2}\right)\right\}=$$

$$\left\{t!\left(\nu_1(1)-\frac{t+1}{2}\right)+\frac{t!}{3}(t-2)\left(\nu_1(1)-\frac{t+1}{2}\right)\right\}=$$

$$\frac{(t+1)}{3}t!\left(\nu_1(1)-\frac{t+1}{2}\right).$$

Other entries are treated similarly. Part (iii) follows directly from (ii).

To show part (iv) it suffices to show

$$\left(\mathbf{T}_K'\mathbf{T}_K\right)^2=\frac{t!}{3}\left(\mathbf{T}_K'\mathbf{T}_K\right)+\frac{4t!}{3}\left(\mathbf{T}_S'\mathbf{T}_S\right). \tag{4.11}$$

In fact, this follows since the rs term of the left-hand side of (4.11) is equal to

$$\sum_{k<l}\sum_{k'<l'}sgn\left[\nu_r(k)-\nu_r(l)\right]sgn\left[\nu_s(k')-\nu_s(l')\right]$$

$$\times\sum_{h=1}^{t!}sgn\left[\nu_h(k)-\nu_h(l)\right]sgn\left[\nu_h(k')-\nu_h(l')\right]=$$

$$\frac{t!}{3}\sum_{k<l}\sum_{k'<l'}sgn\left[\nu_r(k)-\nu_r(l)\right]sgn\left[\nu_s(k')-\nu_s(l')\right]$$

$$+\frac{t!}{3}\sum_{k<l}\sum_{k<l'}sgn\left[\nu_r(k)-\nu_r(l)\right]sgn\left[\nu_s(k)-\nu_s(l')\right]$$

$$=\frac{t!}{3}\sum_{k<l}\sum_{k'<l'}sgn\left[\nu_r(k)-\nu_r(l)\right]sgn\left[\nu_s(k')-\nu_s(l')\right]$$

$$+4\frac{t!}{3}\sum_{k}\left(\nu_r(k)-\frac{t+1}{2}\right)\left(\nu_s(k)-\frac{t+1}{2}\right).$$

□

**Theorem 4.2.** *The asymptotic distribution of the Spearman statistic under the null hypothesis of randomness is given by*

$$(t-1)\{(n-1)\overline{\rho}_n+1\} \Rightarrow_L \chi^2_{t-1}. \tag{4.12}$$

*The left-hand side of (4.12) can also be expressed as*

$$\frac{12n}{t(t+1)}\sum_{i=1}^{t}\left(\bar{R}_i-\frac{t+1}{2}\right)^2$$

*which is the usual Friedman statistic.*

*Proof.* The asymptotic distribution of $(t-1)\{(n-1)\overline{\rho}_n+1\}$ is that of a weighted $\chi^2$ where the weights are determined by the eigenvalues of the idempotent matrix $\frac{(t-1)}{t!}Q_S$. Hence its eigenvalues are 0 or 1. Moreover, the rank of the matrix is $(t-1)$.

The left-hand side of (4.12) is equal to

$$\begin{aligned}(t-1)Z'QZ &= \frac{t-1}{c_S}n\parallel \mathbf{T}\hat{\mathbf{p}}_n\parallel^2\\ &= \frac{12n(t-1)}{t(t^2-1)}\sum_{i=1}^{t}\left(\bar{R}_i-\frac{t+1}{2}\right)^2.\end{aligned}$$

□

**Theorem 4.3.** *The asymptotic distribution of the Kendall statistic under the null hypothesis of randomness is given by*

$$(n-1)\bar{\tau}_n+1 \Rightarrow_L \frac{2}{3t(t-1)}\left\{(t+1)\chi^2_{t-1}+\chi^2_{\binom{t-1}{2}}\right\}-1. \tag{4.13}$$

*The left-hand side of (4.13) can also be expressed as*

$$\frac{\sum(2x_i-n)^2}{n\binom{t}{2}}$$

*where the summation is taken over all $\binom{t}{2}$ pairs of objects and $x_i$ is the number of judges whose ranking of the pair i of objects agrees with the ordering of the same pair in a criterion ranking such as the natural ordering.*

*Proof.* From Lemma 4.2,

$$A = Q_K - \frac{2(t+1)}{3t}Q_S$$

and it follows that

$$A^2 = \frac{2t!}{3t\,(t-1)}A.$$

This implies $\frac{3t(t-1)}{2t!}A$ is an idempotent matrix. Noting that

$$Trace\,(A) = \frac{t!\,(t-2)}{3t} = rank\,(A)\,,$$

we see that $Q_K$ has two distinct nonzero eigenvalues,

$$\lambda_1 = \frac{2t!\,(t+1)}{3t\,(t-1)}, \lambda_2 = \frac{2t!}{3t\,(t-1)}$$

and (4.13) follows.

Now, let $a_{ij} = 1$, if judge j agrees with the ranking in pair i and $= -1$ if he disagrees. Then, setting $a_i = \sum_j a_{ij}$ and noting that if $x_i$ =number of judges who agree with the ranking in pair i and $y_i$ =number who disagree, we have

$$x_i + y_i = n, x_i - y_i = a_i,$$

then $a_i = 2x_i - n$. The left-hand side of (4.13) is equal to

$$\frac{\sum (a_i)^2}{n\binom{t}{2}}$$

and the result follows. □

The preceding theorems did not consider the situation where ties are possible in the rankings. This situation was considered in the literature for the case of the Spearman statistic (Lehmann 1975) wherein the asymptotic distribution is obtained by conditioning on the observed ties. Consider the following example where it may not be desirable to condition on the observed ties only. Suppose that tasters are asked to rank in order of preference each of three varieties of tea. If ties are permitted, the sample space would consist of all possible permutations, including those where either two or all three varieties are tied. Alvo and Cabilio (1985) derived the correction for ties under precisely such situations. This correction for ties is made once and for all. This approach allows for comparisons to be made when the same experiment is repeated. We recall for completeness the definition of a tied ordering, previously given in Chapter 3.

**Definition 4.1.** A tied ordering of $n$ objects is a partition into $e$ sets, $1 \leq e \leq t$, each of which contains $d_i$ objects, $d_1 + d_2 + \ldots + d_e = t$, so that the $d_i$ objects in each set share the rank $i$, $1 \leq i \leq e$. Such a tie pattern is denoted by $\boldsymbol{\delta} = (d_1, d_2, \ldots, d_e)$. The ranking denoted by $\boldsymbol{\nu}_\delta = (\nu_\delta(1), \nu_\delta(2), \ldots, \nu_\delta(t))$, resulting from such an ordering, is a tied ranking and is one of $t!/(d_1!d_2!\ldots d_e!)$ possible permutations.

Let $k_i = \frac{t!}{d_1!\ldots d_e!}$. Then the total number of possible permutations is given by $k = \sum_{i=1}^{2^t-1} k_i$. Define $t_i = \frac{1}{12}\sum_{j=1}^{e}\left(d_{ij}^3 - d_{ij}\right)$, $\theta_i = 1 - \frac{12t_i}{t(t^2-1)}$, $\theta = \sum_{i=1}^{2^t-1} k_i\theta_i$.

**Theorem 4.4.** *(a) The asymptotic distribution of the Spearman statistic under the null hypothesis $H_0 : \mathbf{p}_i = \frac{1}{k}$ is as $n \to \infty$ given by*

$$(t-1)\frac{k}{\theta}\left\{(n-1)\bar{\rho}_n + \frac{\theta}{k}\right\} \to_L \chi^2_{t-1}.$$

*(b) The asymptotic distribution of the Kendall statistic under the null hypothesis $H_0 : \mathbf{p}_i = \frac{1}{k}$ is given by*

$$n\bar{\tau}_n \Rightarrow_L \frac{2}{3t(t-1)}\left\{\frac{\theta}{k}(t+1)\chi^2_{t-1} + \frac{3(\beta-2\gamma)}{k}\chi^2_{\binom{t-1}{2}}\right\} - \frac{\beta}{k}$$

*where the two $\chi^2$ variates are independent and*

$$\beta = \sum_{i=1}^{2^t-1} k_i\beta_i, \beta_i = \left(t^2 - \sum_j d_{ij}^2\right) / (t(t-1))$$

$$\gamma = \sum_{i=1}^{2^t-1} k_i\gamma_i, \gamma_i = \frac{1}{t-2}\left\{\theta_i\frac{t+1}{3} - \beta_i\right\}.$$

*Proof.* See Alvo and Cabilio (1985) for the proof. □

When ties are not allowed, $\theta_i = \beta_i = 1, \theta = \beta = k = t!$, $\gamma_i = \frac{1}{3}$, $\gamma = \frac{k}{3}$.

## 4.2 Tests for Agreement Among Groups

We may wish to compare two groups of patients with respect to how they perceive their hospitalization, those who require bed rest and those who are mobile in their recovery. Each patient is presented with a set of situations and asked to rank them in order of severity of stress. The result is that two sets of rankings are obtained and it is necessary to determine if the groups are responding in a similar manner. In another example Hollander and Sethuraman (1978) considered data of C. Sutton in his/her 1976 thesis on leisure preferences and attitudes on retirement of the elderly for 14 white and 13 black females in the age group 70–79 years. Each individual was asked: with which sex do you wish to spend your leisure? Each female was asked to rank the three responses: male(s), female(s) or both, assigning rank 1 for the most desired and 3 for the least desired. The first object in the ranking corresponds to "male," the second to "female," and the third to "both." It was desired to compare these two groups. The data is reproduced in Table 4.1.

**Table 4.1** Sutton data on leisure preferences

| Rankings | (123) | (132) | (213) | (231) | (312) | (321) |
|---|---|---|---|---|---|---|
| Frequencies for white females | 0 | 0 | 1 | 0 | 7 | 6 |
| Frequencies for black females | 1 | 1 | 0 | 5 | 0 | 6 |

We begin with a general introduction to the concepts of diversity and dissimilarity. These concepts provide a generalization of the classical analysis of variance and are particularly applicable to data in the form of rankings. Consider a set of $g$ populations where the individuals are characterized by a set of rankings chosen from the set of all possible rankings $\mathcal{P}$ in accordance with some distribution.

**Definition 4.2.** The diversity coefficient of the population whose distribution on the set of possible rankings is $\mathbf{p}_i$ is defined to be

$$H_i = \mathbf{p}_i' \Delta \mathbf{p}_i$$

where $\Delta$ is the matrix of pairwise distances between rankings. The diversity coefficient is the average difference between two randomly chosen individuals from the $i$th population.

Similarly, we may define the similarity coefficient when one individual is drawn from the $i$th and another from the $j$th population

$$H_{ij} = \mathbf{p}_i' \Delta \mathbf{p}_j. \tag{4.14}$$

The dissimilarity coefficient or between population diversity is then defined to be the difference

$$H_{ij} - \frac{1}{2}\left(H_i + H_j\right) = -\frac{1}{2}\left(\mathbf{p}_i - \mathbf{p}_j\right)' \Delta \left(\mathbf{p}_i - \mathbf{p}_j\right). \tag{4.15}$$

Suppose now that the individuals are mixed together in accordance with the proportions $\lambda_1, \ldots, \lambda_g$ such that $\sum_{i=1}^{g} \lambda_i = 1$. The convex set generated by the mixture leads to a new population with probability vector $\mathbf{p} = \sum_{i=1}^{g} \lambda_i \mathbf{p}_i$. The notions of diversity and between population diversity can now be formally defined.

**Definition 4.3.** The total diversity, the within population diversity, and the between population diversity are defined respectively to be

(i) $H(\mathbf{p}) = \mathbf{p}' \Delta \mathbf{p}$,
(ii) $H_W = \sum_i \lambda_i \mathbf{p}_i' \Delta \mathbf{p}_i$,
(iii) $H_B = -\sum_{i<j} \lambda_i \lambda_j \left(\mathbf{p}_i - \mathbf{p}_j\right)' \Delta \left(\mathbf{p}_i - \mathbf{p}_j\right)$.

It can be seen that

$$H(\mathbf{p}) = H_B + H_W.$$

The requirement that the between population diversity $H_B$ be positive demands that

$$a'\Delta a \leq 0 \text{ whenever } \sum_{i=1}^{t!} a(i) = 0.$$

It can be seen from (4.15) that this condition is equivalent to requiring that $H$ be a concave function. The concavity requirement imposes certain conditions on the distance function which must be verified for each potential distance measure. It does not follow from the right-invariance property. The following lemma makes this requirement more precise.

**Lemma 4.3.** *If the distance measure $d(\mu, \nu)$ is right invariant on the set of permutations, then there exists a constant $c > 0$ such that*

$$\Delta \mathbf{1} = (ct!)\,\mathbf{1}$$

*and $H(p)$ is concave if and only if*

$$Q^* = cJ - \Delta$$

*is positive semidefinite. Moreover, in this case $H(p)$ has the maximum value $c$ at $u = \frac{1}{t!}\mathbf{1}$.*

*Proof.* The existence of the eigenvalue $ct!$ follows from the right-invariance property of the distance measure. We note that whenever $a'\mathbf{1} = 0$, for any $x = a + b\mathbf{1}$, which includes all points in $R^{t!}$

$$x'Q^*x = cb'Jb - x'\Delta x \geq 0.$$

Writing $\mathbf{p} = u + (\mathbf{p} - u)$ we note that since u is an eigenvector of $\Delta$ orthogonal to $(\mathbf{p} - u)$ we have

$$H(\mathbf{p}) = u'\Delta u + (\mathbf{p} - u)'\,\Delta\,(\mathbf{p} - u) \leq c$$

showing that for right-invariant measures, the uniform distribution over the set of all permutations is most diverse among diversity measures. □

Specializing to the Spearman and Kendall distances, we saw earlier in Chap. 3 that the matrix $cQ$ can be expressed as

$$c_K Q_K = \mathbf{T}_K'\mathbf{T}_K,\ c_S Q_S = \mathbf{T}_S'\mathbf{T}_S.$$

In the next result we establish the link between the characteristic $\mathbf{T}\pi$ and the between population diversity, thus showing that it is this characteristic which forms the basis for inference when comparing populations.

**Lemma 4.4.** *For a right-invariant metric on the set of permutations, the between population diversity is given by*

$$\sum_{i<j} \lambda_i \lambda_j \left\| \mathbf{Tp}_i - \mathbf{Tp}_j \right\|_s^2 = tr\left\{var\left(\mathbf{Tp}_I\right)\right\}$$

*where* $\|.\|_s$ *is the Euclidean norm in* $R^s$ *and* $I$ *has the distribution* $P(I = i) = \lambda_i$.

*Proof.* We note that since $\Delta = cJ - \mathbf{T}'\mathbf{T}$,

$$-\sum_{i<j} \lambda_i \lambda_j \left[\left(\mathbf{p}_i - \mathbf{p}_j\right)' \Delta \left(\mathbf{p}_i - \mathbf{p}_j\right)\right] =$$

$$\sum_{i<j} \lambda_i \lambda_j \left[\left(\mathbf{p}_i - \mathbf{p}_j\right)' \mathbf{T}'\mathbf{T} \left(\mathbf{p}_i - \mathbf{p}_j\right)\right] = \sum_{i<j} \lambda_i \lambda_j \left\| \mathbf{Tp}_i - \mathbf{Tp}_j \right\|_s^2.$$

□

Suppose that we have a random sample of $n_i$ judges from population $i$ each of whom chooses a ranking in accordance with some distribution $\mathbf{p}_i$. Set $N = \sum n_i$. Given that the basis for inference for comparing two or more groups are the characteristics $\mathbf{Tp}$, consider therefore a test of the null hypothesis

$$H_0 : \mathbf{Tp}_1 = \mathbf{Tp}_2 = \ldots = \mathbf{Tp}_g \tag{4.16}$$

against the alternative that at least two among the $T\mathbf{p}_i$ are not equal. We observe for each group $i$, the relative frequency of occurrence of each ranking $\nu_l, l = 1, \ldots, t!$ denoted by $\hat{\mathbf{p}}_i(l), i = 1, \ldots, g$. Set $\hat{\mathbf{p}}_i = \left(\hat{\mathbf{p}}_i(1), \ldots, \hat{\mathbf{p}}_i(t!)\right)'$. A central limit theorem exists for each of the statistics $T\hat{\mathbf{p}}_i$.

**Theorem 4.5.** *Suppose that* $n_i/N \to \lambda_i > 0$ *as* $N \to \infty$. *Then*

*(a)*

$$\sqrt{n_i}\mathbf{T}\left(\hat{\mathbf{p}}_i - \mathbf{p}_i\right) \Rightarrow Z_i$$

*where*

$$Z_i \sim N_s\left(0, \mathbf{T}\Sigma_i\mathbf{T}'\right)$$

*and the* $Z_i$ *are independent with*

$$\Sigma_i = \Pi_i - \mathbf{p}_i\mathbf{p}_i', \Pi_i = diag\left(\mathbf{p}_i(1), \ldots, \mathbf{p}_i(t!)\right).$$

*(b) Under* $H_0$,

$$\sqrt{N}\mathbf{T}\left(\hat{\mathbf{p}}_i - \hat{\mathbf{p}}_j\right) \Rightarrow N_s\left(0, \mathbf{T}\Sigma\mathbf{T}'\right)$$

*where*

$$\Sigma = \frac{\Sigma_i}{\lambda_i} + \frac{\Sigma_j}{\lambda_j}.$$

*Moreover, for a consistent estimator* $\hat{\Sigma}$ *of* $\Sigma$ *and if* $\hat{D}$ *is the Moore–Penrose inverse of* $\mathbf{T}\hat{\Sigma}\mathbf{T}'$, *then*

$$N\left(\hat{\mathbf{p}}_i - \hat{\mathbf{p}}_j\right)'\mathbf{T}'\hat{D}\mathbf{T}\left(\hat{\mathbf{p}}_i - \hat{\mathbf{p}}_j\right) \Rightarrow \chi_r^2$$

*where* $r = rank(\mathbf{T}\Sigma\mathbf{T}')$.

*Proof.* (a) The multivariate central limit theorem applies to multinomial vectors

$$\sqrt{n_i}\left(\hat{\mathbf{p}}_i - \mathbf{p}_i\right) \Rightarrow N_{t!}\left(0, \Sigma_i\right).$$

The result follows.

(b) Using standard multivariate normal theory (Timm 1975), this part follows from the independence of the $Z_i's$ and the null hypothesis.

□

We note that the use of the Moore–Penrose inverse may be circumvented by choosing the matrix $\mathbf{T}$ so that $\mathbf{T}\Sigma\mathbf{T}'$ is of full rank. Hence, in the case of the Spearman distance, we may reduce the matrix $\mathbf{T}_S$ by using only the ranks of the first (t-1) objects. This problem does not immediately arise for the Kendall distance since there is no singularity in $\mathbf{T}_\mathbf{K}$.

An unbiased estimate of the covariance matrix $\Sigma_i$ is given by

$$\hat{\Sigma}_i = \frac{n_i}{n_i - 1}\left(\hat{\Pi}_i - \hat{\mathbf{p}}_i\hat{\mathbf{p}}_i'\right)$$

where

$$\hat{\Pi}_i = diag\left(\hat{\mathbf{p}}_i\left(1\right), \ldots, \hat{\mathbf{p}}_i\left(t!\right)\right).$$

Suppose now that we are interested in the two-sample problem and that we wish to test the null hypothesis that

$$H_o\left(1\right) : \mathbf{T}\mathbf{p}_1 = \mathbf{T}\mathbf{p}_2.$$

Under $H_o\left(1\right)$, it follows that an estimate of the covariance matrix $\Sigma$ in Theorem 4.5 is given by

$$\hat{\Sigma}_{Separate} = N\left(\frac{\hat{\Sigma}_1}{n_1} + \frac{\hat{\Sigma}_2}{n_2}\right).$$

The separate estimation of the covariances is appropriate in this case since the covariances are not assumed to be equal. In the situation when the null hypothesis is given by

$$H_o(2) : \mathbf{p}_1 = \mathbf{p}_2,$$

we may pool the separate estimates as

$$\hat{\Sigma}_{Pooled} = N\left(\frac{1}{n_1} + \frac{1}{n_2}\right)\left(\frac{(n_1 - 1)\,\hat{\Sigma}_1 + (n_2 - 1)\,\hat{\Sigma}_2}{N - 2}\right).$$

Hollander and Sethuraman (1978) actually used the combined estimate

$$\hat{\Sigma}_{combined} = \left(\frac{N-2}{N-1}\right)\hat{\Sigma}_{pooled} + \left(\frac{N}{N-1}\right)(f_1 - f_2)(f_1 - f_2)'$$

where $f_1, f_2$ are the frequency vectors. It should be noted that the estimates of the $(s \times s)$ covariance matrices are based on the observed score vectors $\{t\left(X_{ij}\right)\}$; that is,

$$\mathbf{T}\hat{\Sigma}_i\mathbf{T}' = \frac{\sum_{j=1}^{n_i}\left(t\left(R_{ij}\right) - \bar{t}_{i.}\right)\left(t\left(R_{ij}\right) - \bar{t}_{i.}\right)'}{n_i - 1}$$

where $R_{ij}$ is the observed ranking of judge j in group i and

$$\bar{t}_i = \frac{1}{n_i}\sum_{j=1}^{n_i} t\left(R_{ij}\right).$$

Consequently, the calculations do not require computation of the individual covariance matrices $\hat{\Sigma}_i$.We may apply the methodology to the following example on leisure time preferences.

*Example 4.1.* Sutton data was analyzed in Feigin and Alvo (1986) using both the Spearman and Kendall test statistics. The total diversity was apportioned as indicated in Table 4.2. It can be seen that there is strong evidence that the two groups of females differ significantly.

The hypothesis expressed in (4.16) can alternatively be tested by using general multivariate analysis of variance methods. We do not pursue this further but instead refer the reader to Timm (1975).

**Table 4.2** Analysis of the Sutton data

| | Spearman | Kendall |
|---|---|---|
| Within | 0.88 | 1.51 |
| Between | 0.41 | 0.54 |
| Total | 1.29 | 2.05 |
| | $\chi_2^2$ | $\chi_3^2$ |
| Separate | 28.0 | 28.1 |
| Pooled | 28.5 | 28.5 |

## 4.3 Test for Interaction in a Two-Way Layout

In this section, we consider the general two-factor design with equal numbers of replications in each cell. Such designs are utilized in statistics to test for main effects and for interactions in a variety of experiments. In more recent times, they have been applied in a genetics environment in order to understand the underlying biological mechanisms. See Gao and Alvo (2005b) for an application in a more general situation. In the gene expression data of *Drosoplila melanogaster* (Jin et al. 2001) for example, there are 24 cDNA microarrays, 6 for each combination of two genotypes (Oregon R and Samarkand) and two sexes. As each array used two different dyes, there were in total 48 separate labeling reactions. Focusing on the individual expression level of a gene and its relationship with genotypes and sexes, the objective of the study was to identify genes whose expression levels are affected by the interaction between the two factors. For such data, the assumption of normality for the error terms is not warranted and consequently, nonparametric procedures are needed. We shall consider a nonparametric test for interaction based on the row ranks and column ranks of the data.

We consider the following general two-way layout with interaction

$$X_{ijn} = \mu + \alpha_i + \beta_j + \gamma_{ij} + \epsilon_{ijn}, i = 1, \ldots, I, j = 1, \ldots, J, n = 1, \ldots, N$$

where $X_{ijin}$ is the response, $\{\alpha_i\}$ and $\{\beta_j\}$ are main effects, $\{\gamma_{ij}\}$ are interaction effects, and $\{\epsilon_{ijn}\}$ are independent and identically distributed according to a continuous cumulative distribution $F_{ij}$. We wish to test the null hypothesis of no interaction effects

$$H_0 : \gamma_{ij} = 0 \text{ for all } i, j$$

against the alternative

$$H_1 : \gamma_{ij} \neq 0 \text{ for some } i, j.$$

We propose a test statistic based on both row and column ranks. This statistic is invariant under monotone transformations and therefore can be applied directly on the original data. In order to motivate the test, let $R_{ijn}$ be the rank of $X_{ijn}$ with respect to the entries in the $i$th row. Similarly, let $C_{ijn}$be the rank of $X_{ijn}$ with respect to the entries in the $j$th column. Define the score

$$a_{ijn} = \frac{R_{ijn}}{NJ+1} + \frac{C_{ijn}}{NI+1}. \tag{4.17}$$

Set the indicator function

$$u(x) = \begin{cases} 1 & x \geq 0, \\ 0 & x < 0. \end{cases}$$

It then follows that

$$\begin{aligned} E\left(a_{ijn}\right) &= \frac{1}{NJ+1}\sum_{b=1}^{J}\sum_{n'=1}^{N} Eu\left(X_{ijn} - X_{ibn'}\right) + \frac{1}{NI+1}\sum_{a=1}^{I}\sum_{n'=1}^{N} Eu\left(X_{ijn} - X_{ajn'}\right) \\ &= \frac{1}{NJ+1}\left(N\sum_{b=1}^{J}\int F_{ib}dF_{ij} + \frac{1}{2}\right) + \frac{1}{NI+1}\left(N\sum_{a=1}^{I}\int F_{aj}dF_{ij} + \frac{1}{2}\right). \end{aligned}$$

Under the null hypothesis of no interaction effects

$$\sum_{b}\int F_{ib}dF_{ij} = \int F\left(x - \alpha_i - \beta_0\right)dF\left(x - \alpha_i - \beta_j\right) = \int F\left(x + \beta_j - \beta_0\right)$$

which does not depend on i. Similarly, $\sum_a \int F_{aj}dF_{ij}$ does not depend on j. Setting

$$\overline{a}_{ij.} = \frac{1}{N}\sum_n a_{ijn}, \overline{a}_{i..} = \frac{1}{NJ}\sum_n a_{ijn}, \overline{a}_{.j.} = \frac{1}{NI}\sum_n a_{ijn}, \overline{a}_{...} = \frac{1}{NIJ}\sum_n a_{ijn}$$

it follows that

$$E\left(\overline{a}_{ij.} - \overline{a}_{i..} - \overline{a}_{.j.} + \overline{a}_{...}\right) = 0.$$

The quantity $\overline{a}_{ij.} - \overline{a}_{i..} - \overline{a}_{.j.} + \overline{a}_{...}$ serves as the nonparametric analogue of $\overline{X}_{ij.} - \overline{X}_{i..} - \overline{X}_{.j.} + \overline{X}_{...}$ which is the measure of the interaction effect appearing in the F statistic in the usual normal theory test for interaction.

### 4.3.1 Proposed Row–Column Test Statistic

In light of the motivation for the test statistic, define the sum of the row ranks in the $(i, j)$ cell

$$\mathbf{S}_N(i, j) = \frac{1}{NJ + 1} \sum_{n=1}^{N} R_{ijn}$$

and set

$$\mathbf{S}_N = (\mathbf{S}_N(1, 1), \ldots, \mathbf{S}_N(I, J))'.$$

Similarly, define the sum of the column ranks in the $(i, j)$ cell

$$\mathbf{T}_N(i, j) = \frac{1}{NI + 1} \sum_{n=1}^{N} C_{ijn}$$

and set

$$\mathbf{T}_N = (\mathbf{T}_N(1, 1), \ldots, \mathbf{T}_N(I, J))'.$$

Let $\mathbf{I}_I$ and $\mathbf{I}_J$ represent the I × I and J × J identity matrices, respectively, and let $\mathbf{J}_I$and $\mathbf{J}_J$ represent the I × I and J × J matrices with all elements equal to one, respectively. Set

$$\mathbf{A} = \mathbf{J}_I \otimes \left(-\frac{1}{I}\mathbf{I}_J\right) + \mathbf{I}_I \otimes \mathbf{I}_J,$$

$$\mathbf{B} = \mathbf{I}_J \otimes \left(\mathbf{I}_J - \frac{1}{J}\mathbf{J}_J\right).$$

We note that the $(i, j)$ term of $\frac{1}{N}(\mathbf{AS}_N + \mathbf{BT}_N)$ is $\left(\overline{a}_{ij.} - \overline{a}_{i..} - \overline{a}_{.j.} + \overline{a}_{...}\right)$. The proposed test statistic is then given by

$$W = \frac{1}{N}(\mathbf{AS}_N + \mathbf{BT}_N)' \left(\hat{\sum}\right)^{-} (\mathbf{AS}_N + \mathbf{BT}_N) \tag{4.18}$$

where $\left(\hat{\sum}\right)^{-}$ is the generalized inverse of the estimate of the variance-covariance matrix of $\mathbf{AS}_N + \mathbf{BT}_N$. The covariance matrix is not of full rank since there exist I+J-1 linear combinations of $\mathbf{AS}_N + \mathbf{BT}_N$ which are constants. We may obtain a formal expression for the estimate of the covariance matrix. Let

$$\sum\nolimits_1 = \lim \frac{1}{N} Var(\mathbf{S_N}), \sum\nolimits_2 = \lim \frac{1}{N} Var(\mathbf{T_N}), \sum\nolimits_{12} = \lim \frac{1}{N} cov(\mathbf{S_N}, \boldsymbol{T}_N).$$

Then

$$\hat{\sum} = \mathbf{A}\hat{\sum}_1 \mathbf{A}' + \mathbf{B}\hat{\sum}_2 \mathbf{B}' + 2\mathbf{A}\hat{\sum}_{12} \mathbf{B}'$$

where estimates of the covariances denoted by hats will be given in the next section.

### 4.3.2 Asymptotic Distribution of the Test Statistic Under the Null Hypothesis

The asymptotic distribution of the test statistic $W$ in (4.18) is a consequence of the general theory for linear rank statistics. We begin by recalling some theorems of Hajek.

Let $X_1, \ldots, X_N$ be independent random variables with continuous distribution functions $F_1, \ldots, F_N$, respectively. Let $R_i$ be the rank of $X_i$ among $X_1, \ldots, X_N$ and let $c_i, i = 1, \ldots, N$ be regression coefficients. Let $\alpha_N(x)$ be generated by a real values function $\phi(x)$ having a second derivative as

$$\alpha_N(i) = \phi\left(\frac{i}{N+1}\right).$$

A simple linear rank statistic takes the form

$$\mathbf{S} = \sum_{i=1}^{N} c_i \alpha_N(R_i).$$

Let

$$\overline{c} = \frac{1}{N}\sum c_i,$$

$$\overline{\phi} = \int_0^1 \phi(x)\,dx,$$

$$H(x) = \frac{1}{N}\sum F_i(x),$$

$$\mu = \sum c_i \int \phi(H(x))\,dF_i(x).$$

We quote the following two theorems from Hajek (1968).

**Theorem 4.6.** *Let*

$$L_i(x) = \frac{1}{N}\sum_{j=1}^{N}(c_j - c_i)\int [u(y-x) - F_i(x)]\,\phi'(H(x))\,dF_j(x)$$

*and*

$$\sigma^2 = \sum var\left(L_i\left(X_i\right)\right).$$

*If for every $\varepsilon > 0$, there exists $K_\epsilon$ such that*

$$Var\left(\mathbf{S}\right) > K_\epsilon \max_{1 \le i \le N} \left(c_i - \overline{c}\right)^2,$$

*then*

$$\max_{\infty < x < \infty} \left| P\left(\mathbf{S} - E\mathbf{S} < x\left(var\mathbf{S}\right)^{1/2}\right) - \Phi\left(x\right) \right| < \epsilon$$

*where $\Phi$ denotes the standard normal distribution function. The conclusion still holds if var($\mathbf{S}$) is replaced by $\sigma^2$. If $\sum c_i^2$ is bounded by a multiple of $\sum\left(c_i - \overline{c}\right)^2$, $\mathbf{ES}$ can be replaced by $\mu$ in the conclusion.*

We note that an integration by parts yields

$$\int \left[u\left(y - x\right) - F_i\left(x\right)\right] \phi'\left(H\left(x\right)\right) dF_i\left(x\right) = \int_x^\infty \phi'\left(H\left(y\right)\right) dF_j\left(y\right) + \text{constant}.$$

Moreover, $EL_i\left(X_i\right) = 0$.

The proof of Theorem 4.6 makes use of a projection argument. It is shown that the statistic $\mathbf{S} - E\mathbf{S}$ can be approximated best in the mean square sense by the statistic

$$\hat{\mathbf{S}} = \sum_{i=1}^{N} L_i\left(X_i\right)$$

which is the projection onto the Hilbert space generated by sums of independent square integrable linear functions of the $X_i$. The next result from Hajek makes this notion more precise.

**Theorem 4.7.** *Let $Z_i = L_i\left(X_i\right)$. There exists a constant M independent of N such that*

$$E\left(\mathbf{S} - E\mathbf{S} - \sum_{i=1}^{N} Z_i\right)^2 \le \frac{M}{N} \sum_{i=1}^{N} \left(c_i - \overline{c}\right)^2$$

*and*

$$E\left(\mathbf{S} - \mu\right)^2 \le \frac{M}{N} \sum_{i=1}^{N} c_i^2.$$

The proof of the asymptotic normality of our test statistic rests on extending Hajek's result to the study of composite linear rank statistics. We illustrate this result in the following simple situation whereby $X_1, \ldots, X_{n_1}, \ldots, X_{n_1+n_2}, \ldots, X_N$ are independent random variables. Consider the two simple linear rank statistics

$$\mathbf{S}_1 = \sum_{i=1}^{n_1+n_2} c_i^{(1)} a\left(R_i^{(1)}\right),$$

$$\mathbf{S}_2 = \sum_{i=n_1+1}^{N} c_i^{(2)} a\left(R_i^{(2)}\right),$$

where $R_i^{(1)}$ is the rank of $X_i$ among $\{X_1, \ldots, X_{n_1+n_2}\}$ and $R_i^{(2)}$ is the rank of $X_i$ among $\{X_{n_1+1}, \ldots, X_N\}$. We are interested in the asymptotic normality of the composite linear rank statistic formed by the sum

$$\mathbf{S} = \mathbf{S}_1 + \mathbf{S}_2.$$

This is done by adapting the projection argument. First, $\mathbf{S}_1$ is projected onto the space spanned by linear combinations of $\{X_1, \ldots, X_{n_1+n_2}\}$. Next, $\mathbf{S}_2$ is projected onto the space spanned by linear combinations of $\{X_{n_1+1}, \ldots, X_N\}$. Then the sum is projected onto the combined space $\{X_1, \ldots, X_N\}$. Let

$$W_i = \begin{cases} Z_i & i = 1, \ldots, n_1, \\ Z_i + Z_i^* & i = n_1 + 1, \ldots, n_1 + n_2, \\ Z_i^* & i = n_1 + n_2 + 1, \ldots, N, \end{cases}$$

where $Z_i = L_i(X_i)$ and $Z_i^* = L_i^*(X_i)$ are the respective projections. Here,

$$L_i(x) = \frac{1}{n_1+n_2} \sum_{j=1}^{n_1+n_2} \left(c_j^{(1)} - c_i^{(1)}\right) \int \left[u(y-x) - F_i(x)\right] \phi'(H_1(x))\, dF_j(x),$$

$$L_i^*(x) = \frac{1}{n_2+n_3} \sum_{j=1}^{n_2+n_3} \left(c_j^{(2)} - c_i^{(2)}\right) \int \left[u(y-x) - F_i(x)\right] \phi'(H_2(x))\, dF_j(x),$$

with

$$H_1(x) = \frac{1}{n_1+n_2} \sum_{i=1}^{n_1+n_2} F_i(x),\ H_2(x) = \frac{1}{n_2+n_3} \sum_{i=1}^{n_2+n_3} F_i(x).$$

We note that $EW_i = 0$.

**Theorem 4.8.** *Let $S_1, S_2$ be defined as above. Also let*

$$L_N = \max\left\{\sup\left(c_j^{(1)} - \overline{c}^{(1)}\right)^2, \sup\left(c_j^{(2)} - \overline{c}^{(2)}\right)^2\right\}$$

*and*

$$\sigma_N^2 = Var\left(\sum_{i=1}^{N} W_i\right).$$

*If the following condition holds*

$$\lim \frac{L_N}{\sigma_N^2} = 0, as \ \min\left(n_1 + n_2, n_2 + n_3\right) \to \infty,$$

*then*

$$\frac{\mathbf{S}_1 + \mathbf{S}_2 - E\left(\mathbf{S}_1 + \mathbf{S}_2\right)}{\sigma_N} \Rightarrow_L N\left(0, 1\right).$$

*Proof.* From (4.7), there exist constants $M_1, M_2$ such that

$$E\left(\mathbf{S}_1 - E\mathbf{S}_1 - \sum_{i=1}^{n_1+n_2} Z_i\right)^2 \leq \frac{M_1}{n_1 + n_2} \sum_{i=1}^{n_1+n_2} \left(c_i^{(1)} - \overline{c}^{(1)}\right)^2,$$

$$E\left(\mathbf{S}_2 - E\mathbf{S}_2 - \sum_{i=n_1+1}^{N} Z_i^*\right)^2 \leq \frac{M_2}{n_2 + n_3} \sum_{i=n_1+1}^{N} \left(c_i^{(2)} - \overline{c}^{(2)}\right)^2.$$

Hence,

$$E\left(\mathbf{S}_1 + \mathbf{S}_2 - E\left(\mathbf{S}_1 + \mathbf{S}_2\right) - \sum_{i=1}^{N} W_i\right)^2 \leq 2\left(M_1 + M_2\right) L_M.$$

It remains to show that $\sum_{i=1}^{N} W_i \sigma_N$ is asymptotically normally distributed with mean 0 and variance 1. This follows from the Lindeberg theorem. □

We state the general limiting distribution of the vector $(\mathbf{S}_N, \mathbf{T}_N)$.

**Theorem 4.9.** *Under the assumption that the errors $\{\epsilon_{ijn}\}$ are independent identically distributed in the two-way layout, we have that as $N \to \infty$*

*(i)* $\frac{1}{\sqrt{N}}\begin{pmatrix}\mathbf{S}_N - E(\mathbf{S}_N)\\ \mathbf{T}_N - E(\mathbf{T}_N)\end{pmatrix} \Longrightarrow N_{2IJ}(0, \Sigma)$

*(ii)* $W \Longrightarrow \chi^2_{(I-1)(J-1)}$ *under* $H_0$

*(iii)* $W \Longrightarrow \chi^2_{(I-1)(J-1)}(\delta)$ *under* $H_1$ *where* $\delta$ *is the noncentrality parameter under Pitman alternatives*

*Proof.* The proof makes use of projection arguments and Theorems 4.7 and 4.8 above. We refer the reader to Gao and Alvo (2005a) for details of the proof. □

To estimate the covariance of the test statistics define the following variables involving the empirical distribution functions:

$$C_{abn}^{(i,j)} = \begin{cases} -\frac{1}{NJ}\sum_{n'=1}^{N} u\left(X_{abn} - X_{ajn'}\right) & a = i, b \neq j, \\ \frac{1}{NJ}\sum_{j\neq j'}\sum_{n'=1}^{N} u\left(X_{abn} - X_{aj'n'}\right) & a = i, b = j, \\ 0 & a \neq i. \end{cases}$$

The fact that $C_{abn}^{(i,j)} - W_{abn}^{(i,j)} \to 0$ almost surely leads to the following consistent estimator

$$\hat{\sigma}_1^2\left(i, j, i', j'\right) = \sum_{a,b} \frac{1}{N} \sum_{n=1}^{N} \left(C_{abn}^{(i,j)} - \overline{C_{ab.}^{(i,j)}}\right)\left(C_{abn}^{(i',j')} - \overline{C_{ab.}^{(i',j')}}\right).$$

Similarly, defining

$$G_{abn}^{(i,j)} = \begin{cases} -\frac{1}{NI}\sum_{n'=1}^{N} u\left(X_{abn} - X_{ajn'}\right) & a \neq i, b = j, \\ \frac{1}{NI}\sum_{i\neq i'}\sum_{n'=1}^{N} u\left(X_{abn} - X_{aj'n'}\right) & a = i, b = j, \\ 0 & b \neq j, \end{cases}$$

we may construct consistent estimators for $\Sigma_2$ and $\Sigma_{12}$, respectively,

$$\hat{\sigma}_2^2\left(i, j, i', j'\right) = \sum_{a,b} \frac{1}{N} \sum_{n=1}^{N} \left(G_{abn}^{(i,j)} - \overline{G_{ab.}^{(i,j)}}\right)\left(G_{abn}^{(i',j')} - \overline{G_{ab.}^{(i',j')}}\right),$$

$$\hat{\sigma}_{12}^2\left(i, j, i', j'\right) = \sum_{a,b} \frac{1}{N} \sum_{n=1}^{N} \left(C_{abn}^{(i,j)} - \overline{C_{ab.}^{(i,j)}}\right)\left(G_{abn}^{(i',j')} - \overline{G_{ab.}^{(i',j')}}\right).$$

Gao and Alvo (2005a) report the result of some simulation studies which compare the proposed row–column statistic with the aligned test as well as the rank transform test. It is shown that the row–column test performs very well under a variety of underlying distributions including the normal, contaminated normal, and Cauchy. The following example was also considered.

*Example 4.2.* Consider the gene expression data of *D. melanogaster* of Jin et al. (2001). The gene fs(1) $k10$ is known to be expressed in reproductive systems and its expression level was reportedly affected by the gender and genotype interaction. The row–column statistic was applied to this data to account for the genotype, the gender, and the genotype-gender interaction. It was found that the interaction effect was statistically significant with a p-value equal to 0.004. The parametric F statistic and the aligned rank transform using the residuals yielded similar results. In order to illustrate the robustness of the nonparametric procedures, the analyses were redone with the first observation changed to an arbitrarily large number. The performance of the F statistic was severely affected and yielded a nonsignificant result. On the other hand, the nonparametric procedures were unaffected.

Next, we recall that in an example of a $3 \times 4$ factorial design considered by Box and Cox (1964) it was claimed that only after application of a nonlinear transformation can the error term be stabilized and the data made suitable for standard statistical analysis. We applied the row–column procedure to the untransformed data and obtained a p-value of 0.44. Thus the hypothesis of no interaction was not rejected, a finding that concurs with Box and Cox. The aligned test on the other hand yielded a p-value of 0.02 which indicates the presence of interaction. However, for the transformed data, the aligned test with a p-value of 0.45 did not reject the null hypothesis.

## Chapter Notes

Alvo et al. (1982) developed a new approach to test for randomness. This allowed the consideration of various distance functions including Kendall's distance. Theorems 4.2 and 4.3 provide the asymptotic distributions of the Spearman and Kendall test statistics in the complete randomized block design. Iman and Davenport (1980) describe the F distribution approximation to the Friedman statistic which is used later in Chap. 10.

One question of interest in connection with the asymptotic results is how accurate are the asymptotic distributions. Alvo and Cabilio (1984) considered the accuracy of the asymptotic distribution of Kendall's test statistic and compared it to other approximations for small values of $t$ and $n$. In addition, tables of the exact distribution were computed for $t = 3, n = 3, \ldots, 19; t = 4, n = 3, \ldots, 9$; and $t = 5, n = 3, 4, 5$. Some exact calculations are made of the Bahadur efficiency where it is demonstrated that the Kendall tau is more efficient.

Feigin and Alvo (1986) considered the two-group problem by placing it in the context of diversity and described an extensive discussion of the literature on the subject. Bu et al. (2009) developed an extension of the two-sample situation to the case where there are missing data. Although not discussed in this book, it may be of interest to consider the problem of paired comparisons whereby a judge ranks a set of objects before and after a treatment.

Gao and Alvo (2005b) provide a brief historical look at the analysis of unbalanced two-way layout with interaction effects. Using the notion of a weighted rank, they present tests for both main effects as well as for interaction effects. In addition, there is a discussion of the asymptotic relative efficiency of the proposed tests relative to the parametric F test. Various simulations further exemplify the power of the proposed tests. In a specific application, it is shown that the test statistic is the most robust in the presence of extreme outliers compared to other procedures.

Gao et al. (2008) also consider nonparametric multiple comparison procedures for unbalanced one-way factorial designs whereas Gao and Alvo (2008) treat nonparametric multiple comparison procedures for unbalanced two-way layouts.

# Chapter 5
# Block Designs

In the previous chapter, we were concerned with the study of complete randomized block designs. In biological studies involving animals, however, it is not always possible to compare several treatments within litters since the size of the litter will be a function of the particular species used. In such cases, it is then necessary to consider various types of incomplete experimental designs. The methodology presented here rests on the concept of compatibility and the extended notion of distance between rankings. This approach provides a natural extension of the well-known Friedman and Durbin statistics to some partially balanced incomplete designs. The tests developed are also applicable to general block designs with ties and multiple observations per cell.

We also address the general problem of what are the best choices of functions of the rankings based on considerations of efficiency.

## 5.1 Incomplete Block Designs

Consider the situation in which $t$ objects are ranked $k_h$ at a time, $2 \leq k_h \leq t$ by $b$ judges (blocks) independently and in such a way that each object is presented to $r_i$ judges and each pair of objects $(i, j)$ is compared by $\lambda_{ij}$ judges, $h = 1, \ldots, b$, $i, j = 1, \ldots, t$. We would like to test the hypothesis of no treatment effect, that is,

> $H_0$: *each judge, when presented with the specified $k_h$ objects, picks the ranking at random from the space of $k_h!$ permutations of* $(1, 2, \ldots, k_h)$.

In the study of the asymptotic behavior of various statistics for such problems, we consider $n$ replications of such basic designs. In the complete ranking case $k_h = t, b = 1$ for each block, so that the design becomes a randomized block. An example of a test in such a situation is the Friedman test with test statistic studied in Theorem 4.2



M. Alvo, P.L.H. Yu, *Statistical Methods for Ranking Data*, Frontiers in Probability and the Statistical Sciences, DOI 10.1007/978-1-4939-1471-5_5

$$G_S = \sum_{i=1}^{t} \left( R_i - \frac{n\,(t+1)}{2} \right)^2$$

where $R_i$ = sum of the ranks assigned by the judges to object $i$.

Under $H_o$, as $n \to \infty$,

$$\frac{1}{n} G_S \to_{\mathcal{L}} \left\{ \frac{t\,(t+1)}{12} \right\} \chi^2_{t-1}.$$

An interpretation of the Friedman statistic is that it is essentially the average of the Spearman correlations between all pairs of judges' rankings.

In the balanced incomplete block design (BIBD), we have $k_h = k, r_i = r$, $\lambda_{ij} = \lambda, bk = rt$, and $\lambda\,(t-1) = r\,(k-1)$. An example of a test in such a case is the Durbin test with test statistic

$$G_S = \sum_{i=1}^{t} \left( \frac{(t+1)}{(k+1)} R_i - \frac{nr\,(t+1)}{2} \right)^2.$$

Under $H_o$, as $n \to \infty$,

$$\frac{1}{n} G_S \to_{\mathcal{L}} \left\{ \frac{\lambda t (t+1)^2}{12\,(k+1)} \right\} \chi^2_{t-1}.$$

In order to measure the level of concordance of the judges, distance functions and their related measures of similarity can be defined on the $t!$-dimensional space of all complete rankings. The measure of similarity $\mathcal{A}\,(\mu_1, \mu_2)$ between the rankings $\mu_1$ and $\mu_2$ previously defined is an unstandardized rank correlation, which in many cases can be expressed as the inner product

$$\mathcal{A}\,(\mu_1, \mu_2) = \mathbf{t}\,(\mu_1)' \cdot \mathbf{t}\,(\mu_2),$$

where $\mathbf{t}\,(\mu)$ is a column vector whose components are scores which characterize the ranking $\mu$. We recall the Spearman, Kendall, and Hamming measures, respectively,

$$\mathcal{A}_S\,(\mu_1, \mu_2) = \sum_{i=1}^{t} \left( \mu_1\,(i) - \frac{t+1}{2} \right) \left( \mu_2\,(i) - \frac{t+1}{2} \right),$$

$$\mathcal{A}_K\,(\mu_1, \mu_2) = \sum_{i<j} sgn\,(\mu_1(j) - \mu_1\,(i))\, sgn\,(\mu_2(j) - \mu_2\,(i)),$$

$$\mathcal{A}_H\,(\mu_1, \mu_2) = \sum_{i=1}^{t} \sum_{j=1}^{t} \left( I\,[\mu_1\,(i) = j] - \frac{1}{t} \right) \left( I\,[\mu_2\,(i) = j] - \frac{1}{t} \right),$$

where $I\,[\cdot]$ is the indicator function, which is 1 or 0 depending on whether the statement in brackets holds or not. The dimensions of the $\mathbf{t}\,(\mu)$ vector depend on the measure chosen, so that for Spearman it is $t$, while for Kendall it is $\binom{t}{2}$ and for

Hamming $t^2$. Under $H_0$, the expected values of the components of the score vector $\mathbf{t}(\mu)$ are zero. As before, the collection of the score vectors $\mathbf{t}(\mu)$, as $\mu$ ranges over all its $t!$ possible values, is represented by the matrix $\mathbf{T}$. The $(t! \times t!)$ matrix $\mathbf{T}'\mathbf{T}$ has components $\mathcal{A}(\mu_1, \mu_2)$, with $\mu_1$ and $\mu_2$ ranging over all $t!$ permutations of $(1, 2, \ldots, t)$.

We also recall that in the incomplete ranking situation, the distance between $\mu_1^*$ and $\mu_2^*$ is defined as the *average of the distances between all the complete rankings compatible with each of* $\mu_1^*$ *and* $\mu_2^*$. Thus, with each set of incomplete $k_h$-rankings with a given pattern, we associate a $(t! \times k_h!)$ matrix of compatibility $\mathbf{C}_h$, whose column vectors are indicators that identify which of the $t!$ complete rankings are compatible with the particular $k_h$-permutation indexed by the column.

For a given pattern of $t - k_h$ missing observations, each permutation of the $k_h$ objects has its own distinct set of $t!/k_h!$ compatible $t$-rankings, so that each column of $\mathbf{C}_h$ contains exactly $t!/k_h!$ 1's and each row exactly one 1. For any incomplete $k_h$-ranking $\mu^*$, this definition can be shown to lead to an analogue of $\mathbf{T}$ given by

$$\mathbf{T}_h^* = \frac{k_h!}{t!}\mathbf{T}\mathbf{C}_h, \tag{5.1}$$

whose columns are the score vectors of

$$\mu^* = \left(\mu^*(1), \mu^*(2), \ldots, \mu^*(k_h)\right)'$$

as $\mu^*$ ranges over each of the $k_h!$ permutations.

*Example 5.1.* Let $t = 3$, $k_h = 2$. The complete rankings associated with the rows are in the order (123), (132), (213), (231), (312), (321). For the incomplete rankings (12_), (21_) indexing the columns, the associated matrix $\mathbf{C}_h$ is

$$\mathbf{C}_h = \begin{bmatrix} 1\,0 \\ 1\,0 \\ 0\,1 \\ 1\,0 \\ 0\,1 \\ 0\,1 \end{bmatrix}.$$

In the Spearman case, for the incomplete rankings (12_), (21_)

$$\frac{k_h!}{t!}\mathbf{T}_S\mathbf{C}_h = \frac{1}{3}\cdot \begin{bmatrix} -1 & -1 & 0 & 0 & 1 & 1 \\ 0 & 1 & -1 & 1 & -1 & 0 \\ 1 & 0 & 1 & -1 & 0 & -1 \end{bmatrix} \begin{bmatrix} 1\,0 \\ 1\,0 \\ 0\,1 \\ 1\,0 \\ 0\,1 \\ 0\,1 \end{bmatrix} = \begin{bmatrix} -\frac{2}{3} & \frac{2}{3} \\ \frac{2}{3} & -\frac{2}{3} \\ 0 & 0 \end{bmatrix}.$$

If for any such fixed incomplete ranking $\mu^*$ we denote by $\mathcal{C}(\mu^*)$ the class of complete rankings compatible with $\mu^*$, then, under $H_0$, the columns of $\mathbf{T}_h^*$ are the conditional expected scores $E(\mathbf{t}(\mu)|\mathcal{C}(\mu^*))$.

For a specific pattern of missing observations for each of the $b$ blocks, the matrix of scores is given by

$$\mathbf{T}^* = \left(\mathbf{T}_1^* \mid \mathbf{T}_2^* \mid \ldots \mid \mathbf{T}_b^*\right) = \mathbf{T}\left(\frac{k_1!}{t!}\mathbf{C}_1 \mid \frac{k_2!}{t!}\mathbf{C}_2 \mid \ldots \mid \frac{k_b!}{t!}\mathbf{C}_b\right),$$

where the index of each block identifies the pattern of missing observations, if any, in that block. Patterns may or may not be different from block to block, so that the $\mathbf{T}_i^*$'s need not be distinct. In this setting the null hypothesis $H_0$ becomes that each judge, presented with $k_h$ objects, picks a ranking of these uniformly from the set of $k_h!$ possible rankings. Because of interblock independence, the matrix $(\mathbf{T}^*)'\,\mathbf{T}^*$ for this specified design contains, under $H_0$, components of the form

$$A\left(\mu_1^*, \mu_2^*\right) = E\left(\mathcal{A}(\mu_1, \mu_2)\,|\mathcal{C}\left(\mu_1^*\right), \mathcal{C}\left(\mu_2^*\right)\right),$$

and defines a type of unstandardized rank correlation between incomplete rankings. Recall that $\mathbf{T}'\mathbf{T}$ is the matrix of similarities $\left[\mathcal{A}\left(\mu_i, \mu_j\right)\right]$. In order to detail the elements of $(\mathbf{T}^*)'\,\mathbf{T}^*$, consider the case of say $\left(\mathbf{T}_1^*\right)'\mathbf{T}_2^* = \frac{k_1!}{t!}\mathbf{C}_1'\left(\mathbf{T}'\mathbf{T}\right)\mathbf{C}_2\frac{k_2!}{t!}$. The first row of $\mathbf{C}_1'\left(\mathbf{T}'\mathbf{T}\right)$ gives row elements

$$\left[\sum_m \mathcal{A}(\mu_m, \mu_1), \sum_m \mathcal{A}(\mu_m, \mu_2), \ldots, \sum_m \mathcal{A}(\mu_m, \mu_{t!})\right],$$

where the summation is over all complete rankings compatible with the permutation $\mu_1^*$ identified with the first columns of $\mathbf{C}_1$.

If we now multiply this row by any column of $\mathbf{C}_2$, the resulting element will be

$$\sum_\ell \sum_m \mathcal{A}(\mu_m, \mu_\ell)$$

where the outer summation is over all complete rankings compatible with the permutation $\mu_2^*$ identified with that column of $\mathbf{C}_1$. The proposed test rejects for large values of

$$G \equiv \left(\mathbf{T}^*\mathbf{f}\right)'\left(\mathbf{T}^*\mathbf{f}\right) = \mathbf{f}'\left(\mathbf{T}^*\right)'\mathbf{T}^*\mathbf{f}, \tag{5.2}$$

where the $\sum_{h=1}^{b} k_h!$-dimensional vector of frequencies $\mathbf{f}$ is given by

$$\mathbf{f} = \left(\mathbf{f}_1 \mid \mathbf{f}_2 \mid \ldots \mid \mathbf{f}_b\right)'$$

and $\mathbf{f}_h$ is the $k_h!$-dimensional vector of the observed frequencies of each of the $k_h!$ ranking permutations for the incomplete pattern $h = 1, 2, \ldots, b$.

## 5.2 Asymptotic Distribution

The asymptotic distribution of the various test statistics is given in the following:

**Theorem 5.1.** *Under $H_0$ as $n \to \infty$,*

$$\frac{1}{\sqrt{n}}\mathbf{T}^*\mathbf{f} \to_{\mathcal{L}} \mathcal{N}(\mathbf{0}, \mathbf{\Gamma}), \tag{5.3}$$

*where $\mathcal{N}(\mathbf{0}, \mathbf{\Gamma})$ is a multivariate normal with mean 0 and covariance*

$$\mathbf{\Gamma} = \sum_{h=1}^{b} \frac{1}{k_h!}\left(\mathbf{T}_h^*\right)\left(\mathbf{T}_h^*\right)' = \sum_{h=1}^{b} \frac{1}{k_h!}\left(\frac{k_h!}{t!}\mathbf{TC}_h\right)\left(\frac{k_h!}{t!}\mathbf{TC}_h\right)'. \tag{5.4}$$

*Thus*

$$n^{-1}G = n^{-1}\left(\mathbf{T}^*\mathbf{f}\right)'\left(\mathbf{T}^*\mathbf{f}\right) \to_{\mathcal{L}} \sum \alpha_i z_i^2, \tag{5.5}$$

*where $\{z_i\}$ are independent identically distributed normal variates and $\{\alpha_i\}$ are the eigenvalues of $\mathbf{\Gamma}$.*

*Proof.* The proof is a straightforward application of standard multivariate normal theory. □

Note that the covariance of $\mathbf{T}^*\mathbf{f}$ can be written in an explicit form, and often its eigenvalues can be found analytically. If necessary, the eigenvalues of $\mathbf{\Gamma}$ may readily be calculated numerically since the dimension of $\mathbf{\Gamma}$ is not very large, being $(t \times t)$ in the Spearman case and $\left(\binom{t}{2} \times \binom{t}{2}\right)$ in the Kendall case.

In the following, we will need to make use of notation to differentiate between the elements of the space of possible incomplete rankings with a particular pattern of missing observations and a ranking picked from this space. For a given pattern of missing observations $h = 1, \ldots, b$, the permutations of $(1, 2, \ldots, k_h)$ comprise the space of possible rankings, and these permutations may be indexed by $s = 1, 2, \ldots, k_h!$. Let $\mu_{h(s)}^*(i)$ denote the rank of object $i$ for the permutation indexed by $s$ in block pattern $h$, while $\mu_h^*(i)$ will denote the observed rank of object $i$ for this block.

## 5.3 Spearman Case

We shall be interested in determining the eigenvalues of the $\mathbf{\Gamma}$ matrix. It will be shown that there is a direct link to the information matrix encountered in the study of experimental designs.

**Lemma 5.1.** *The* $t \times t$ *matrix*

$$\Gamma = \sum_{h=1}^{b} \frac{1}{k_h!}\left(\frac{k_h!}{t!}\mathbf{T}_S\mathbf{C}_h\right)\left(\frac{k_h!}{t!}\mathbf{T}_S\mathbf{C}_h\right)' = \sum_{h=1}^{b} \gamma_h^2 \mathbf{A}_h$$

*where*

$$\mathbf{A}_h = \begin{cases} \frac{k_h-1}{k_h}\delta_h(j) & \text{on the diagonal} \\ -\frac{1}{k_h}\delta_h(j)\,\delta_h(j') & \text{off the diagonal} \end{cases} \tag{5.6}$$

*and* $\gamma_h^2 = \frac{1}{(k_h-1)}\sum_{j=1}^{k_h}\left(\frac{(t+1)}{(k_h+1)}j - \frac{(t+1)}{2}\right)^2$.

*Proof.* Consider the matrix $\mathbf{T}_h^* = \frac{k_h!}{t!}\mathbf{T}_S\mathbf{C}_h$ , where $\mathbf{T}_S$ is the $(t \times t!)$ matrix of Spearman score vectors. The vectors $\mathbf{t}(\mu)$ are vectors of mean adjusted ranks. For a given permutation of $(1, 2, \ldots, k_h)$, indexed by $s = 1, 2, \ldots, k_h!$, the corresponding $i$th row element of column $s$ of $\frac{k_h!}{t!}\mathbf{T}_S\mathbf{C}_h$ is found to be

$$\frac{(t+1)}{(k_h+1)}\left(\mu_{h(s)}^*(i) - \frac{k_h+1}{2}\right)\delta_h(i) = \left(\frac{(t+1)}{(k_h+1)}\mu_{h(s)}^*(i) - \frac{t+1}{2}\right)\delta_h(i); \tag{5.7}$$

$\delta_h(i)$ is either 1 or 0 depending on whether the object $i$ is, or is not, ranked in block $h$, and $\mu_{h(s)}^*(i)$, as defined above, is the rank of object $i$ for the permutation indexed by $s$ for block pattern $h$. An $(i, j)$ element of $\left(\frac{k_h!}{t!}\mathbf{T}_S\mathbf{C}_h\right)\left(\frac{k_h!}{t!}\mathbf{T}_S\mathbf{C}_h\right)'$ is thus of the form

$$\sum_{s=1}^{k_h!}\left(\frac{(t+1)}{(k_h+1)}\mu_{h(s)}^*(i) - \frac{t+1}{2}\right)\left(\frac{(t+1)}{(k_h+1)}\mu_{h(s)}^*(j) - \frac{t+1}{2}\right)\delta_h(i)\delta_h(j),$$

and a direct evaluation of this yields the result. □

The elements of $\Gamma$ are thus

$$\begin{cases} (t+1)^2\left[\frac{1}{12}\sum_{h=1}^{b}\frac{k_h-1}{k_h+1}\delta_h(j)\right] & \text{on the diagonal,} \\ (t+1)^2\left[-\frac{1}{12}\sum_{h=1}^{b}\frac{1}{k_h+1}\delta_h(j)\delta_h(j')\right] & \text{off the diagonal.} \end{cases} \tag{5.8}$$

Note that the elements of each row of $\Gamma$ sum to 0, so that $rank\,(\Gamma) \leq t - 1$.

In the case of a basic design of $b$ blocks, so that the number of replications is $n = 1$, the test statistic is found as follows. Since

$$\mathbf{T}_S^*\mathbf{f} = \sum_{h=1}^{b}\frac{k_h!}{t!}\mathbf{T}_S\mathbf{C}_h\mathbf{f}_h,$$

then its $i$th component is given by

$$\sum_{h=1}^{b}\frac{(t+1)}{(k_h+1)}\left(\mu_h^*(i)-\frac{k_h+1}{2}\right)\delta_h(i)=\sum_{h=1}^{b}\left(\frac{(t+1)}{(k_h+1)}\mu_h^*(i)-\frac{t+1}{2}\right)\delta_h(i), \tag{5.9}$$

where $\mu_h^*(i)$ is the observed rank of object $i$ in block $h$. In terms of $n \geq 2$ replications of the basic design of $b$ blocks, the statistic is

$$n^{-1}G_S = n^{-1}\left(\mathbf{T}_S^*\mathbf{f}\right)'\left(\mathbf{T}_S^*\mathbf{f}\right) = n^{-1}\sum_{i=1}^{t}\left(S_{nb}^*(i) - nr_i\frac{(t+1)}{2}\right)^2, \tag{5.10}$$

where

$$S_{nb}^*(i) = \sum_{h=1}^{nb}\frac{(t+1)}{(k_h+1)}\mu_h^*(i)\,\delta_h(i) \tag{5.11}$$

is a weighted sum of the ranks given to treatment $i$ and $r_i = \sum_{h=1}^{nb}\delta_h(i)$. Under $H_0$, $n^{-1/2}\mathbf{T}^*\mathbf{f}$ has mean 0 and covariance matrix $\mathbf{\Gamma}$, so that an alternative form of the statistic is

$$n^{-1}\left(\mathbf{T}_S^*\mathbf{f}\right)\mathbf{\Gamma}^-\left(\mathbf{T}_S^*\mathbf{f}\right) = n^{-1}\left(\mathbf{S}_{nb} - n\frac{(t+1)}{2}\mathbf{r}\right)'\mathbf{\Gamma}^-\left(\mathbf{S}_{nb} - n\frac{(t+1)}{2}\mathbf{r}\right) \tag{5.12}$$

where

$$\mathbf{S}_{nb} = \left(S_{nb}^*(1), S_{nb}^*(2), \ldots, S_{nb}^*(t)\right)', \mathbf{r} = (r_1, r_2, \ldots, r_t)',$$

and $\mathbf{\Gamma}^-$ is a generalized inverse of $\mathbf{\Gamma}$. This statistic has an asymptotic $\chi^2$-distribution with degrees of freedom equal to $rank\,(\mathbf{\Gamma}) \leq t-1$.

The matrix $\mathbf{\Gamma}$ with elements given in Lemma 5.1 is closely related to the *information matrix* of a block design. John and Williams (1995) detail how this matrix occurs in the least squares estimation of treatment effects and the role its eigenvalues play in determining optimality criteria for choosing between different designs. This information matrix $\mathbf{A}$ has components

$$\begin{cases} \sum_{h=1}^{b}\frac{k_h-1}{k_h}\delta_h(i) & \text{on the diagonal,} \\ -\sum_{h=1}^{b}\frac{1}{k_h}\delta_h(i)\delta_h(j) & \text{off the diagonal.} \end{cases} \tag{5.13}$$

Note that $\mathbf{A}$ and $\mathbf{\Gamma}$ share the same rank.

In some designs there may be one or more sets of objects which have the property that there is no block in which objects of one set are ever compared to objects of another set. These are known as "non-compared" sets in Benard and van Elteren

(1953), and their existence leads to a "disconnected" design. A design which is not disconnected is said to be *connected*. The rank of $\mathbf{\Gamma}$ in a connected design will be $t-1$.

There are various important connected designs for which the number of objects ranked in each block is $k_h = k$, and the number of judges who rank object $i$ is $r_i = r$.

In such cases

$$\mathbf{\Gamma} = \frac{k\,(t+1)^2}{12\,(k+1)}\mathbf{A},$$

so that we may make use of the known eigenvalues of $\mathbf{A}$ for various incomplete block designs in order to apply Theorem 5.1 to the statistic $n^{-1}G_S$.

### 5.3.1 Applications

*Example 5.2 (Complete Case and the BIBD).* In the case of a complete randomized block design, every treatment appears in every block $k = t, b = 1$. The statistic is the Friedman test,

$$n^{-1}G_S = n^{-1}\sum_{i=1}^{t}\left(S_n\,(i) - n\frac{(t+1)}{2}\right)^2$$

where $S_n\,(i) =$ sum of the ranks assigned by the judges to object $i$. The eigenvalues of $\mathbf{\Gamma}$ are

$$\alpha_1 = t\,(t+1)\,/12,$$

with multiplicity $t-1$, so that under $H_o$, as $n \to \infty$,

$$n^{-1}G_S \to_{\mathcal{L}} \{t\,(t+1)\,/12\}\,\chi^2_{t-1}.$$

Similarly in the BIBD, $\lambda_{ij} = \lambda, bk = rt$, and $\lambda\,(t-1) = r\,(k-1)$. The statistic is the well-known Durbin test. The eigenvalues of $\mathbf{\Gamma}$ are

$$\alpha_1 = \frac{1}{12}\lambda t(t+1)^2\,(k+1)^{-1},$$

with multiplicity $t-1$, so that under $H_o$, as $n \to \infty$,

$$n^{-1}G_S \to_{\mathcal{L}} \left\{\frac{1}{12}\lambda t(t+1)^2\,(k+1)^{-1}\right\}\chi^2_{t-1}.$$

**Table 5.1** Example of a group divisible design

| Treatments | 1 | 2 | 3 | 4 | 5 | 6 |
|---|---|---|---|---|---|---|
| Block 1 | X | | X | X | | |
| Block 2 | | X | | X | X | |
| Block 3 | | | X | | X | X |
| Block 4 | | | | X | X | |
| Block 5 | X | | | X | | X |
| Block 6 | X | X | | | X | |
| Block 7 | | X | X | | | X |

In the parametric case, the efficiency factor for the BIBD relative to the randomized complete block design is given as the ratio of the variances between two treatment means. It is given here as

$$\frac{\lambda t}{rk} = \frac{r(k-1)t}{rk(t-1)}$$
$$= \frac{t}{(t-1)}\left(1 - \frac{1}{k}\right)$$

which shows that the efficiency increases with increasing block size. The BIBD will be more precise if the ratio of the variance of the BIBD to the complete randomized block design is smaller than $\frac{t}{(t-1)}\left(1 - \frac{1}{k}\right)$.

*Example 5.3 (Group Divisible Designs).* It may not always be possible to construct BIBDs. For example, the smallest number of replications for a BIBD is $r = \frac{\lambda(t-1)}{(k-1)}$. When $t = 6, k = 4$, the balanced design would require $r = 10$ replications and hence $rt = 60$ experimental units. Such a large number of units may either not be available or too costly for the researcher to acquire. The partially balanced group divisible design helps to reduce the number of experimental units required at the cost of forcing some comparisons between treatments to be more precise than others. In this design, the $t$ objects occur in $g$ groups of $d$ objects, $t = gd$. Within each group pairs of objects are compared by $\lambda_1$ judges, whereas each pair of objects between groups is compared by $\lambda_2$ judges. Such designs must satisfy the additional conditions

$$bk = rt, r(k-1) = (d-1)\lambda_1 + d(g-1)\lambda_2$$

If $\lambda_1 = \lambda_2$, then we have a BIBD.

We illustrate such a design in Table 5.1 for $t = 6, k = 4, r = 2, g = 2, \lambda_1 = 1, \lambda_2 = 2$. Treatments $(1,4)$, $(2,5)$, $(3,6)$ are compared $\lambda_2 = 2$ times whereas all the other pairs are compared only once. The number of experimental units required here is $rt = 12$ compared to the 60 for a BIBD.

In general, the construction of such designs and, the form of $\mathbf{A}$ and its eigenvalues are all given in John and Williams (1995). The eigenvalues of $\mathbf{\Gamma}$ are

**Table 5.2** Example of a cyclic design

| Treatments | 1 | 2 | 3 | 4 | 5 | 6 |
|---|---|---|---|---|---|---|
| Block 1 | X | X | | X | X | |
| Block 2 | | X | X | | X | X |
| Block 3 | X | | X | X | | X |

$$\begin{cases} \alpha_1 = \frac{(t+1)^2}{12} \frac{\{r(k-1)+\lambda_1\}}{(k+1)} & \text{with multiplicity } g\,(d-1)\,, \\ \alpha_2 = \frac{(t+1)^2}{12} \frac{t\lambda_2}{(k+1)} & \text{with multiplicity } (g-1)\,. \end{cases}$$

In the parametric case, the efficiency for the comparison of two treatments for the first group is

$$\frac{(\lambda_1 + rk - r)}{rk}$$

and for the second group

$$\frac{t\lambda_2\,(\lambda_1 + rk - r)}{rk\,(\lambda_1 + \lambda_2 t - \lambda_2)}.$$

For our example, these efficiencies become 0.875 for the first group and 0.955 for the second group. On the other hand, for the corresponding BIBD, the efficiency is 0.9.

*Example 5.4 (Cyclic Designs).* Incomplete block designs often require the use of tables which may not always be available in the field. Moreover, care must be taken to record data correctly for such designs. Cyclic designs on the other hand are easily constructed from an initial block and the treatments can be easily assigned. Such designs are obtained by cyclic development of an initial block or combinations of such sets. Let $\lambda_0 = r$ and $\lambda_{j-1} = \lambda_{1j}, j = 2, \ldots, t$, be the number of judges that compare object 1 with object $j$. The matrix $\mathbf{A}$ is a circulant related to the matrix derived by the cyclic development of $(\lambda_0, \lambda_1, \ldots, \lambda_{t-1})$ and its eigenvalues are given in John and Williams (1995). The eigenvalues of $\mathbf{\Gamma}$ are

$$\alpha_i = \frac{(t+1)^2}{12} \left\{ \frac{r\,(k-1)}{(k+1)} - \frac{1}{(k+1)} \sum_{h=1}^{t-1} \lambda_h \cos\left( \frac{2\pi i h}{t} \right) \right\}, \quad i = 1, 2, \ldots, t-1.$$

As an example of a cyclic design, consider Table 5.2 with $t = 6, k = r = 3, b = 6$. Note that this is not a BIBD since, for example, $\lambda_{14} = 2 \neq \lambda_{13} = 1$.

Finally, we note that the canonical efficiency factors of an incomplete block design in the parametric case are given in terms of the eigenvalues of a matrix related to the information matrix (John and Williams 1995). As we have seen, that matrix also arises in the nonparametric context. Consequently, we may by analogy define in exactly the same manner the efficiency factors for the nonparametric case. We do not pursue this subject further in this book.

## 5.4 Kendall Case

Recall that the Kendall scores are given by the $\left(\binom{t}{2} \times t!\right)$ matrix $\mathbf{T}_K$ where the row elements of each column of $\mathbf{T}_K$ are $sgn\,(\mu_s(j) - \mu_s\,(i))$ for each object pair $(i, j)$, $1 \leq i < j \leq t$.

The element of $\mathbf{T}_h^* = \frac{k_h!}{t!}\mathbf{T}_K\mathbf{C}_h$ in the row which corresponds to the given object pair $(i, j)$, $i < j$, and the column $s = 1, 2, \ldots, k_h!$, which indexes a given permutation of $(1, 2, \ldots, k_h)$, is

$$a_{h(s)}(i,j) = \begin{cases} sgn(\mu^*_{h(s)}(j) - \mu^*_{h(s)}(i)) \text{ if } \delta_h(i) = \delta_h(j) = 1, \\ 1 - \dfrac{2\mu^*_{h(s)}(i)}{(k_h + 1)} & \text{if } \delta_h(i) = 1, \delta_h(j) = 0, \\ \dfrac{2\mu^*_{h(s)}(j)}{(k_h + 1)} - 1 & \text{if } \delta_h(i) = 0, \delta_h(j) = 1, \\ 0 & \text{otherwise,} \end{cases} \tag{5.14}$$

where, as previously, for the block pattern $h$, $\mu^*_{h(s)}(i)$ is the rank of object $i$ in the permutation indexed by $s$ and $\delta_h(i)$ is either 1 or 0 depending on whether the object $i$ is, or is not, ranked in block $h$.

We consider a simple example. For the incomplete rankings (12_), (21_)

$$\frac{k_h!}{t!}\mathbf{T}_K\mathbf{C}_h = \frac{1}{3}\begin{bmatrix} 1 & 1 & -1 & 1 & -1 & -1 \\ 1 & 1 & 1 & -1 & -1 & -1 \\ 1 & -1 & 1 & -1 & 1 & -1 \end{bmatrix}\begin{bmatrix} 1 & 0 \\ 1 & 0 \\ 0 & 1 \\ 1 & 0 \\ 0 & 1 \\ 0 & 1 \end{bmatrix} = \begin{bmatrix} 1 & -1 \\ \frac{1}{3} & -\frac{1}{3} \\ -\frac{1}{3} & \frac{1}{3} \end{bmatrix}.$$

In general, the component of $\mathbf{T}^*\mathbf{f} = \sum_{h=1}^{b} \frac{k_h!}{t!}\mathbf{T}_K\mathbf{C}_h\mathbf{f}_h$ corresponding to object pair $(i, j)$ is given by $\sum_{h=1}^{b} a_h(i, j)$, where, for a given block pattern $h$, $a_h(i, j)$ is that value corresponding to the observed ranking $\mu_h^*$. Thus the test statistic in this case is

$$G_K = \sum_{i<j}\left\{\sum_{h=1}^{b} a_h(i,j)\right\}^2, \tag{5.15}$$

which in terms of $n$ replications of a basic design would be altered by taking the inner sum over all $nb$ blocks, with the statistic becoming $n^{-1}G_K$. The elements of $\left(\frac{k_h!}{t!}\mathbf{T}_K\mathbf{C}_h\right)\left(\frac{k_h!}{t!}\mathbf{T}_K\mathbf{C}_h\right)'$ (and thus of the covariance $\boldsymbol{\Gamma}$) are somewhat more complicated to write down in general than in the Spearman case and are given in Alvo and Cabilio (1996) for the case where $k_h = k$ for all blocks. As an alternative to the statistic $n^{-1}G_K$, we may again proceed as in the Spearman case and define the statistic

$$n^{-1}K = n^{-1}\left(\mathbf{T}^*\mathbf{f}\right)'\boldsymbol{\Gamma}^-\left(\mathbf{T}^*\mathbf{f}\right). \tag{5.16}$$

For the complete and balanced incomplete designs the matrix $\boldsymbol{\Gamma}$ is shown to be of full rank $\binom{t}{2}$, so that in such cases the inverse of $\boldsymbol{\Gamma}$ replaces $\boldsymbol{\Gamma}^-$ in Brunden and Mohberg (1976). In the Spearman situation the statistics are the same for such balanced designs, but this is not the case with the Kendall-based statistics.

In the complete case $b = 1, t = k, r = r_i = 1, \lambda_{ij} = \lambda = 1$. Let the index

$$q(i,j) = (i-1)(t-\frac{i}{2}) + (j-i), 1 \leq i < j \leq t,$$

correspond to the object pair $(i,j)$.

The $(q(i,j), q(\ell,m))$ element of $\boldsymbol{\Gamma}$ is

$$\begin{cases} 1 & i = \ell, j = m \text{ (diagonal)}, \\ \frac{1}{3} & i = \ell, j \neq m \text{ or } i \neq \ell, j = m, \\ -\frac{1}{3} & i = m, j \neq \ell \text{ or } i \neq m, j = \ell, \\ 0 & i, j, \ell, m \text{ all different} \end{cases} \tag{5.17}$$

while the Kendall scores become

$$a_s(i,j) = sgn(\mu_s(j) - \mu_s(i)) = 2I\left[\mu_s(j) > \mu_s(i)\right] - 1, i < j.$$

A test for this design of the form Brunden and Mohberg (1976) was introduced by Wormleighton (1959). The differences between the two forms of such Kendall-based tests are explored in the following example.

*Example 5.5.* Consider once again the case with $t = 3$. Although in practice the statistic $n^{-1}G_K$ would be calculated directly, it is useful in this example to write it in general as a function of the observed frequencies through the form $(\mathbf{Tf})'(\mathbf{Tf})$. In this case we index the $t! = 6$ possible permutations (123), (132), (213), (231), (312), (321) in this order from $s = 1$ to 6, so that $f_s$ refers to the observed frequency in the $n$ trials of the ranking indexed by $s$. Using the fact that in this case $\boldsymbol{\Gamma} = \frac{1}{6}\mathbf{TT}'$ gives

$$\boldsymbol{\Gamma} = \begin{bmatrix} 1 & -\frac{1}{3} & \frac{1}{3} \\ -\frac{1}{3} & 1 & -\frac{1}{3} \\ \frac{1}{3} & -\frac{1}{3} & 1 \end{bmatrix},$$

so that

$$\boldsymbol{\Gamma}^{-1} = \frac{3}{4}\begin{bmatrix} 2 & -1 & 1 \\ -1 & 2 & -1 \\ 1 & -1 & 2 \end{bmatrix}.$$

Direct computation of Brunden and Mohberg (1976) with $\mathbf{T}$ replacing $\mathbf{T}^*$ shows that

$$K = 3\left\{(f_1 - f_6)^2 + (f_2 - f_5)^2 + (f_3 - f_4)^2\right\},$$

while a calculation of $(\mathbf{Tf})'(\mathbf{Tf})$ gives

$$G_K = K + 2\{(f_1 - f_6)(f_2 - f_5) + (f_1 - f_6)(f_3 - f_4) - (f_2 - f_5)(f_3 - f_4)\}.$$

Note that while $K$ and $G_K$ both compare conjugate rankings, the statistic $G_K$ also includes interactions between these comparisons.

The null distributions of these statistics are quite different. For example, in the case where $n = 3$, $K$ does not differentiate between

| Object | 1 2 3 |
|---|---|
| Judge 1 | 3 2 1 |
| Judge 2 | 3 2 1 |
| Judge 3 | 3 1 2 |

| Object | 1 2 3 |
|---|---|
| Judge 1 | 3 2 1 |
| Judge 2 | 3 2 1 |
| Judge 3 | 1 3 2 |

assuming the same value of 15 for both. On the other hand, $G_K = 19$ for the first and 11 for the second. It may be argued that the judges in the first realization share a higher level of agreement than do those in the second.

The null distribution of $G_K$ is much richer than that of $K$. In the case $n = 6$, using the lower-tail probabilities in Table I in Wormleighton (1959), the upper-tail probabilities of $K$ with probability 0.2 or less are

| $u$ | 108 | 78 | 60 | 54 | 48 | 42 | 36 |
|---|---|---|---|---|---|---|---|
| $P(K \geq u)$ | 0.0001 | 0.0032 | 0.0109 | 0.0315 | 0.0400 | 0.1017 | 0.1172 |

while use of Table II in Alvo and Cabilio (1984), with the quantity there labeled $x$ transformed to $u = 18(5x + 1)$, gives us the upper-tail probabilities of $G_K$

| $u$ | 108 | 88 | 76 | 72 | 68 | 56 | 52 | 48 |
|---|---|---|---|---|---|---|---|---|
| $P(G_K \geq u)$ | 0.0001 | 0.0017 | 0.0055 | 0.0081 | 0.0135 | 0.0367 | 0.0521 | 0.0606 |

| $u$ | 44 | 40 | 36 | 32 |
|---|---|---|---|---|
| $P(G_K \geq u)$ | 0.0760 | 0.0914 | 0.1519 | 0.1905 |

## 5.5 Hamming Case

We introduced the Hamming distance in Chap. 3. As noted, this distance has found applications in coding theory for binary strings. In this section we consider its applications in the analysis of block designs. We recall that the distance is defined as

$$d_H(\mu,\nu) = t - \sum_{i=1}^{t}\sum_{j=1}^{t} I(\mu(i)=j)\, I(\nu(i)=j) \tag{5.18}$$

where $I(.)$ is the indicator function which is 1 or 0 depending on whether the statement in brackets holds or not. The following lemmas provide information on the test statistic based on the Hamming distance. Define

$$\phi_h(j,l) = E\left(I\left(\mu(i)=j\right)|C\left(\mu^*\right)\right).$$

It is seen

$$\phi_h(j,l) = \begin{cases} \binom{j-1}{l-1}\binom{t-j,}{k_h-l}\binom{t}{k}^{-1} & 1 \le l \le j, 0 \le k_h \le t-j \\ 0 & otherwise. \end{cases}$$

Then, from (5.1), the element of $\mathbf{T}_h^* = \frac{k_h!}{t!}\mathbf{T}_H C_h$ in row $q(i,j)$ and column $s = 1,\ldots,k_h!$ which indexes a given permutation of $(1,2,\ldots,k_h)$ is

$$\left(\phi_h\left(j,\mu^*_{h(s)}(i)\right) - \frac{1}{t}\right)\delta_h(i)$$

where $\mu^*_{h(s)}(i)$ is the rank of object $i$ in the permutation indexed by $s$ and $\delta_h(i)$ is either 1 or 0 depending on whether object i is or is not ranked in the $h$ block.

The component $q(i,j)$ of $\mathbf{T}^* f = \sum \frac{k_h!}{t!}\mathbf{T}_H C_h f_h$ becomes

$$\sum_{h=1}^{b}\sum_{l=1}^{k_h}\left\{\phi_h(j,l) - \frac{1}{t}\right\}\delta_h(i)\, I\left[\mu_h^*(i) = l\right]$$

leading to the test statistic

$$G_H = \sum_{i=1}^{t}\sum_{j=1}^{t}\left\{\sum_{h=1}^{b}\sum_{l=1}^{k_h}\left\{\phi_h(j,l) - \frac{1}{t}\right\}\delta_h(i)\, I\left[\mu_h^*(i) = l\right]\right\}^2. \tag{5.19}$$

**Lemma 5.2.** *In the special case where each judge ranks exactly k out of t, the test statistic in (5.19) is given by*

$$G_H = \sum_{i=1}^{t}\sum_{j=1}^{t}\left\{\sum_{l=1}^{k}\left\{\phi(j,l)\,D_i(l) - \frac{1}{t}\right\}\right\}^2 - \frac{1}{t}\sum_{i=1}^{t} r_i^2$$

*where $D_i(l)$ is the number of judges that assign rank $l$ to object $i$ and $r_i$ is the number of judges who consider object $i$.*

*Proof.* We note that

$$D_i(l) = \sum_{h=1}^{b} \delta_h(i)\, I\left[\mu_h^*(i) = l\right]$$

and

$$r_i = \sum_{l=1}^{k} D_i(l)\,.$$

Hence

$$G_H = \sum_{i=1}^{t}\sum_{j=1}^{t}\left\{\sum_{l=1}^{k}\left\{\phi(j,l)\,D_i(l) - \frac{r_i}{t}\right\}\right\}^2.$$

Since

$$\sum_{j=l}^{t-k+l} \phi(j,l) = 1$$

the cross term above becomes

$$-\frac{2}{t}\sum_{i=1}^{t} r_i^2$$

and the result follows. □

We note that in the context of n replications of a basic design of b blocks, $r_i$ in Lemma 5.2 would be replaced by $nr_i$, $D_i(l)$ would be taken over all $nb$ blocks, and the statistic used would be $G_H/n$. In the next lemma, we obtain a specific form for the elements of the covariance matrix $\boldsymbol{\Gamma}$.

**Lemma 5.3.** *The $(q(i,j), q(l,m))$ element of the $t^2 \times t^2$ matrix*

$$\frac{1}{k_h!}\left(\frac{k_h!}{t!}\mathbf{T}_H C_h\right)\left(\frac{k_h!}{t!}\mathbf{T}_H C_h\right)'$$

*with index* $q(i,j) = (i-1)t + j, 1 \le i, j \le t$ *is*

$$\begin{cases} \frac{1}{k_h}\left\{\psi_h(j,j) - \frac{k_h}{t^2}\right\}\delta_h(i) & i = l, j = m, \\ \frac{1}{k_h}\left\{\psi_h(j,m) - \frac{k_h}{t^2}\right\}\delta_h(i) & i = l, j \neq m, \\ -\frac{1}{k_h(k_h-1)}\left\{\psi_h(j,j) - \frac{k_h}{t^2}\right\}\delta_h(i)\,\delta_h(l) & i \neq l, j = m, \\ -\frac{1}{k_h(k_h-1)}\left\{\psi_h(j,m) - \frac{k_h}{t^2}\right\}\delta_h(i)\,\delta_h(l) & i \neq l, j \neq m, \end{cases}$$

*where for* $u = \max(j,m) - t + k_h, v = \min(j,m)$

$$\psi_h(j,m) = \begin{cases} \sum_{x=u}^{v} \binom{j-1}{x-1}\binom{t-j}{k_h-x}\binom{m-1}{x-1}\binom{t-m}{k_h-x}\binom{t}{k_h}^{-2} & u \le v, \\ 0 & otherwise. \end{cases}$$

*Proof.* The proof proceeds by considering a typical element of the product

$$\sum_{s=1}^{k_h!}\left\{\phi_h\left(j, \mu^*_{h(s)}\right) - \frac{1}{t}\right\}\left\{\phi_h\left(m, \mu^*_{h(s)}\right) - \frac{1}{t}\right\}\delta_h(i)\,\delta_h(l)$$

and noting that

$$\sum_{x=1}^{k_h}\phi_h(j,x) = \frac{k_h}{t}.$$

□

We shall now consider the special case when each judge ranks exactly k out t objects. Define the $(t \times k)$ matrix $H = (\phi(j,l))$ and set $E = HH' - \frac{k}{t^2}J$ to be a $t \times t$ matrix. We may then write the Kronecker product

$$\mathbf{\Gamma} = \frac{1}{k-1}\boldsymbol{A} \otimes E$$

where $\boldsymbol{A} \otimes E$ is the matrix formed by multiplying each element of $E$ by the previously defined matrix $\boldsymbol{A}$ with

$$\boldsymbol{A} = \sum \boldsymbol{A_h}.$$

Hence, the eigenvalues of $\mathbf{\Gamma}$ are $\frac{1}{k-1}\alpha_i\eta_j$ where the $\alpha_i$'s and $\eta_j$'s are the eigenvalues of $\boldsymbol{A}$ and $E$, respectively.

### 5.5.1 Applications

*Example 5.6 (Complete Case).* In the complete case $k = t, r_i = \lambda_{il} = 1, H = I$, and $E = A = \left(I - \frac{J}{t}\right)$, so that

$$\mathbf{\Gamma} = \frac{1}{t-1}\left(I - \frac{J}{t}\right) \otimes \left(I - \frac{J}{t}\right).$$

In view of the fact that $\left(I - \frac{J}{t}\right)$ is idempotent and has rank $(t-1)$, it follows that $\mathbf{\Gamma}$ has a distinct eigenvalue $\frac{1}{t-1}$ with multiplicity $(t-1)^2$ . Hence

$$\frac{t-1}{n} G_H \Rightarrow \chi^2_{(t-1)^2}.$$

In that case,

$$\begin{aligned} G_H &= \sum\sum\left\{D_i(l) - \frac{n}{t}\right\}^2 \\ &= \sum\sum D_i^2(l) - n^2. \end{aligned}$$

This test was first introduced by Anderson (1959) and rediscovered by Kannemann (1976) who claimed that it was more sensitive to small differences in treatment location than the Friedman test.

*Example 5.7 (General Case).* Let $J_{i\times j}$ be the $i \times j$ matrix of ones. Since $HJ'_{t\times k} = \frac{k}{t} J_{t\times t}$, it follows that

$$\left(H - \frac{J_{t\times k}}{t}\right)\left(H - \frac{J_{t\times k}}{t}\right)' = E. \tag{5.20}$$

Moreover, $\left(H - \frac{J_{t\times k}}{t}\right)$ is orthogonal to $J_{t\times k}$ so that its rank is $(k-1)$. In the next lemma we state a result for the eigenvalues of the matrix $E$.

**Lemma 5.4.** *In the Hamming case with $k_h = k$, the $(k-1)$ eigenvalues of the matrix E given in (5.20) are*

$$\eta_j = \binom{t-j-1}{k-j-1}\binom{t+j}{k+j}\binom{t}{k}^{-2}, j = 1, \ldots, k-1.$$

*Proof.* The proof appears in Alvo and Cabilio (1998) where it is shown that the corresponding eigenvectors are given by the Chebyshev polynomials. □

*Example 5.8 (The BIBD).* In the BIBD design, $r_i = r$ and the design matrix $\mathbf{A} = \frac{\lambda t}{k}\left(I - \frac{J}{t}\right)$ has a distinct eigenvalue $\frac{\lambda t}{k}$ with multiplicity $(t-1)$. Thus the eigenvalues of $\mathbf{\Gamma}$ are

$$\frac{\lambda t}{k(k-1)}\eta_j, i = 1, \ldots, k-1,$$

each with multiplicity $(t-1)$. In that case,

$$G_H = \sum_{i=1}^{t}\sum_{j=1}^{t}\left\{\sum_{l=1}^{k}\phi(j,l)\,D_i(l)\right\}^2 - r^2.$$

The Anderson statistic was also studied by Shach (1979) who modified it instead and showed that as n→ ∞

$$\left(\frac{t-1}{t}\right)\left[\frac{1}{nb}\sum_{i=1}^{t}\sum_{l=1}^{k}\left\{D_i(l) - \frac{nb}{t}\right\}^2\right] \Rightarrow_L \chi^2_{(k-1)(t-1)}.$$

## 5.6 Score Statistics

The statistics that have been considered thus far were based directly on the ranks themselves without consideration of whether they are optimal in any sense. In this section we shall consider a generalization of such statistics by replacing the ranks assigned by each judge by real valued functions $a(j, k_i)$, $1 \le j \le k_i \le t$, which we shall call scores. We wish to test the hypothesis

> $H_0$: *each judge, when presented with the specified $k_i$ objects, picks the ranking at random from the space of $k_i!$ permutations of* $(1, 2, \ldots, k_i)$.

For a given function $a(j,t)$, $1 \le j \le t$, and a complete ranking $\mu$, define the vector of adjusted scores

$$\mathbf{a}(\mu) = (a(\mu(1), t) - \overline{a}_t, a(\mu(2), t) - \overline{a}_t, \ldots, a(\mu(t), t) - \overline{a}_t)' \quad (5.21)$$

where under $H_0$,

$$\overline{a}_t = t^{-1}\sum_{r=1}^{t} a(r,t),$$

For a random ranking $\mu$, it can be shown that the covariance matrix of $\mathbf{a}(\mu)$ is given by

$$\boldsymbol{\Sigma}_0 = \frac{t}{t-1}\sigma_0^2\left(\mathbf{I} - \frac{1}{t}\mathbf{J}\right) \quad (5.22)$$

where

$$\sigma_0^2 \equiv Var(a(\mu(j), t)) = t^{-1}\sum_{r=1}^{t}(a(r,t) - \overline{a}_t)^2,$$

**I** is the identity $(t \times t)$ matrix, and **J** is the $(t \times t)$ matrix of 1's. Let

$$S_n(j) = \sum_{i=1}^{n} a(\mu_i(j), t),$$

where $\mu_i(j), i = 1, \ldots, n$, represents the observed rankings of object $j$. In analogy with the Spearman measure of similarity between the complete rankings $\mu_1$ and $\mu_2$ define

$$\mathcal{A}(\mu_1, \mu_2) = \mathbf{a}(\mu_1)' \mathbf{a}(\mu_2) = \sum_{j=1}^{t} (a(\mu_1(j), t) - \overline{a}_t)(a(\mu_2(j), t) - \overline{a}_t). \tag{5.23}$$

Let the $(t \times t!)$ matrix **T** represent the collection of adjusted score vectors $\mathbf{a}(\mu)$ as $\mu$ ranges over all its $t!$ possible values, and let **f** be the $t!$ vector of frequencies of the observed rankings. The $(t! \times t!)$ matrix $\mathbf{T}'\mathbf{T}$ has components $\mathcal{A}(\mu_1, \mu_2)$, with $\mu_1$ and $\mu_2$ ranging over all $t!$ permutations of $(1, 2, \ldots, t)$. Let

$$\mathbf{S}_n = (S_n(1), S_n(2), \ldots, S_n(t))',$$

so that $\mathbf{Tf} = (\mathbf{S}_n - n\overline{a}_t\mathbf{1})$, where $\mathbf{1}$ is the $t$-vector of 1's. Proceeding as before, the proposed statistic is the quadratic form

$$n^{-1}(\mathbf{Tf})'(\mathbf{Tf}) = n^{-1} \sum_{j=1}^{t} (S_n(j) - n\overline{a}_t)^2. \tag{5.24}$$

Large values of this statistic are inconsistent with $H_0$. Under $H_0$, as $n \to \infty$, $n^{-1/2}\mathbf{Tf}$ is asymptotically normal with mean vector $\mathbf{0}$ and covariance matrix $\mathbf{\Sigma}_0$. Consequently, since $\mathbf{I} - \frac{1}{t}\mathbf{J}$ is an idempotent matrix of rank $t-1$, the statistic

$$Q_n = \frac{t-1}{nt\sigma_0^2} \sum_{j=1}^{t} (S_n(j) - n\overline{a}_t)^2$$

has asymptotically as $n \to \infty$ a $\chi^2$ distribution with $t-1$ degrees of freedom.

Consider now the incomplete block situation in which the $i$th judge ranks $2 \leq k_i \leq t$ objects. In this setup, a basic design of $b$ blocks is defined and then replicated $n$ times, so that a total of $nb$ judges rank the specified subsets of the $t$ objects. Recall that $r_j$ is the total number of blocks that include object $j$. For an incomplete ranking pattern let $\mu_{h(s)}$ $s = 1, 2, \ldots, k_h!$ denote the possible $k_h$-rankings, that is, the permutations of $(1, 2, \ldots, k_h)$. The analogue of the adjusted matrix **T**, the collection of adjusted score vectors $\mathbf{a}(\mu)$ is given by

$$\mathbf{T}^* = \left(\mathbf{T}_1^* \mid \mathbf{T}_2^* \mid \ldots \mid \mathbf{T}_b^*\right) \tag{5.25}$$

where

$$\mathbf{T}_h^* = \frac{k_h!}{t!}\mathbf{TC}_h$$

and $\mathbf{C}_h$ is the compatibility matrix associated with $\mu_{h(s)}, s = 1, 2, \ldots, k_h!$. Denote by $\mathcal{C}(\mu_{h(s)})$ the class of complete rankings compatible with the specified $k_h$-permutation $\mu_{h(s)}$ indexed by column $s$ of $\mathbf{C}_h$. Under $H_0$, the columns of $\mathbf{T}_h^*$ are the conditional expected scores $E\left(\mathbf{a}(\mu)\left|\mathcal{C}(\mu_{h(s)})\right.\right)$. We define the notion of a weighted score as follows.

**Definition 5.1.** For a given score function $a(j,t)$, $1 \leq j \leq t$, if object $j$ is ranked in a given block $1 \leq h \leq b$ and if $\mu_h(j) = r$, the weighted score is given by

$$a^*(r, k_h) = \sum_{q=r}^{t-k_h+r} a(q,t)\binom{q-1}{r-1}\binom{t-q}{k_h-r}\binom{t}{k_h}^{-1}, r = 1, 2, \ldots, k_h. \quad (5.26)$$

**Theorem 5.2.** *For an incomplete ranking with* $\mu_{h(s)}$ $s = 1, 2, \ldots, k_h!$ *denoting the permutations of* $(1, 2, \ldots, k_h)$ *within the specified incomplete pattern, the row* $j$*, column* $s$ *element of* $\mathbf{T}_h^* = \frac{k_h!}{t!}\mathbf{TC}_h$ *is*

$$\left(a^*\left(\mu_{h(s)}(j), k_h\right) - \overline{a}_t\right)\delta_h(j)$$

*Proof.* See Alvo and Cabilio (2005) for details of the proof which follows by computing the conditional expectation the score onto the space of compatible rankings. □

Under $H_0$, for an incomplete ranking $\mu_h$, we have

$$\begin{aligned} k_h^{-1}\sum_{r=1}^{k_h} a^*(r, k_h) &= E\left(a^*(\mu_h(j), k_h)\right) \\ &= E\left(a(\mu(j), t)\right) \\ &= \overline{a}_t \end{aligned}$$

so that the vector of adjusted weighted scores,

$$\begin{aligned} \mathbf{a}(\mu_h) = &\left[\left(a^*(\mu_h(1), k_h) - \overline{a}_t\right)\delta_h(1), \left(a^*(\mu_h(2), k_h) - \overline{a}_t\right)\delta_h(2), \ldots,\right. \\ &\times \left.\left(a^*(\mu_h(t), k_h) - \overline{a}_t\right)\delta_h(t)\right]', \end{aligned}$$

has the covariance matrix $\mathbf{\Sigma}_h = \gamma_h^2\mathbf{A}_h$, where

$$\gamma_h^2 = (k_h - 1)^{-1}\sum_{r=1}^{k_h}\left(a^*(r, k_h) - \overline{a}_t\right)^2$$

and the $(t \times t)$ matrix

$$\mathbf{A}_h = \begin{cases} \frac{k_h-1}{k_h}\delta_h(j) & \text{on the diagonal,} \\ -\frac{1}{k_h}\delta_h(j)\,\delta_h(j') & \text{off the diagonal.} \end{cases} \tag{5.27}$$

In order to extend the notation for the weighted scores to all the $nb$ rankings, define

$$a^*(r, k_i) = a^*(r, k_h) \text{ if } i \equiv h \,(\text{mod } b) \; for \; b+1 \le h \le nb.$$

The test statistic in this more general setting is given by

$$n^{-1}(\mathbf{T}^*\mathbf{f})'(\mathbf{T}^*\mathbf{f}) = n^{-1}\sum_{j=1}^{t}\left(\sum_{i=1}^{nb}\left(a^*(\mu_i(j), k_i) - \overline{a}_t\right)\delta_i(j)\right)^2 \tag{5.28}$$

Set $\mathbf{S}_{nb} = \left(S^*_{nb}(1), S^*_{nb}(2), \ldots, S^*_{nb}(t)\right)'$, $\mathbf{r} = (r_1, r_2, \ldots, r_t)'$, where

$$S^*_{nb}(j) = \sum_{i=1}^{nb} a^*(\mu_i(j), k_i)\,\delta_i(j),$$

so that the test statistic is

$$G_n = n^{-1}(\mathbf{S}_{nb} - n\bar{a}_t\mathbf{r})'(\mathbf{S}_{nb} - n\bar{a}_t\mathbf{r}) = n^{-1}\sum_{j=1}^{t}\left(S^*_{nb}(j) - nr_j\bar{a}_t\right)^2 \tag{5.29}$$

which is asymptotically distributed as $\sum \alpha_i z_i^2$, where $\{z_i\}$ are independent identically distributed normal variates and $\{\alpha_i\}$ are the eigenvalues of $\boldsymbol{\Sigma}_0$. Under $H_0$, $n^{-1/2}\mathbf{T}^*\mathbf{f}$ has covariance matrix

$$\boldsymbol{\Sigma}_0 = \sum_{i=1}^{b}\gamma_i^2\mathbf{A}_i, \tag{5.30}$$

so that an alternative form of the statistic is

$$n^{-1}(\mathbf{T}^*\mathbf{f})'\,\boldsymbol{\Sigma}_0^-\mathbf{T}^*\mathbf{f} = n^{-1}(\mathbf{S}_{nb} - n\bar{a}_t\mathbf{r})'\,\boldsymbol{\Sigma}_0^-(\mathbf{S}_{nb} - n\bar{a}_t\mathbf{r}) \tag{5.31}$$

where $\boldsymbol{\Sigma}_0^-$ is a generalized inverse of $\boldsymbol{\Sigma}_0$.

We shall require the following assumption.

**Assumption A.** *The random variables $(X_{i1}, X_{i2}, \ldots, X_{it})$, which may or may not be observable, underlie the rankings*

$$\mu_i = (\mu_i(1), \mu_i(2), \ldots, \mu_i(t))', i = 1, \ldots, b.$$

*These random variables are assumed independent with absolutely continuous distribution functions* $F_{ij}(x) = F_i(x - \tau_j)$, *where* $\sum \tau_j = 0$ *and* $F_i(x)$ *have continuous densities* $f_i(x)$, *for which* $\int_{-\infty}^{\infty} f_i^2(x)\,dx < \infty$.

The null hypothesis of random uniform selection of rankings becomes

$$H_0' : \tau = 0,$$

where $\boldsymbol{\tau}' = (\tau_1, \tau_2, \ldots, \tau_t)$. If the asymptotics of interest are simply that the number of blocks $b$ becomes large, the definitions and notation used earlier may be modified by setting $n = 1$ as appropriate. The test statistic may be rewritten as

$$G_b^* = (\mathbf{S}_b - \bar{a}_t \mathbf{r})' \boldsymbol{\Sigma}_0^- (\mathbf{S}_b - \bar{a}_t \mathbf{r}),$$

and, as $b \to \infty$, if the design is connected, $G_b^*$ has an asymptotic $\chi^2$-distribution with $t-1$ degrees of freedom.

When the score function is $a(j,t) = j$, the Wilcoxon score, the weighted score becomes

$$\begin{aligned} a^*(r, k_i) &= r \sum_{q=r}^{t-k_h+r} \binom{q}{r} \binom{t-q}{k_h - r} \binom{t}{k_h}^{-1} \\ &= r \binom{t+1}{t-k_i} \binom{t}{k_h}^{-1} \\ &= r\,\frac{t+1}{k_i+1}. \end{aligned}$$

The measure of similarity is simply the Spearman measure and our previous results hold. We may now consider the asymptotic distribution of the test statistic under the alternative:

$$H_{1n} : \tau = \tau_n = n^{-1/2}\theta, \theta = (\theta_1, \theta_2, \ldots, \theta_t)'.$$

The following theorem provides the basis for determining optimal score functions.

**Theorem 5.3.** *Under the alternative,* $n^{-1} (\mathbf{S}_{nb} - n\bar{a}_t \mathbf{r})' \boldsymbol{\Sigma}_0^- (\mathbf{S}_{nb} - n\bar{a}_t \mathbf{r})$ *has for* $n \to \infty$ *a noncentral chi-square distribution with* $t-1$ *degrees of freedom and a specified noncentrality parameter* $\delta_n(a^*)$. *The form of this noncentrality parameter is simplified if* $k_h = k$ *for all* $h = 1, 2, \ldots, b$, *and, in such a case, this parameter is maximized when*

$$a^*(r,k) - \bar{a}_t = c(\beta_{r-2} - \beta_{r-1}). \tag{5.32}$$

*where* $\beta_s = \binom{k-2}{s} \int_{-\infty}^{\infty} F(x)^s (1 - F(x))^{k-2-s} f^2(x)\,dx,\ s = 0, 1, \ldots, k-2.$

*Proof.* See Alvo and Cabilio (2005) for the proof. □

### 5.6.1 Special Score Functions

We consider weighted scores which are associated with certain score functions $a_f(r,t)$ detailed for the complete ranking case in Sen (1968). In the complete case, under **Assumption A**, and with the additional assumption that block effects are additive so that the underlying density is $f(x)$, the score functions which optimize the test statistics satisfy the relationship

$$a_f(r,t) - \overline{a}_t = c(\beta_{r-2} - \beta_{r-1}) \tag{5.33}$$

where $\beta_s$ is defined above with $t$ replacing $k_h$. The corresponding weighted score for an incomplete $k_i$-block is

$$a^*(r,k_i) = \sum_{q=r}^{t-k_i+r} a_f(q,t)\binom{q-1}{r-1}\binom{t-q}{k_i-r}\binom{t}{k_i}^{-1}. \tag{5.34}$$

1. For the Wilcoxon score, $a_f(j,t) = j$, and $Q_n$ is optimal in the case that the rankings result from samples from the logistic distribution with density $f(x) = e^{-x}/(1+e^{-x})^2, -\infty < x < \infty$. The weighted score may be written as $a^*(r,k_i) = \left(\frac{t+1}{k_i+1}\right) a_f(r,k_i)$.
2. With the score $a_f(1,t) = 1, a_f(t,t) = -1, a_f(r,t) = 0$ otherwise, $Q_n$ is optimal when sampling from the uniform distribution with density $f(x) = 1; 0 \le x \le 1$. Direct substitution into (5.32) gives $a^*(r,k_i) = \frac{k_i}{t} a_f(r,k_i)$.
3. With the score $a_f(1,t) = 1, a_f(r,t) = 0$ otherwise, $Q_n$ is optimal when sampling from the exponential distribution with density $f(x) = e^{-x}; 0 \le x < \infty$. Again, direct substitution into (5.32) gives $a^*(r,k_i) = \frac{k_i}{t} a_f(r,k_i)$.
4. When sampling from the double exponential distribution with density $f(x) = e^{-|x|}, -\infty < x < \infty$, $Q_n$ associated with the score function $a_f(r,t) = 1 - 2\sum_{i=0}^{r-1}\binom{t}{i}2^{-t}$ is shown to be optimal. It may be shown that $a^*(r,k_i) = a_f(r,k_i)$.

All the weighted scores associated with these score functions have the property that

$$a^*(r,k_i) = K(k_i,t)\, a_f(r,k_i),$$

where $K(k_i,t)$ depends only on the number of objects ranked in the block. Note that since

$$k_i^{-1}\sum_{r=1}^{k_i} a^*(r,k_i) = \overline{a}_t,$$

it follows that for such scores, $\overline{a}_t = K(k_i,t)\,\bar{a}_{k_i}$. If the design is such that the number of objects ranked in each block is constant, that is, $k_i = k$ for all $i$, these weighted scores $a^*(r,k)$ satisfy the relation (5.32). Thus the statistic

$$n^{-1}(\mathbf{S}_{nb} - n\bar{a}_t\mathbf{r})'\,\boldsymbol{\Sigma}_0^-\,(\mathbf{S}_{nb} - n\bar{a}_t\mathbf{r})$$

is optimal in the cases enumerated above.

## Chapter Notes

Alvo and Cabilio (1996) consider non-doubly balanced incomplete block designs (DBIBD), whereby each triplet of objects is not necessarily presented to the same number of judges and they obtain the asymptotic distribution for the Kendall statistic in this case. Although we have concentrated on the measures of Spearman and Kendall, the methods presented here for the analysis of block designs are quite general and applicable to other measures of similarity between rankings, particularly to those that can be written as inner products of score vectors. These methods have been applied in Alvo and Cabilio (1998) to the case of Hamming measure. On the whole, unlike various other approaches to such problems, the resulting tests have forms which are easily calculated immediate extensions of their versions in the complete block situation.

The asymptotic distributions are linear combinations of independent chi squares, with coefficients that are given analytically for many designs based on the Spearman or Kendall measures and which can in any case be calculated quite simply. Once the coefficients are determined, the critical values can be approximated using a procedure such as that in Jensen and Solomon (1972).

The statistics may be modified in order to simplify their asymptotic distributions to chi square, but this comes at the cost of making the statistics more complex. The example given previously for the Kendall case further shows that such statistics may have exact distributions whose support is less dense than that of the forms derived here.

Caution should be exercised in the use of the large sample critical values in conducting a small sample test. Various studies indicate that at least for complete block and BIBDs, other approximations to the small sample critical values may be a great deal more accurate (see for example Alvo and Cabilio 1995b). One approach which may have some value in dealing with small samples and with unbalanced designs is to generate the p-values of the test by simulation methods.

Ties for Hamming distance are also discussed in Alvo and Cabilio (1998). The discussion on the choice of scores follows closely the development in Alvo and Cabilio (2005) for the incomplete case where some simulation results are reported. This in turn was motivated by the work of Sen (1968)

A companion result to Lemma 5.4 showing a further use of Chebyshev polynomials appears in Alvo and Cabilio (2000). In particular, it is shown that one can compute values of the hypergeometric distributions recursively.

# Chapter 6
# General Theory of Hypothesis Testing

The notion of distance was fruitfully utilized in previous chapters in order to develop tests of hypotheses for both complete and incomplete rankings. In this chapter we consider a more general framework for constructing tests of hypotheses. We begin by defining two sets of rankings: one set consists of all the rankings which are most in agreement with the observed ranking while the second set contains all the rankings which are most in agreement with the alternative hypothesis. A distance function is then defined between those two sets of rankings. The notion of distance between sets is well known in mathematics and is often taken to be the minimum distance between pairs of elements, one from each set. In the present statistical context however, the more workable definition of distance is chosen to be the average of all pairwise distances between pairs of rankings, one from each set. Then the test rejects the null hypothesis whenever the distance between the two sets is small. Following a description of the basic construction of the sets, we then consider some general hypothesis testing problems. We begin with the multi-sample location problem with ordered alternatives and then consider tests with umbrella alternatives. The general theory is further exemplified in Chap. 7.

## 6.1 The Construction

The construction of the test statistic follows a simple procedure. Let $H_0, H_1$ denote the null and alternative hypotheses, respectively, in a typical testing situation. Let $\mathcal{P}_{\mathrm{n}} = \{\mu : [\mu(1), \ldots, \mu(n)]\}$ be the set of all permutations of the integers $1, 2, \ldots, n$ and let $d(\mu, \nu)$ be a distance function between the two permutations $\mu, \nu$.

Step 1: Let $X_1, \ldots, X_n$ be a collection of random variables from continuous distributions and let $\pi(i)$ be the rank of $X_i$ among the $X's$. The continuity assumption ensures that with probability one there can be no ties in the permutation



M. Alvo, P.L.H. Yu, *Statistical Methods for Ranking Data*, Frontiers in Probability and the Statistical Sciences, DOI 10.1007/978-1-4939-1471-5_6

$$\pi = [\pi (1), \ldots, \pi (n)]'.$$

Step 2: Define $\{\pi\}$ to be the subclass of permutations which are "equivalent" to the observable permutation $\pi$ in the sense that ranks occupied by identically distributed random variables are exchangeable.

Step 3: Define $E$ to be the subclass of extremal permutations consisting of all permutations which are most "in agreement" with $H_1$.

Step 4: Define the distance between the subclasses $\{\pi\}$ and $E$ to be the sum of pairwise distances between permutations $\mu, \nu$ with $\mu \epsilon \{\pi\}, \nu \epsilon E$:

$$d (\{\pi\}, E) = \sum_{\mu \epsilon \{\pi\}} \sum_{\nu \epsilon E} d (\mu, \nu).$$

It is seen consequently that small values of $d (\{\pi\}, E)$ are consistent with the alternative and lead to the rejection of the null hypothesis. It should be pointed out that the extremal set $E$ is not identical to the entire critical region of the test but rather consists of those permutations exhibiting the strongest evidence in favor of $H_1$. In the next sections we consider various examples of tests of hypotheses and develop corresponding test statistics. Before proceeding we consider an example to illustrate the computation.

*Example 6.1.* Consider the two sample problem and suppose we wish to test the hypothesis $H_0 : F_1 (x) = F_2 (x)$ against the alternative $H_1 : F_1 (x) \geq F_2. (x)$ for some $x$. Suppose that we observe $\mathrm{x}_1 = 2, x_2 = 4.5$ from $F_1$ and $\mathrm{x}_3 = 4, x_4 = 7$ from $F_2$. Then, using the convention that the rankings of the first population are placed first, the observed ranking $\pi = [1, 3, 2, 4]$. The subclass $\{\pi\} = \{[1324], [3124], [1342], [3142]\}$. The extremal set $E$ consists of those permutations which allocate ranks $3, 4$ to population 1 and ranks $1, 2$ to population 2. Hence, $E = \{[3412], [4312], [3421], [4321]\}$. The sum of the pairwise distances using the Spearman distance is then given by

$$\begin{aligned} d_S (\{\pi\}, E) &= \sum_{\mu \epsilon \{\pi\}} \sum_{\nu \epsilon E} d_S (\mu, \nu) \\ &= 112. \end{aligned}$$

## 6.2 The Multi-Sample Location Problem with Ordered Alternatives

We consider now the general location problem with ordered alternatives. Let $F_1 (x), \ldots, F_r (x)$ be r continuous distributions and suppose that we wish to test

$$H_0 : F_r (x) = \ldots = F_1 (x)$$

against the ordered alternative

$$H_1 : F_r(x) \leq \ldots \leq F_1(x)$$

with strict inequality for some $x$. Let $X_{N_{k-1}+1}, \ldots, X_{N_k}$ be a random sample of size $n_k$ from $F_k(x)$ where $N_0 = 0$ and $N_k = n_1 + \ldots + n_k, k = 1, \ldots, r$. Rank the $N_r$ observations among themselves and write the permutation $\pi$ as

$$\pi = [\pi(1), \ldots, \pi(N_1) | \ldots, |\pi(N_{r-1}+1), \ldots, \pi(N_r)]'$$

where ranks from the same distribution are placed together. The equivalent subclass $\{\pi\}$ corresponding to $\pi$ may be written as

$$\{\pi\} = \Big\{ [\pi(i_{11}), \ldots, \pi(i_{1n_1}) | \ldots, |\pi(N_{r-1}+i_{r1}), \ldots, \pi(N_r + i_{m_r})] : \\ (i_{k1}, \ldots, i_{kn_k}) \epsilon \mathcal{P}_{n_k}, 1 \leq k \leq r \Big\}'$$

and it consists of all permutations of the integers $1, \ldots, N_r$ which assign the same set of ranks to the populations as $\pi$ does. On the other hand, the extremal set consists of all permutations which assign ranks $N_{k-1}+1, \ldots, N_k$ to population $k$. Formally,

$$E = \{[i_{11}, \ldots, i_{1n_1} | \ldots, |N_{r-1}+i_{r1}, \ldots, N_r + i_{m_r}] : (i_{k1}, \ldots, i_{kn_k}) \epsilon \mathcal{P}_{n_k}, 1 \leq k \leq r\}.$$

The main result is the following theorem.

**Theorem 6.1.** *The test statistics corresponding to various distances for the location problem with ordered alternatives are indicated below. The null hypothesis is rejected in all cases whenever the distances are small.*

*(i) Spearman*

$$d_S(\{\pi\}, E) = \Pi_{k=1}^r \frac{(n_k!)^2 N_r (N_r+1)(4N_r-1)}{12} \\ - \frac{1}{2}\left[\Pi_{k=1}^r (n_k!)^2 \sum_{k=1}^r (N_{k-1}+N_k) \sum_{i=N_{r_{k-1}}+1}^{N_k} \pi(i)\right].$$

*(ii) Kendall*

$$d_K(\{\pi\}, E) = \Pi_{k=1}^r \frac{(n_k!)^2 N_r (N_r-1)}{2} \\ + \Pi_{k=1}^r (n_k!)^2 \sum_{k=1}^{r-1} \sum_{t=k}^{r-1} \sum_{i=N_{k-1}+1}^{N_k} \sum_{j=N_k+1}^{N_{k+1}} (sgn(\pi(i) - \pi(j))).$$

*(iii) Spearman Footrule*

$$d_F(\{\pi\}, E) = 2\Pi_{k=1}^r (n_k!)^2 \sum_{k=1}^r \sum_{j_1=N_{k-1}+1}^{N_k} \sum_{j_2=N_{k-1}+1}^{N_k} \frac{[\pi(j_1) - \pi(j_2)] I_{[\pi(j_1) > \pi(j_2)]}}{n_k}.$$

*(iv) Hamming*

$$d_H(\{\pi\}, E) = \Pi_{k=1}^{r}(n_k!)^2 \left(N_r - \sum_{k=1}^{r} \frac{Y_k}{n_k}\right)$$

*where $Y_k$ is the cardinality of $\{N_{k-1}+1, \ldots, N_k\}$ and $\{\pi(N_{k-1}+1), \ldots, \pi(N_k)\}$.*

*Proof.* The proofs are the result of direct calculations. □

In all cases, the null hypothesis is rejected whenever the distance between sets is large. The Spearman statistic is a linear rank statistic involving the sums of the rankings observed for each population. The Kendall statistic measures the disagreements between pairs of objects, one from each population. The Spearman Footrule test statistic is a function of the differences by which an observed ranking from a given population exceeds each of the rankings prescribed by the alternative for that population. The Hamming statistic counts for each population the number of common rankings between the observed and the alternative.

Specializing to the two-sample case, $r = 2$ we see that the test statistics associated with Spearman and Kendall are equivalent. However, such is not the case for $r > 2$. Nonetheless, they are asymptotically equivalent.

**Theorem 6.2.** *Under the null hypothesis, the standardized Spearman and Kendall test statistics in the multi-sample location problem with ordered alternatives are asymptotically equivalent when $min(n_1, \ldots, n_r) \to \infty$.*

*Proof.* The proof appears in Alvo and Pan (1997). It consists of obtaining expressions for the variances of the Spearman and Kendall test statistics and then showing that the standardized statistics and equivalent in mean square. □

In the next theorem, we state the result on the asymptotic distribution of the test statistics under the null hypothesis.

**Theorem 6.3.** *Consider the multi-sample location problem with an ordered alternative. Let $n_k/N_r \to w_k > 0$ as $min(n_1, \ldots, n_r) \to \infty$ and set $W_k = w_1 + \ldots + w_k$. Then, under the null hypothesis,*

*(i) Spearman*

$$S_r = \sum_{i=1}^{N_k} c(i)\frac{\pi(i)}{N_r+1}$$

$$\approx Normal\left(\frac{N_r}{2}, \sigma_S^2\right)$$

*where*

$$c(i) = (N_{k-1} + N_k), N_{k-1} < i \leq N_{k,}, 1 \leq k \leq r$$

*and*

$$\sigma_S^2 = \frac{N_r^3}{12} \sum w_k W_k W_{k-1}.$$

*The test rejects for large values of* $S_r$.

*(ii) Spearman Footrule*

$$\begin{aligned} F_r &= \sum_{i=1}^{N_r} c_{i\pi(i)} \\ &\approx Normal\left(\frac{N_r^2}{6}, \sigma_F^2\right) \end{aligned}$$

*where*

$$c_{ij} = \sum_{s=1}^{N_r} a_{is} (j - s) I_{(j>s)}$$

*and*

$$a_{is} = \begin{cases} \frac{1}{n_k} & i, s \epsilon \{N_{k-1} + 1, \ldots, N_k\}, 1 \le k \le r, \\ 0 & otherwise. \end{cases}$$

*Moreover, the variance*

$$\begin{aligned} \sigma_F^2 &= \frac{1}{N_r - 1} \sum_{i,j}^{N_r} d^2(i, j) \\ &\approx M N_r^3 + o\left(N_r^3\right) \end{aligned}$$

*where* $d(i, j) = c_{ij} - \overline{c}_{i.} - \overline{c}_{.j} + \overline{c}_{..}$. *The test rejects for small values of* $F_r$.

*(iii) Hamming*

$$\begin{aligned} H_r &= \sum_{i=1}^{N_r} a_{i\pi(i)} \\ &\approx Normal\left(1, \frac{r-1}{N_r - 1}\right). \end{aligned}$$

*The test rejects for large values of* $H_r$.

*Proof.* The proofs make direct use of Hoeffding's general result and appear in Alvo and Pan (1997). □

### 6.2.1 Asymptotic Distribution Under the Alternative

In this section we consider the asymptotic distribution of the test statistics under contiguous alternatives. This will permit us to compute the relative efficiencies of the test statistics. Let $F(x)$ be a continuous distribution function with density $f(x)$. Define the inverse of $F(x)$

$$F^{-1}(u) = inf\{x : F(x) \geq u\}.$$

Set the information function, assumed finite,

$$I(f) = \int_{-\infty}^{\infty} \left[\frac{f'(x)}{f(x)}\right]^2 f(x)\,dx$$

and let

$$\varphi(u, f) = \frac{f'\left(F^{-1}(u)\right)}{f\left(F^{-1}(u)\right)}, 0 < u < 1.$$

We rephrase the hypothesis testing problem as follows:

$$H_0 : F_1(x) = \ldots = F_r(x) = F\left(x - \bar{d}\right)$$

against the ordered alternative

$$H_1 : F_k(x) = F(x - d_k), 1 \leq k \leq r$$

with $\bar{d} = \frac{1}{N_r}\sum_{i=1}^{N_r} d_i$ and $d_1 < \ldots < d_r$. We shall assume as $min(n_1, \ldots, n_r) \to \infty$

$$N_r^{1/2} d_k \to \delta_k, 1 \leq k \leq r.$$

We quote the following theorem which provides the asymptotic normality of linear rank statistics under the alternative.

**Theorem 6.4 (Theorem VI.2.4 Hájek and Sidak 1967).** *Let* $S_n = \sum_{I=1}^{n}(c_i - \bar{c})\, a_n(\pi(i))$ *where the scores satisfy*

$$lim \int_0^1 \{a_n(1 + [un] - \varphi(u))\}^2\, du = 0$$

*and*

$$\frac{\sum(c_i - \bar{c})^2}{max_{1 \leq i \leq n}(c_i - \bar{c})^2} \to \infty.$$

*Then if* $max_{1 \le i \le n} \left(d_i - \bar{d}\right)^2 \to 0$ *and* $I(f) \sum \left(d_i - \bar{d}\right)^2 \to b^2$, a finite constant, we have that $S_n$ is asymptotically normal with mean

$$m = \sum_{i=1}^{n} (c(i) - \bar{c}) \left(d_i - \bar{d}\right) \int_0^1 \varphi(u) \varphi(u, f) \, du$$

and variance

$$\sigma^2 = \sum_{i=1}^{n} (c(i) - \bar{c})^2 \int_0^1 (\varphi(u) - \bar{\varphi})^2 \, du.$$

We are now ready for the main result.

**Theorem 6.5.** *Consider the multi-sample location problem with an ordered alternative. Let* $n_k / N_r \to w_k > 0$ *as* $min(n_1, \ldots, n_r) \to \infty$ *and set* $W_k = w_1 + \ldots + w_k$. *Then, under the alternative hypothesis,*

*(i) Spearman*

$$S_r = \sum_{i=1}^{N_k} c(i) \frac{\pi(i)}{N_r + 1}$$
$$\approx Normal\left(m_S^*, \sigma_S^2\right)$$

*where*

$$m_S^* = \frac{N_r^2}{2} + \sum_{i=1}^{N_r} (c(i) - \bar{c}) \left(d_i - \bar{d}\right) \int_0^1 u\varphi(u, f) \, du$$
$$c(i) = (N_{k-1} + N_k), \; N_{k-1} < i \le N_{k,}, 1 \le k \le r, N_0 = 0$$

*and*

$$\sigma_S^2 = \frac{N_r^3}{12} \sum w_k W_k W_{k-1}.$$

*(ii) Spearman Footrule*

$$F_r = \sum_{i=1}^{N_r} c_{i\pi(i)}$$
$$\approx Normal\left(\frac{N_r^2}{6} + \nu_F, \sigma_F^2\right)$$

*where*

$$\nu_F = N_r^{3/2} \sum \left( \delta_k - \sum w_j \delta_j \right) \times \left\{ \int_{W_{k-1}}^{W_k} \frac{(u - W_{k-1})^2}{2} \left[ \varphi(u, f) - \bar{\varphi}(f) \right] du + \int_{W_k}^{1} w_k \left( u - \frac{W_k + W_{k-1}}{2} \right) \left[ \varphi(u, f) - \bar{\varphi}(f) \right] du \right\}.$$

*(iii) Hamming*

$$H_r = \sum_{i=1}^{N_r} a_{i\pi(i)} \approx Normal\left(1 + \nu_H, \frac{r-1}{N_r - 1}\right)$$

*where*

$$\nu_H = N_r^{-1/2} \sum \delta_k \int_{W_{k-1}}^{W_k} \left[ \varphi(u, f) - \bar{\varphi}(f) \right] du.$$

*Proof.* The proofs make direct use of Theorem 6.4 and Hoeffding's general result and appear in Alvo and Pan (1997). □

The asymptotic distributions of the test statistics under both the null and the alternative hypotheses permit us now to compute the asymptotic power efficiencies. The latter depend on the underlying distributions. Alvo and Pan (1997) have shown that for the test statistics considered, the asymptotic power is of the form

$$1 - \Phi(k_\alpha - B)$$

where $k_\alpha$ denotes the $(1 - \alpha)$ quantile of the standard normal distribution and $B$ is a constant depending on $F$ and on the test statistic. The asymptotic power efficiency is then defined to be $B^2$. (See Chap. VII.1.3 in Hájek and Sidak 1967.) The following examples were considered.

*Example 6.2.* Let $r = 2, n_1 = n_2$.

(a) Let $f(x)$ be a standard normal density. Then

$$B_S^2 = 0.9554,$$
$$B_F^2 = 0.8808,$$
$$B_H^2 = 0.6369.$$

(b) Let $f(x) = \frac{1}{2} exp(-|x|)$ be the double exponential density. Then

$$B_S^2 = 0.7500,$$
$$B_F^2 = 0.8333,$$
$$B_H^2 = 1.$$

(c) Let $f(x) = e^{-x}(1 + e^{-x})^2$ be the logistic density. Then

$$B_S^2 = 1,$$
$$B_F^2 = 0.9765,$$
$$B_H^2 = 0.7500.$$

We see that the Spearman (or equivalently the Kendall) statistic achieves higher asymptotic power than either the Spearman Footrule or the Hamming statistic when the underlying distributions are normal or logistic but lower if the distribution is double exponential. In all cases, it worthy to note that the Spearman Footrule is robust.

## 6.3 Tests Under Umbrella Alternatives

The general theory of hypothesis testing may also be applied in the case of an umbrella alternative. Consider the following example on intelligence scores.

*Example 6.3.* The Wechsler Adult Intelligence Scale scores shown in Table 6.1 were recorded on 12 males listed by age groups. If we assume that the peak is located in the 35–54 age group, we would like to test the null hypothesis that there is no difference due to age against the alternative that the scores rise monotonically prior to the ages 35–54 and decrease thereafter. More generally, we may not want to specify the location of the peak age group.

Formally, let $X_{i(1)}, \ldots, X_{i(m_i)}, i = 1, \ldots, k$, be $k$ independent random samples with $X_{i(l)}, l = 1, \ldots, m_i$ having an absolutely continuous distribution function $F_i(x)$. In the parametric case, we may have $F_i(x) \equiv F(x - \theta_i)$, where $F$ has median zero. We shall be concerned with testing the hypothesis of no treatment

**Table 6.1** Wechsler adult intelligence scale scores

| Age group | | | | |
|---|---|---|---|---|
| 16–19 | 20–34 | 35–54 | 55–69 | >70 |
| 8.62 | 9.85 | 9.98 | 9.12 | 4.80 |
| 9.94 | 10.43 | 10.69 | 9.89 | 9.18 |
| 10.06 | 11.31 | 11.40 | 10.57 | 9.27 |

effect against the alternative that there is a monotone treatment effect subject to a change in direction. Alternatives of this type arise in situations where the treatment effect changes in direction after reaching a peak. For example, the effectiveness of a drug may change with time, the effectiveness in learning as a function of age may peak at a certain stage, the reaction to increasing levels of a drug dosage may peak at a certain point and decrease thereafter, etc. Formally, letting $F_p$ be the distribution where the turning point occurs, the null and the alternative hypothesis are respectively

$$H_0 : F_1(x) = \ldots = F_k(x),$$

$$H_1 : F_1(x) \geq \ldots \geq F_{p-1}(x) \geq F_p(x) \leq F_{p+1}(x) \leq \ldots \leq F_k(x),$$

with at least one strict inequality for some $x$. Equivalently, in the parametric case, the hypotheses become

$$H_0 : \theta_1 = \ldots = \theta_k,$$

$$H_1 : \theta_1 \leq \ldots \leq \theta_{p-1} \leq \theta_p \geq \theta_{p+1} \geq \ldots \geq \theta_k,$$

with at least one strict inequality in $H_1$. We note that an umbrella alternative contains as special cases the ordered alternatives corresponding to $p = k$ or $p = 1$.

We may use the general theory for constructing test statistics based on the ranks of the observations for the situation when the location of the peak is known. Test statistics will be determined using both the Spearman and Kendall distances. We shall then obtain their asymptotic null distributions under the assumption that the minimum of the sample sizes gets large. Finally, we shall propose an algorithm to estimate the location of peak when it is unknown.

### 6.3.1 *The Construction of the Test Statistics*

In arriving at test statistics, we will assume that there is a single peak and that its location is known.

We will also adopt the following notation. Let

$$n_i = \sum_{h=1}^{i} m_h, i = 1, \ldots, k, n = n_k, \hat{n} = n - m_p, n_0 = 0.$$

In keeping with the general approach, we propose the following steps.

Step 1: Rank all the observations together so that the smallest gets rank 1, the next smallest rank 2, etc. Let the $n$-dimensional vector

$$\pi = (\pi(1), \ldots, \pi(m_1) | \pi(m_1 + 1), \ldots, \pi(m_1 + m_2) | \ldots, \pi(n))$$

represent the ranks of the $\{X_{i(l)}\}, i = 1, \ldots, k,\ l = 1, \ldots, m_i$ and grouped by populations. In view of the continuity assumption on the distributions, ties among the observations occur with probability zero.

Step 2: Define $\{\pi\}$ to be the subclass of permutations $\pi$ in the sense that ranks occupied by identically distributed random variables are exchangeable. This subclass consists of all the permutations $\pi$ where the rankings within each population are permuted among themselves only. The cardinality of $\{\pi\}$ is given by the product $(\Pi m_h!)$.

Step 3: In the present context, permutations in the extremal set $E$ are such that ranks occupied by observations from $F_i$ are always less than those from $F_{i'}$, if $i < i' \leq p$, whereas the reverse is true if $p \leq i < i'$. Moreover, ranks attributed to a distribution consist of consecutive integers. Hence the cardinality of $E$ is equal to $c\ (\Pi m_i!)$ where $c = \binom{k-1}{p-1}$.

The enumeration of the extremal set $E$ is a two-stage procedure. First, choose the relative order of the $(p-1)$ populations $F_1, \ldots, F_{p-1}$ among $F_1, .., F_{p-1}, F_{p+1}, \ldots, F_k$. This can be done in $c = \binom{k-1}{p-1}$ ways. Then partition the integers $1, \ldots, n$ in accordance with the prescribed ordering of the populations while taking into account corresponding sample sizes. The extremal set $E$ is finally obtained by permuting the integers within each population. Population $F_p$ is always allocated the last $m_p$ integers, namely $\hat{n}+1, \ldots, n$. The cardinality of $E$ is therefore equal to $c\ (\Pi m_i!)$.

Step 4: Let $d(\mu, \nu)$ be a distance function between two permutations $\mu, \nu$ and define the distance between the two sets $\{\pi\}$ and an extremal set $E$

$$d(\{\pi\}, E) = \sum_{\mu \epsilon \{\pi\}} \sum_{\nu \epsilon E} d(\mu, \nu). \tag{6.1}$$

Small values of $d(\{\pi\}, E)$ are inconsistent with the null hypothesis and consequently lead to rejection of $H_0$. In what follows, we shall consider the Spearman and Kendall distances between permutations adapted to the present context:

Spearman:

$$\begin{aligned} d_S(\mu, \nu) &= \frac{1}{2} \sum_{i=1}^{k} \sum_{l=1}^{m_i} [\mu(i(l)) - \nu(i(l))]^2 \\ &= \frac{n(n^2-1)}{12} - \sum_{i=1}^{k} \sum_{l=1}^{m_i} [\nu(i(l))] \left[\mu(i(l)) - \frac{n+1}{2}\right]. \end{aligned}$$

Kendall:

$$d_K(\mu, \nu) = \sum_{i_1(l) < i_2(l')} \{1 - sgn\left[\mu(i_1(l)) - \mu\left(i_2(l')\right)\right] sgn\left[\nu(i_1(l)) - \nu\left(i_2(l')\right)\right]\}$$

$$= \frac{n(n-1)}{2} - \sum_{i_1(l)<i_2(l')} \left\{ sgn\left[\mu(i_1(l)) - \mu\left(i_2(l')\right)\right] sgn\left[\nu(i_1(l)) - \nu\left(i_2(l')\right)\right]\right\}.$$

### 6.3.2 *The Test Statistic Corresponding to Spearman Distance*

In this section, we derive the test statistic corresponding to Spearman distance under the extremal set $E$ when the location of the peak is known. Throughout, we shall assume that permutations defined by the extremal sets are arranged in columns indexed by $1 \le i(l) \le n$ in such a way that ranks are in increasing order for populations $F_i, i \le p$ and in decreasing order when $i \ge p$.

Suppose that $F_i, i < p$ is in relative position $j$ and hence populations $F_1, \ldots, F_{i-1}$ are in relative positions chosen from among the first $(j-1)$ positions. Populations $F_{i+1}, \ldots, F_{p-1}$ are then assigned positions chosen from $(j+1), \ldots, (k-1)$. This can happen with frequency $\binom{j-1}{i-1}\binom{k-1-j}{p-1-i}$. The positions of the remaining populations are then automatically determined. Together populations $F_1, \ldots, F_{i-1}$ and $F_{k+i-j+1}, \ldots, F_k$ are assigned the first $a_{ij}$ integers where

$$\begin{aligned} a_{ij} &= \left(\sum_{h=1}^{i-1} m_h\right) + \left(\sum_{h=k+i-j+1}^{k} m_h\right) \\ &= n_{i-1} + n - n_{k+i-j}. \end{aligned}$$

Population $F_i$ is assigned integers $a_{ij}+1, \ldots, a_{ij}+m_i$ whose sum is equal to $\left(a_{ij} + \frac{m_i+1}{2}\right) m_i$. The process of permuting ranks within $F_i$ implies that each entry will contribute the sum $(m_i - 1)!$ times. Hence, for each entry taking into account the permutations, we have

$$(\Pi m_i!) \left(a_{ij} + \frac{m_i+1}{2}\right), i < p.$$

Finally, summing over each $j$, we have

$$(\Pi m_i!) \sum_{j=i}^{k+i-p} \left(a_{ij} + \frac{m_i+1}{2}\right)\binom{j-1}{i-1}\binom{k-1-j}{p-1-i}.$$

On the other hand for the data vector we have for each entry in $F_i$

$$(\Pi m_i!) \left[\bar{\pi}_i - \frac{n+1}{2}\right]$$

where $\bar{\pi}_i$ represents the average of the ranks for the $i$th population. Similarly for $i > p$, we may define

$$\begin{aligned} b_{ij} &= \left(\sum_{h=i+1}^{k} m_h\right) + \left(\sum_{h=1}^{j+i-k-1} m_h\right) \\ &= n - n_i + n_{j+i-k-1}. \end{aligned}$$

The calculation of (6.1) then yields

$$d_S(\{\pi\}, E) = \frac{n(n^2-1)}{12} c(\Pi m_i!)^2 - c(\Pi m_i!)^2 S$$

where $S = \sum_{i=1}^{k} m_i v_i \left[\bar{\pi}_i - \frac{n+1}{2}\right]$ and

$$v_i = \begin{cases} c^{-1} \sum_{j=i}^{k+i-p} \left(a_{ij} + \frac{1+m_i}{2}\right) \binom{j-1}{i-1} \binom{k-1-j}{p-1-i} & i < p, \\ \left(\hat{n} + \frac{1+m_p}{2}\right) & i = p, \\ c^{-1} \sum_{j=k-i+1}^{k-i+p} \left(b_{ij} + \frac{1+m_i}{2}\right) \binom{j-1}{k-i} \binom{k-1-j}{i-1-p} & i > p. \end{cases} \tag{6.2}$$

It is instructive to consider the special case $m_i = m$ where an equal number of observations is taken from each population. In that case, $a_{ij} = b_{ij} = (j-1)m$ and

$$v_i = \begin{cases} c^{-1} \sum_{j=i}^{k+i-p} \left(jm + \frac{1-m}{2}\right) \binom{j-1}{i-1} \binom{k-1-j}{p-1-i} & i < p, \\ km + \left(\frac{1-m}{2}\right) & i = p, \\ c^{-1} \sum_{j=k-i+1}^{k-i+p} \left(jm + \frac{1-m}{2}\right) \binom{j-1}{k-i} \binom{k-1-j}{i-p-1} & i > p \end{cases} \tag{6.3}$$

and hence

$$v_i = \begin{cases} n\frac{i}{p} + \left(\frac{1-m}{2}\right) & i \le p, \\ n\frac{(k+1-i)}{k+1-p} + \left(\frac{1-m}{2}\right) & i > p. \end{cases} \tag{6.4}$$

It follows that

$$S = mn\left\{\sum_{i\le p}\frac{i}{p}\left[\bar{\pi}_i - \frac{n+1}{2}\right] + \sum_{i>p}\frac{(k+1-i)}{(k+1-p)}\left[\bar{\pi}_i - \frac{n+1}{2}\right]\right\}. \tag{6.5}$$

### 6.3.3 The Test Statistic Corresponding to Kendall Distance

In this section, we derive the test statistic corresponding to the Kendall distance. Consider for now the situation when there is only one observation per population. Fix $1 \le i_1 < p < i_2 \le k$. Suppose that integer $j$ is assigned to $F_{i_1}$ and integer $j_2$ is assigned to $F_{i_2}$, with $j_2 > j$. Then the frequency with which this can happen is given by

$$\binom{j-1}{i_1-1}\binom{j_2-j-1}{j_2+i_2-i_1-k-1}\binom{k-j_2}{i_2-p}, j_2 > j. \tag{6.6}$$

In fact, from the point of view of the $(p-1)$ populations $F_1, \ldots, F_{p-1}$, the number of ways of choosing $(i_1-1)$ integers to be less than $j$ is $\binom{j-1}{i_1-1}$. If $q$ is the number of populations among $F_{(i_1+1)}, \ldots, F_{(p-1)}$ which are assigned ranks greater than $j$ but less than $j_2$, then we must have

$$q + i_1 + (k - i_2) = j_2 - 1. \tag{6.7}$$

Their ranks are chosen from $(j+1), \ldots, (j_2-1)$.

Summing over $j_2$ we obtain the total number of negatives

$$H(i_1, i_2) = \sum_{j_2>j}\binom{j-1}{i_1-1}\binom{j_2-j-1}{q}\binom{k-j_2-1}{i_2-p-1} \tag{6.8}$$

$$= \sum_{j=i_1}^{k+i_1-i_2}\binom{j-1}{i_1-1}\binom{k-1-j}{k+i_1-j-p} \tag{6.9}$$

$$= \sum_{j=i_1}^{k+i_1-i_2}\binom{j-1}{i_1-1}\binom{k-1-j}{p-1-i_1}. \tag{6.10}$$

Alternatively, we may first sum over $j$

$$H(i_1, i_2) = \sum_{j_2=k+i_1-i_2}^{k+p-i_2}\ \sum_{j=i_1}^{k+i_1-i_2}\binom{j-1}{i_1-1}\binom{j_2-j-1}{q}\binom{k-j_2-1}{i_2-p-1} \tag{6.11}$$

$$= \sum_{j_2=k+i_1-i_2}^{k+p-i_2} \binom{j_2-1}{k-i_2}\binom{k-j_2-1}{i_2-p-1}. \tag{6.12}$$

It follows that the sum over the signs is given by the positives less the negatives

$$c - 2H(i_1, i_2). \tag{6.13}$$

Considering only the second term in the Kendall distance and setting

$$W(i_1, i_2) = \begin{cases} -c & 1 \le i_1 < i_2 \le p, \\ c - 2H(i_1, i_2) & 1 \le i_1 < p < i_2 \le k, \\ c & p \le i_1 < i_2 \le k, \end{cases} \tag{6.14}$$

it follows that the Kendall test statistic becomes

$$\sum_{i_1<i_2} W(i_1, i_2)\{sgn[\pi(i_1) - \pi(i_2)]\}. \tag{6.15}$$

We now consider the more general case with unequal observations when the extremal set is $E_2$. In that case, the extremal set is determined by first specifying the ordering of the populations and then permuting within populations. It follows that the weight function will be a function only of the indices $i_1, i_2$ since the sign of the difference $[\pi(i_1(l)) - \pi(i_2(l'))]$ is determined entirely by the ordering of the populations. The data set on the other hand consists of permuting the ranks occupied by the ranks within populations. This yields the double sum

$$\sum_{l}\sum_{l'} sgn\left[\left[\pi(i_1(l)) - \pi\left(i_2(l')\right)\right]\right].$$

There is no contribution to the sum from permutations within each population. Set

$$U(i_1, i_2) = \sum_{l=1}^{m_{i_1}}\sum_{l'=1}^{m_{i_2}} \left\{sgn\left[\left[\pi(i_1(l)) - \pi\left(i_2(l')\right)\right]\right]\right\}. \tag{6.16}$$

Hence the Kendall test statistic for unequal numbers of observations when the extremal set is $E$ becomes

$$\tau = \sum_{i_1<i_2} W(i_1, i_2)\, U(i_1, i_2). \tag{6.17}$$

### 6.3.4 The Asymptotic Distribution of the Test Statistics Under the Null Hypothesis

In this section, we consider the asymptotic distribution of the Spearman and Kendall test statistics under the null hypothesis.

**Theorem 6.6.** *Assume that* $\min(m_i) \to \infty$ *in such a way as*

$$\frac{m_i}{n} \to \lambda_i > 0.$$

*Define*

$$\sigma_p^2 = (n+1)^2 \frac{\sum_{i=1}^k m_i (\nu_i - \bar{\nu})^2}{12}.$$

*Then the test statistic* $S = \sum_{i=1}^{k} m_i \nu_i \left[\bar{\pi}_i - \frac{n+1}{2}\right]$ *corresponding to the Spearman distance is asymptotically normal with mean* 0 *and variance equal to* $\sigma_p^2$.

*Proof.* We note that the Spearman test statistic $S$ can be represented as a normalized linear rank statistic

$$(n+1)\sum_{i=1}^{k} \nu_i \sum_{l=1}^{m_i} \left[\frac{R_{i(l)}}{n+1} - \frac{1}{2}\right].$$

Since $w_i/n$ converges to a constant, the result follows since

$$\max_i \frac{(\nu_i - \bar{\nu})^2}{\sum m_i (\nu_i - \bar{\nu})^2} \le \frac{1}{\min m_i} \max_{i \le k} \frac{(\nu_i - \bar{\nu})^2}{\sum_{i=1}^k (\nu_i - \bar{\nu})^2}$$
$$\to 0 \text{ as } \min(m_i) \to \infty.$$

It can be shown that

$$\bar{\nu} = \frac{n+1}{2},$$

and in the equal sample size case the variance is given by

$$\sigma_p^2 = \frac{[n(n+1)]^2 m}{12}\left[\sum_{i=1}^{p}\left(\frac{i}{p} - \frac{k+1}{2k}\right)^2 + \sum_{i=p+1}^{k}\left(\frac{k+1-i}{k+1-p} - \frac{k+1}{2k}\right)^2\right].$$

□

**Theorem 6.7.** *The projection of*

$$\tau = \sum_{i_1 < i_2} W(i_1, i_2) U(i_1, i_2)$$

*onto the space of linear rank statistics is given by*

$$\hat{\tau} = \frac{4c}{n} S.$$

*Proof.* The proof of the asymptotic distribution of the Kendall statistic makes use of the following projection adapted from Hájek and Sidak (1967):

$$E\left\{sgn\left[\pi\left(i_1(l)\right) - \pi\left(i_2\left(l'\right)\right)\right] | \pi(i) = j\right\} = \begin{cases} \frac{1}{n-1}(2j - (n+1)) & i = i_1(l), 1 \leq j \leq n, \\ \frac{1}{n-1}((n+1) - 2j) & i = i_2\left(l'\right), 1 \leq j \leq n, \\ 0 & \text{Otherwise.} \end{cases}$$

We refer the reader to Alvo (2008) for further details of the proof. □

In the next theorem, we state the result showing that the Kendall and Spearman statistics are asymptotically equivalent.

**Theorem 6.8.** *Assume that* $\min(m_i) \to \infty$ *in such a way as*

$$\frac{m_i}{n} \to \lambda_i > 0.$$

*Then* $Var\ \tau / Var\ \hat{\tau} \to 1$ *and the Kendall and Spearman test statistics are asymptotically equivalent.*

*Proof.* It was shown in Alvo (2008) that the ratio of the variances

$$\frac{Var(\tau)}{Var(\hat{\tau})} = 1 + \frac{\sum \frac{m_{i_1}}{n} \frac{m_{i_2}}{n} W^2(i_1, i_2)}{n \sum \frac{m_i}{n} \left(\frac{u_{1i}}{n} - \frac{u_{2i}}{n}\right)^2} \to 1 \text{ as } n \to \infty.$$

Hence the result follows. □

The asymptotic distribution under the alternative hypothesis can be obtained in a similar way as in Alvo and Pan (1997). Assume that $F(x)$ is a continuous distribution function with density $f(x)$. Let $F^{-1}(u) = \inf\{x : F(x) \geq u\}$ and define

$$\varphi(u, f) = \frac{f'\left(F^{-1}(u)\right)}{f\left(F^{-1}(u)\right)}, 0 < u < 1$$

and

$$I(f) = \int_{-\infty}^{\infty} \left[ \frac{f'(x)}{f(x)} \right]^2 f(x)\,dx.$$

It can be shown that the asymptotic power efficiency is given by the expression

$$\frac{12\left[\int_0^1 u\varphi(u, f)\,du\right]^2}{I(f)}.$$

We may thus conclude that the asymptotic power efficiency is greater when the underlying distribution is logistic as compared to either the normal or the double exponential.

### 6.3.5 The Test Statistics When the Location of the Peak is Unknown

In the case when the location of the peak is unknown, we may construct a test statistic as follows. Allowing p to vary, let

$$S_p^* = \frac{S_p}{\sigma_p}, D_p^* = \frac{D_p}{\sqrt{Var(D_p)}}, p = 1, \ldots, k$$

be the standardized statistics and let

$$S_{max} = max_p S_p^*, D_{max} = max_p D_p^*.$$

The test based on Spearman distance rejects the null hypothesis in favor of an umbrella alternative whenever $S_{max}$ is large. Similarly, the test based on Kendall distance rejects whenever $D_{max}$ is large. The asymptotic distribution of the respective test statistics under the null hypothesis is given in the next theorem.

**Theorem 6.9.** *Let the vector* $S = \left(S_1^*, \ldots, S_k^*\right)'$ *and let* $cov(S) = BB'$. *Under the null hypothesis, if* $\min(m_i) \to \infty$ *in such a way as*

$$\frac{m_i}{n} \to \lambda_i > 0, i = 1, \ldots, k,$$

*then* $S$ *has asymptotically the distribution of* $BZ$ *where* $Z$ *has a standard multivariate normal with mean* 0 *and covariance matrix* $I$. *Consequently,* $S_{max}$ *has asymptotically the distribution of maxBZ.*

**Table 6.2** Critical values for the Spearman test statistic

| k\% | 1 | 1.5 | 2 | 2.5 | 3 | 3.5 | 4 | 4.5 | 5 | 10 |
|---|---|---|---|---|---|---|---|---|---|---|
| 3 | 2.67 | 2.54 | 2.44 | 2.36 | 2.29 | 2.23 | 2.18 | 2.14 | 2.09 | 1.80 |
| 4 | 2.77 | 2.63 | 2.53 | 2.45 | 2.39 | 2.33 | 2.28 | 2.23 | 2.19 | 1.89 |
| 5 | 2.82 | 2.67 | 2.59 | 2.51 | 2.44 | 2.38 | 2.33 | 2.29 | 2.24 | 1.95 |
| 6 | 2.86 | 2.73 | 2.63 | 2.55 | 2.48 | 2.42 | 2.37 | 2.32 | 2.28 | 1.98 |
| 7 | 2.89 | 2.74 | 2.65 | 2.57 | 2.50 | 2.44 | 2.39 | 2.24 | 2.30 | 2.00 |

*Similarly, let* $D = \left(D_1^*, \ldots, D_k^*\right)'$*; then cov* $(D)$ *and cov* $(S)$ *are asymptotically equivalent and* $D_{max}$ *has asymptotically the distribution of maxBZ.*

*Proof.* See Alvo (2008) for the proof.

The distribution of *maxBZ* cannot be explicitly determined although it can be easily simulated. The approximate p-value of the Spearman-based test is given by

$$P\left(maxBZ \geq s_{max}\right)$$

where $s_{max}$ is the observed value of the $S_{max}$. The asymptotic critical values for the Spearman statistic were very stable for $m \geq 3$. They are displayed in Table 6.2 for $m = 10$. □

*Example 6.4.* Returning to Example 6.3 on adult intelligence scores on males by age groups, we calculate $S = 155.79, \sigma_p^2 = 4486.464$, and a standardized test statistic $\frac{S}{\sigma_p} = 2.326$. The resulting p-value is 0.01. On the other hand, using Kendall tau, we calculate that $\tau = 256$, $Var\hat{\tau} = 11485.348$, $Var\hat{\tau} = 12229.348$. The standardized test statistic becomes $\frac{\tau}{\sqrt{Var\ \tau}} = \frac{256}{110.586} = 2.315$ which gives a p-value of 0.01. It can be seen that the statistics based on the Spearman and Kendall distances provided very close results.

### 6.3.6 Simulation Study

In this section, we report results on a limited simulation study. Four families of distributions were considered: normal, double exponential, logistic, and exponential. Let $k = 5, p = 3$.

In Table 6.3, when the location of the peak is assumed to be known, we report the probability of rejecting the null hypothesis when it is true. In that case the critical value for rejection from Table 6.2 is 2.24. It is seen that the probabilities are very close to the nominal value chosen to be 5 %. For each value of the sample size, we used 10,000 repetitions.

The power function was then determined for distributions $F_1, \ldots, F_5$ whose location parameters were set equal to $0, \frac{1}{2}, 1, \frac{1}{2}, 0$, respectively, under the alternative and variances all equal to 1. The powers reported in Table 6.4 show that the test

**Table 6.3** Significance level for the Spearman and Kendall statistics for the case k = 5, p = 3; N, normal; D, double exponential; L, logistic; E, exponential

| Spearman | | | | | Kendall | | | | |
|---|---|---|---|---|---|---|---|---|---|
| m | N | D | L | E | m | N | D | L | E |
| 5 | 0.048 | 0.045 | 0.047 | 0.043 | 5 | 0.044 | 0.049 | 0.041 | 0.042 |
| 6 | 0.050 | 0.044 | 0.045 | 0.047 | 6 | 0.051 | 0.047 | 0.047 | 0.058 |
| 7 | 0.052 | 0.047 | 0.048 | 0.047 | 7 | 0.052 | 0.049 | 0.054 | 0.041 |
| 8 | 0.045 | 0.047 | 0.049 | 0.043 | 8 | 0.047 | 0.048 | 0.051 | 0.055 |
| 9 | 0.049 | 0.050 | 0.052 | 0.049 | 9 | 0.052 | 0.050 | 0.048 | 0.046 |
| 10 | 0.049 | 0.048 | 0.048 | 0.048 | 10 | 0.048 | 0.051 | 0.050 | 0.058 |
| 15 | 0.045 | 0.047 | 0.050 | 0.045 | 15 | 0.052 | 0.049 | 0.050 | 0.051 |

Tests reject when the statistics exceeds 2.24

**Table 6.4** Power function for the Spearman and Kendall statistics for the case k = 5, p = 3; N, normal; D, double exponential; L, logistic; E, exponential

| Spearman | | | | | Kendall | | | | |
|---|---|---|---|---|---|---|---|---|---|
| m | N | D | L | E | m | N | D | L | E |
| 5 | 0.344 | 0.448 | 0.386 | 0.558 | 5 | 0.352 | 0.459 | 0.383 | 0.602 |
| 6 | 0.408 | 0.539 | 0.463 | 0.656 | 6 | 0.421 | 0.555 | 0.462 | 0.700 |
| 7 | 0.478 | 0.619 | 0.519 | 0.744 | 7 | 0.481 | 0.627 | 0.536 | 0.763 |
| 8 | 0.533 | 0.677 | 0.578 | 0.800 | 8 | 0.539 | 0.691 | 0.592 | 0.823 |
| 9 | 0.589 | 0.737 | 0.635 | 0.844 | 9 | 0.598 | 0.740 | 0.640 | 0.867 |
| 10 | 0.648 | 0.778 | 0.696 | 0.883 | 10 | 0.649 | 0.793 | 0.696 | 0.900 |
| 15 | 0.833 | 0.929 | 0.870 | 0.979 | 15 | 0.831 | 0.932 | 0.877 | 0.981 |

Tests reject when the statistics exceeds 2.24

statistics perform well with increasing values of m. Although not reported here, additional simulations showed that the Spearman statistic performs well also when the location parameters under the alternative are not equally spaced.

A further simulation can be done to determine how well one can choose the peak distribution. The following algorithm may be used. Compute the distance $d(\{\pi\}, E)$ to the extremal set for all possible values of $p$ and then choose the corresponding distribution $F_p$ that minimizes that distance. We report in Table 6.5 the results of that simulation for both the normal distributions and the double exponential. It can be seen that the probability of choosing the correct peak location at $p = 3$ increases rapidly as m gets large while the probability of choosing the other locations gets quite small. Moreover, the probability of correct selection is higher for the double exponential than it is for the normal. This is natural since the double exponential is more peaked.

**Table 6.5** Probability of choosing the peak

| Normal | | | | | | Double exponential | | | | | |
|---|---|---|---|---|---|---|---|---|---|---|---|
| m | $F_1$ | $F_2$ | $F_3$ | $F_4$ | $F_5$ | m | $F_1$ | $F_2$ | $F_3$ | $F_4$ | $F_5$ |
| 5 | 0.07 | 0.17 | 0.52 | 0.17 | 0.07 | 5 | 0.05 | 0.16 | 0.59 | 0.15 | 0.05 |
| 6 | 0.06 | 0.16 | 0.57 | 0.16 | 0.06 | 6 | 0.04 | 0.14 | 0.65 | 0.14 | 0.03 |
| 7 | 0.04 | 0.15 | 0.61 | 0.15 | 0.04 | 7 | 0.03 | 0.13 | 0.69 | 0.12 | 0.03 |
| 8 | 0.04 | 0.15 | 0.63 | 0.15 | 0.04 | 8 | 0.03 | 0.12 | 0.72 | 0.12 | 0.02 |
| 9 | 0.03 | 0.14 | 0.67 | 0.13 | 0.03 | 9 | 0.02 | 0.11 | 0.74 | 0.11 | 0.01 |
| 10 | 0.02 | 0.13 | 0.70 | 0.12 | 0.03 | 10 | 0.01 | 0.10 | 0.78 | 0.10 | 0.01 |
| 15 | 0.01 | 0.10 | 0.79 | 0.09 | 0.01 | 15 | 0.00 | 0.06 | 0.86 | 0.07 | 0.00 |

## Chapter Notes

Alvo and Pan (1997) also discussed the situation when the alternatives are unordered by considering the union of the r! possible ordered alternatives.

For the problem of testing for umbrella alternatives, we refer the reader to Alvo (2008) for additional references and for a brief history of the subject. The Spearman statistic considers the data on either side of the peak separately whereas the Kendall statistic (6.15) considers, in addition, the relationship of the data between both sides of the peak. For small sample sizes this may increase the sensitivity of that statistic. The approach presented may have potential applications in the study of isotonic regression.

# Chapter 7
# Testing for Ordered Alternatives

In this chapter, we shall consider a randomized block experiment given by the model

$$X_{ij} = b_i + \tau_j + e_{ij,}\ i = 1, .., n, j = 1, \ldots, t.$$

where $b_i$ is the $i$th block effect, $\tau_j$ is the $j$th treatment effect, and the $e_{ij}$ are independent identically distributed error terms having a continuous distribution. We wish to test the hypothesis of no treatment effect

$$H_0 : \tau_1 = \tau_2 = \ldots = \tau_t$$

against the ordered alternative

$$H_1 : \tau_1 \leq \tau_2 \leq \ldots \leq \tau_t$$

with at least one inequality strict. This problem has been considered in the literature by Page (1963) and by Jonckheere (1954) who proposed different test statistics in the case where there is complete data in each block.

In the next section, we shall use the general theory of hypothesis testing to derive the appropriate test statistics when the data contains no missing values. It will be seen that the approach leads to the Page statistic when using the Spearman distance. However, the use of the Kendall distance does not lead to the Jonckheere test. We obtain the asymptotic distributions in each case. In Sect. 7.2, we consider the more general situation when one or more observations are missing from one or more blocks and we obtain a generalization of the Page and Jonckheere tests. We will apply these tests on the data in Example 7.2.



M. Alvo, P.L.H. Yu, *Statistical Methods for Ranking Data*, Frontiers in Probability and the Statistical Sciences, DOI 10.1007/978-1-4939-1471-5_7

## 7.1 Test Statistics When the Data Is Complete in Each Block

We proceed to obtain the test statistics using the method outlined in Chap. 6. For each i, let

$$R_i = [R_{i1}, \ldots, R_{it}]$$

denote the permutation of the integers $1, \ldots, t$ which preserves the relative order of the original data $X_{i1}, \ldots, X_{it}$.

Step 1: The observed permutation $\pi$ is the ranking of all the $nt$ data. We adopt the convention that object $(i-1)t+j$ is from block i and treatment j, for $i = 1, .., n$ and $j = 1, \ldots, t$. The rank given to this object is then

$$\pi\left((i-1)t+j\right).$$

Step 2: The equivalence class $\{\pi\}$ then consists of all permutations $\mu\epsilon\mathcal{P}$ such that

$$\mu = [\mu_1(1), \ldots, \mu_1(t) \,|\, \ldots \,|\mu_n(1), \ldots, \mu_n(t)]$$

where $(\mu_i(1), \ldots, \mu_i(t))$ preserves the same relative order as $(X_{i1}, \ldots, X_{it})$ for $i = 1, \ldots, n$.

Step 3: Define $E$ to be the subclass of extremal permutations consisting of all permutations $\nu\epsilon\mathcal{P}$ such that

$$\nu = [\nu_1(1), \ldots, \nu_1(t) \,|\, \ldots \,|\nu_n(1), \ldots, \nu_n(t)]$$

where $(\nu_i(1), \ldots, \nu_i(t))$ preserves the same relative order as $(1, \ldots, t)$ for $i = 1, \ldots, n$.

Step 4: Define the distance between the subclasses $\{\pi\}$ and $E$ to be the sum of pairwise distances between permutations $\mu, \nu$ with $\mu\epsilon\{\pi\}, \nu\epsilon E$;

$$d(\{\pi\}, E) = \sum_{\mu\epsilon\{\pi\}} \sum_{\nu\epsilon E} d(\mu, \nu).$$

*Example 7.1.* In order to illustrate the subclass E, suppose that $t = 3, n = 2$. Hence, there are $tn! = 720$ possible rankings of which $\binom{6}{3}\binom{3}{3} = 20$ preserve the monotone increasing ordering in each block. For instance, (123 | 456) and (234 | 156) are such rankings.

In the next theorem we obtain the test statistics corresponding to the various distance functions.

**Theorem 7.1.** *The test statistics for the two-way layout with an ordered alternative are equivalently given by*
*Spearman:*

$$d_S\left(\{\pi\}, E\right) \equiv -\sum_{i=1}^{n}\sum_{j=1}^{t} i R_{ij}.$$

*Kendall:*

$$d_K\left(\{\pi\}, E\right) \equiv -\frac{1}{\binom{2t}{t}^2}\sum_{1\le j_1<j_2\le n}\sum_{1\le i_1<i_2\le t} a\left[R_{i_1 j_1}, R_{i_2 j_2}\right] a\left[i_1, i_2\right]$$
$$-\sum_{j=1}^{n}\sum_{1\le i_1<i_2\le t} sgn\left[R_{i_2 j} - R_{i_1 j}\right]$$

*where*

$$a\left[i, j\right] = \binom{2t}{t} - \sum_{l=0}^{t-i}\binom{l+i+j-1}{i-1}\binom{2t-l-i-j}{t-i}.$$

*Spearman Footrule:*

$$d_F\left(\{\pi\}, E\right) \equiv \sum_{j=1}^{n}\sum_{i=1}^{t} f\left(i, R_{ij}\right)$$

*where*

$$f\left(i, j\right) = \sum_{s=i}^{t(n-1)+i}\sum_{l=j}^{t(n-1)+j}\binom{s-1}{i-1}\binom{tn-1}{t-1}\binom{l-1}{j-1}\binom{tn-l}{t-j} max\left(s, l\right).$$

*Hamming:*

$$d_H\left(\{\pi\}, E\right) \equiv -\sum_{j=1}^{n}\sum_{i=1}^{t} h\left(i, R_{ij}\right)$$

*with*

$$h\left(i, j\right) = \sum_{l=max(i,j)}^{(n-1)t+min(i,j)}\binom{l-1}{j-1}\binom{tn-l}{t-j}\binom{l-1}{i-1}\binom{tn-l}{t-i}.$$

*Proof.* We sketch the proof for the Spearman case and refer the reader to Alvo and Pan (1997) for the others.

Define compatibility $(tn)! \times 1$ vectors $C_{\{\pi\}}, C_E$ such that the $(t(j-1)+i)$ th component is equal to 1 if $\mu(t(j-1)+i)$ is in the set and 0 otherwise. Let $\{\pi_j\}$ for $1 \leq j \leq n$ the set of permutations which keeps the same relative order as $\mathrm{R}_j$ only and similarly $\mathrm{E}_j$ the set which keeps the same relative order for block j only. Then, it can be shown that

$$T_S C_{\{\pi\}} = \frac{1}{(t!)^{n-1}} \sum_{j=1}^{n} T_S C_{\{\pi_j\}}$$

and

$$\left(T_S C_{\{\pi_p\}}\right)' \left(T_S C_{\{\pi_s\}}\right) = 0, p \neq s.$$

A combinatorial argument shows that for the component representing the $i$th treatment and $j$th block,

$$\left(T_S C_{\{\pi\}}\right)_{ij} = \frac{(tn+1)!}{(t+1)!\,(t!)^{n-1}} \left[R_{ij} - \frac{t+1}{2}\right],$$

whereas

$$\left(T_S C_{\{\pi_l\}}\right)_{ij} = \begin{cases} \frac{(tn+1)!}{(t+1)!}\left[R_{ij} - \frac{t+1}{2}\right] & l = j, \\ 0 & otherwise. \end{cases}$$

Finally,

$$\begin{aligned} d_S\left(\{\pi\}, E\right) &= C_E' \Delta_S C_{\{\pi\}} \\ &= C_E' \left[\frac{tn\,(tn+1)\,(tn-1)}{12} J - T_S' T_S\right] C_{\{\pi\}} \\ &= \frac{tn\,(tn+1)\,(tn-1)}{12} \left[\binom{tn}{t} \cdots \binom{2t}{t}\right]^2 \\ &\quad - \frac{\sum_{j=1}^{n} \left(T_S C_{\{\pi_j\}}\right)' \left(T_S C_{\{E_j\}}\right)}{(t!)^{2(n-1)}}. \end{aligned}$$

The result follows. □

We also note that the Spearman and Hamming test statistics are functions of the relative ranks within each block and are sums of independent random variables. As a result, their asymptotic distributions will follow from the usual central limit theorems. The Kendall statistic, although not a sum of independent random variables, can be shown to be asymptotically equivalent to a nondegenerate U-statistic as $n \to \infty$ and its asymptotic distribution will follow from the theory for U statistics. We state these results in the next theorem.

**Theorem 7.2.** *The asymptotic distributions as $n \to \infty$ of the test statistics under the null hypothesis are as follows:*
*Spearman:*

$$\sum_{i=1}^{n}\sum_{j=1}^{t} iR_{ij} \Rightarrow_L N\left(m_S, \sigma_S^2\right)$$

*where*

$$m_S = \frac{nt\,(t+1)^2}{4}$$

*and*

$$\sigma_S^2 = \frac{n}{t-1}\left[\frac{t\left(t^2-1\right)}{12}\right]^2.$$

*Kendall:*

$$-\frac{1}{\binom{2t}{t}^2}\sum_{1\le j_1<j_2\le n}\;\sum_{1\le i_1<i_2\le t} a\left[R_{i_1 j_1}, R_{i_2 j_2}\right] a\left[i_1, i_2\right]$$
$$-\sum_{j=1}^{n}\sum_{1\le i_1<i_2\le t} sgn\left[R_{i_2 j} - R_{i_1 j}\right] \Rightarrow_L N\left(m_K, \sigma_K^2\right)$$

*where*

$$m_K = 0$$

*and*

$$\sigma_K^2 = \frac{n^3 t\,(t+1)^3\left(13t^2-2t+1\right)}{144\,(t-1)}.$$

*Spearman Footrule:*

$$\sum_{j=1}^{n}\sum_{i=1}^{t} f\left(i, R_{ij}\right) \Rightarrow_L N\left(m_F, \sigma_F^2\right)$$

*where*

$$m_F = \frac{n}{t}\sum_{j=1}^{n}\sum_{i=1}^{t} f(i,j)$$

*and*

$$\sigma_F^2 = \frac{n}{t-1}\sum_{j=1}^{n}\sum_{i=1}^{t} F_{ij}^2$$

$$F_{ij} = f(i,j) - \bar{f}(i,.) - \bar{f}(.,j) + \bar{f}(.,.),$$

*with dots indicating averaging over that index.*
*Hamming:*

$$\sum_{j=1}^{n}\sum_{i=1}^{t} f\left(i, R_{ij}\right) \Rightarrow_L N\left(m_H, \sigma_H^2\right)$$

*where*

$$m_H = \frac{n}{t}\sum_{j=1}^{n}\sum_{i=1}^{t} h(i,j)$$

$$\sigma_H^2 = \frac{n}{t-1}\sum_{j=1}^{n}\sum_{i=1}^{t} H_{ij}$$

*and*

$$H(i,j) = h(i,j) - \bar{h}(i,.) - \bar{h}(.,j) + \bar{h}(.,.)$$

*with dots indicating averaging over that index.*

## 7.2 The Incomplete Case

It often happens in experiments, however, that one or more observations are missing from one or more blocks. As an example, we consider the data in Table 7.1 which displays lymph heart pressure measurements in mm Hg taken over 24 h at 6 h intervals on 8 toads during an induced dehydration period. The data shows that some cells are empty. If we identify the 6 h intervals as treatments and the toads with the blocks, then we have $n = 8, t = 4, k_1 = k_2 = k_6 = k_7 = k_8 = 4, k_3 = k_5 = 3, k_4 = 2$. We wish to test against the alternative $H_1 : \tau_1 \geq \tau_2 \geq \tau_3 \geq \tau_4$ with at least one inequality strict. The data shows that some cells are empty.

**Table 7.1** Lymph heart pressure in mm Hg taken over a 24 h period at 6 h intervals on 8 toads during dehydration

| | | Time | | | |
|---|---|---|---|---|---|
| Toad ID | Block | 6 h | 12 h | 18 h | 24 h |
| 21 | 1 | 11.865 | 9.832 | 7.567 | 10.168 |
| 22 | 2 | 5.601 | 4.892 | 4.032 | 3.126 |
| 23 | 3 | | 14.415 | 14.185 | 7.800 |
| 24 | 4 | 13.267 | | | 9.953 |
| 25 | 5 | 8.006 | 7.793 | | 7.582 |
| 27 | 6 | 17.692 | 16.644 | 15.327 | 11.573 |
| 28 | 7 | 9.027 | 7.973 | 11.855 | 6.820 |
| 29 | 8 | 9.789 | 7.967 | 7.758 | 7.849 |

In what follows, we shall obtain a generalization of the Page and Jonckheere test statistics. To this end, define $o_{il}$ to be the label of the $l$th object ranked in the $i$th block and set

$$A_S^*(i) = \frac{(t+1)}{(k_i+1)} \sum_{l=1}^{k_i} \left(o_{il} - \frac{t+1}{2}\right)\left(\mu_i^*(o_{il}) - \frac{k_i+1}{2}\right)$$

where $k_i$ is the number of objects ranked in block $i$. This statistic represents a measure of the association between the observed ranking $\mu_i^*$ for the treatments labeled $\{o_{il}\}$ and the complete criterion ranking specified by the alternative chosen to be $\{1, 2, \ldots, k\}$. It can be interpreted as the sum of a naive statistic which ignores gaps and a correction term:

$$A_S^*(i) = \frac{(t+1)}{(k_i+1)} \sum_{l=1}^{k_i} \left(l - \frac{t+1}{2}\right)\left(\mu_i^*(o_{il}) - \frac{k_i+1}{2}\right) +$$
$$\frac{4}{(k_i+1)} \sum_{l=1}^{k_i} (o_{il} - l)\left(\mu_i^*(o_{il}) - \frac{k_i+1}{2}\right).$$

The term $(o_{il} - l)$ represents the number of missing observations to the left of $o_{il}$ and is a weighting for the observed rank. The proposed test statistic is the sum

$$L_S^* = \sum_{i=1}^{n} A_S^*(i).$$

A further informative expression may be had by defining

$$u_{ij} = \mu_i^*(j)\,\delta_{ij} + \frac{k_i+1}{2}(1-\delta_{ij})$$
$$V_i = \sum_i j\,u_{ij}$$

where $\delta_{ij} = 1$ if treatment $j$ is ranked in block $i$ and $= 0$ otherwise. Note that for each block $i$,

$$\sum_{j=1}^{t} \frac{(t+1)}{(k_i+1)} u_{ij} = \frac{t\,(t+1)}{2}.$$

Then,

$$A_S^*(i) = \frac{t+1}{k_i+1} V_i - \frac{t\,(t+1)^2}{4}$$

and

$$L_S^* = \sum_{i=1}^{n} \frac{t+1}{k_i+1} V_i - \frac{nt\,(t+1)^2}{4}.$$

In this form, we see that instead of taking the sum of the ranks assigned to treatment $j$ as we would in the complete case, we calculate a weighted sum of the scores assigned to that treatment, which is the sum of the ranks in a block in the complete case. Large values of $L_S^*$ are consistent with the alternative.

Turning attention to the Kendall statistic, let

$$a_i(l,m) = \begin{cases} sgn(\mu_i^*(o_{im}) - \mu_i^*(o_{il})) & \text{if } \delta_{il} = \delta_{im} = 1, \\ 1 - \frac{2\mu_i^*(o_{il})}{(k_i+1)} & \text{if } \delta_{il} = 1, \delta_{im} = 0, \\ \frac{2\mu_i^*(o_{im})}{(k_i+1)} - 1 & \text{if } \delta_{il} = 0, \delta_{im} = 1, \\ 0 & \text{otherwise,} \end{cases}$$

where $\delta_{il} = 1$ or 0 according to whether or not the $l$th treatment response is observed in the $i$th block. We may now write the Kendall statistic

$$\begin{aligned} A_K^*(i) &= \sum_{l<m}^{k_i} a_i(l,m) \\ &= \sum_{l<m}^{k_i} sgn\left(\mu_i^*(o_{im}) - \mu_i^*(o_{il})\right) + \\ &\quad \frac{4}{(k_i+1)} \sum_{l=1}^{k_i} (o_{il} - l)\left(\mu_i^*(o_{il}) - \frac{k_i+1}{2}\right). \end{aligned}$$

The extended Jonckheere statistic can then be defined to be

$$L_K^* = \sum_{i=1}^{n} A_K^* (i) .$$

Alternatively, setting $U_i^* (l, m) = \frac{1}{2} (1 + a_i (l, m))$, we obtain the extended Jonckheere statistic

$$L_K{}^* = 2 \sum_{i=1}^{n} \sum_{l<m} U_i^* (l, m) - \frac{nt (t-1)}{2} .$$

The statistics $L_S^*, L_K^*$ are respectively generalizations of the Page and Jonckheere statistics for ordered alternatives. Under $H_0$ and conditional on the pattern of missing observations, the following theorems can be shown to be true.

**Theorem 7.3.** $\frac{L_S^*}{\sigma_S} \Rightarrow N(0,1)$ *as* $n \to \infty$, $k_i \geq 2$, *where* $\sigma_S^2 = \sum_{i=1}^{n} \sigma_S^2 (i)$,

$$\sigma_S^2 (i) = \frac{k_i (t+1)^2}{12 (k_i + 1)} \sum_{l=1}^{k_i} (o_{il} - \overline{o}_l)^2 ,$$

*and* $\overline{o}_l = \sum_{l=1}^{k_i} o_{il} / k_i$.

*Proof.* The proof consists of demonstrating the Lyapunov condition holds and then applying the Lindeberg-Feller central limit theorem. See Alvo and Cabilio (1995b) for the details. □

**Theorem 7.4.** $\frac{L_K^*}{\sigma_K} \Rightarrow N(0,1)$ *as* $n \to \infty$ $k_i \geq 2$, *where* $\sigma_K^2 = \sum_{i=1}^{n} \sigma_K^2 (i)$ *and*

$$\sigma_K^2 (i) = \frac{16}{(t+1)^2} \sigma_S^2 (i) + \frac{5k_i (k_i - 1)}{18} + \frac{8}{3 (k_i + 1)} \sum_{l=1}^{k_i} (o_{il} - l) \left( l - \frac{k_i + 1}{2} \right) .$$

*Proof.* The proof consists once again of demonstrating the Lyapunov condition holds and then applying the Lindeberg-Feller central limit theorem. It is much more involved than the proof of Theorem 7.3. It requires expressions found in Quade (1972) for the moments of the Kendall correlation between the criterion ranking with blanks removed and the incomplete ranking in each block. See Alvo and Cabilio (1995b) for the details. □

*Example 7.2 (Application).* We may illustrate the use of these methods in the following example. We shall use the statistic $L^*$ by first ranking from largest to smallest. Table 7.2 contains the scores $u_{il}$, weights $\frac{t+1}{k_i+1}$, and sums $\sum_{i=1}^{n} \frac{t+1}{k_i+1} u_{il}$.

**Table 7.2** Scores $u_{il}$, weights $\frac{t+1}{k_i+1}$, and sums $\sum_{i=1}^{n} \frac{t+1}{k_i+1} u_{il}$

| Toad ID | Block | Time | | | | |
|---|---|---|---|---|---|---|
| | | 6h | 12 h | 18 h | 24 h | $\frac{t+1}{k_i+1}$ |
| 21 | 1 | 1 | 3 | 4 | 2 | 1 |
| 22 | 2 | 1 | 2 | 3 | 4 | 1 |
| 23 | 3 | 2 | 1 | 2 | 3 | 1.25 |
| 24 | 4 | 1 | 1.5 | 1.5 | 2 | 1.66 |
| 25 | 5 | 1 | 2 | 2 | 3 | 1.25 |
| 27 | 6 | 1 | 2 | 3 | 4 | 1 |
| 28 | 7 | 2 | 3 | 1 | 4 | 1 |
| 29 | 8 | 1 | 2 | 4 | 3 | 1 |
| | $\sum_{i=1}^{n} \frac{t+1}{k_i+1} u_{il}$ | 11.416 | 18.250 | 22.500 | 27.833 | |

**Table 7.3** Null distribution of $V_i$

| $V_1$ none missing | | $V_3$ missing (1) | | $V_4$ missing (2, 3) | | $V_5$ missing (3) | |
|---|---|---|---|---|---|---|---|
| Value | Prob. | Value | Prob. | Value | Prob. | Value | Prob. |
| 20 | 0.0416 | 18 | 0.1666 | 13.5 | 0.5 | 17 | 0.1666 |
| 21 | 0.1250 | 19 | 0.3333 | 16.5 | 0.5 | 18 | 0.1666 |
| 22 | 0.0416 | 21 | 0.3333 | | | 19 | 0.1666 |
| 23 | 0.1666 | 22 | 0.1666 | | | 21 | 0.1666 |
| 24 | 0.0833 | | | | | 22 | |
| 25 | 0.0833 | | | | | | |
| 26 | 0.0833 | | | | | | |
| 27 | 0.1666 | | | | | | |
| 28 | 0.0416 | | | | | | |
| 29 | 0.1250 | | | | | | |
| 30 | 0.0416 | | | | | | |

The value of $L_S^* = 26.75$. Table 7.3 gives the null distributions of the relevant $V_i's$. Note that in this case, $V_1, V_2, V_6, V_7$, and $V_8$ are identically distributed. Note further that in general, the distributions depend on the values of $k_i, t$ as well as the pattern of missing values. The distribution of $L^*$ can now be generated by implementing a program to calculate the distribution of the sum of the independent variables $\frac{t+1}{k_i+1} V_i$. As a result, we find that $P(L_S^* \geq 26.75) = 0.00006$ is the p-value of the test. A naive approach to the problem consists of deleting the incomplete blocks and conducting a Page test on the remaining 5 blocks. This yields an exact p-value of 0.002. It is not difficult to imagine examples where the naive approach would not be viable. Table 7.4 lists other selected upper probabilities for this distribution, specifically the critical values with corresponding probabilities close to but not exceeding .1, .05, .025, .01, .005, and .001.

**Table 7.4** Selected critical values of $L^*$ and corresponding normal probabilities for the block pattern in Table 7.1

| $l$ | $\mathbf{P}\left(L_S^* \geq l\right)$ | $\mathbf{P}\left(Z \geq \frac{l}{7.6376}\right)$ |
|---|---|---|
| 10 | 0.0999 | 0.0952 |
| 12.75 | 0.0489 | 0.0475 |
| 15 | 0.0248 | 0.0248 |
| 17.75 | 0.0092 | 0.0101 |
| 19.5 | 0.0046 | 0.0053 |
| 22.75 | 0.0009 | 0.0014 |

### *7.2.1 Asymptotic Efficiency*

Missing observations in blocks should result in a loss of power and it is interesting to determine for a given value of $t$ how this loss depends on $k_i$. We assume for this exercise that we have $n$ replications of a BIBD of $t$ treatments in $b$ blocks and $k_0$ treatments per block, with each treatment occurring $r$ times in the basic design. Thus there are $nb$ blocks observed. For a basic design, $\lambda$ denotes the number of times each pair of treatments occurs together and

$$bk_0 = tr, \lambda = \frac{r\left(k_o - 1\right)}{\left(t - 1\right)}.$$

As a basis for comparisons, one generally assumes that the number of times each treatment appears is the same for the two designs under consideration. In our case, this requires that for each replication of $b$ blocks in the incomplete design, we have $\frac{bk_0}{t}$ blocks in the complete design. Restricting attention to the extended Page statistic, we can show that the asymptotic relative efficiency in the Pitman sense of the extended Page test to the Page test is

$$\frac{\left(k_0 - 1\right)\left(t + 1\right)}{\left(k_0 + 1\right)\left(t - 1\right)}.$$

## 7.3 Tests for Trend in Proportions

There are several instances in practice when one is interested in testing for a trend in proportions. For instance, one may be interested in the trend in birth rates, in mortality rates, or in the incidence of a certain disease. As an example, we will consider the mortality statistics in South Africa during the period 2000–2008. One may ask if there is an increasing trend in the mortality rates.

In such problems, one usually observes at time values $t_1 < t_2 < \ldots < t_k$ a random sample of $n_i$ binary variables $\{y_{ij}\}$ with $y_{ij}$ taking values 1 or 0 with unknown probabilities $\theta_i$ and $1 - \theta_i$, respectively. The observed data may be viewed as

| Time | $t_1$ | $t_2$ | ... | $t_k$ | Total |
|---|---|---|---|---|---|
| | $y_1$ | $y_2$ | ... | $y_k$ | $y$ |
| | $n_1 - y_1$ | $n_2 - y_2$ | ... | $n_k - y_k$ | $n - y$ |
| Total | $n_1$ | $n_2$ | ... | $n_k$ | $n$ |

where $y_i = \sum_j y_{ij}$, $y = \sum_i y_i$. Let $\bar{\theta}_i = y_i/n_i$ and $\bar{\theta} = y/n$. Without loss in generality, we shall be interested in detecting a monotone increasing trend in $\theta_i$.

Let $\{x_i\}$ represent arbitrary preselected scores, $x_1 < x_2 < \ldots < x_k$, which mimic a monotone increasing time trend. The linear regression model of $y_{ij}$ can be expressed by

$$\theta_i = \alpha + \beta\left(x_i - \bar{x}_k\right) \tag{7.1}$$

and subject to the constraint that $\sum_i n_i\left(x_i - \bar{x}_k\right) = 0$ where $\bar{x}_k = \sum n_i x_i / n$ yields estimates

$$\hat{\alpha} = \frac{\sum_i n_i \bar{\theta}_i}{\sum_i n_i} = \bar{\theta}, \hat{\beta} = \frac{\sum_i n_i\left(x_i - \bar{x}_k\right)\left(\bar{\theta}_i - \bar{\theta}\right)}{\sum_i n_i\left(x_i - \bar{x}_k\right)^2}.$$

The hypothesis of homogeneity is

$$H_0 : \theta_i = \theta, i = 1, ..k$$

(or equivalently $\beta = 0$). Possible alternatives are

$$H_1 : \theta_1 \leq \theta_2 \leq \ldots \leq \theta_k \text{ with at least one strict inequality,}$$

$$H_2 : \theta_i \neq \theta_j, \text{ for at least one pair } (i, j).$$

Hypothesis $H_2$ can also be expressed as

$$\sum_{i<j}\left(\theta_i - \theta_j\right) = \sum_i (k + 1 - 2i)\,\theta_i \neq 0. \tag{7.2}$$

Under the null hypothesis of homogeneity, the estimate of variance of $\hat{\beta}$ is given by

$$V\left(\hat{\beta}\right) = \frac{\bar{\theta}\left(1 - \bar{\theta}\right)}{\sum_i n_i\left(x_i - \bar{x}_k\right)^2}$$

and consequently we reject $H_0$ in favor of $H_1$ (or equivalently $\beta > 0$) for large values of the statistic

$$\hat{\beta}/\sqrt{V\left(\hat{\beta}\right)} = \frac{\sum_i n_i\left(x_i - \bar{x}_k\right)\left(\bar{\theta}_i - \bar{\theta}\right)}{\sqrt{\bar{\theta}\left(1-\bar{\theta}\right)}\sqrt{\sum_i n_i\left(x_i - \bar{x}_k\right)^2}} \tag{7.3}$$

which for large samples has a standard normal. If we suppose further that $\bar{\theta}$ is small (of the order of 1 %–2 %) so that $\bar{\theta}^2$ is negligible, then the statistic becomes

$$\frac{\sum_i n_i\left(x_i - \bar{x}_k\right)\left(\bar{\theta}_i - \bar{\theta}\right)}{\sqrt{\sum_i n_i\bar{\theta}\left(x_i - \bar{x}_k\right)^2}}.$$

The difference between observed and expected frequencies $n_i\left(\bar{\theta}_i - \bar{\theta}\right)$ is multiplied by the score effect $(x_i - \bar{x}_k)$ in the numerator whereas the square of the score effects is weighted by the expected frequencies in the denominator. The statistic provides a test of trend in frequencies as opposed to a test of trend on the proportions. Since the sample correlation between $\left\{\bar{\theta}_i - \bar{\theta}\right\}$ and $\{x_i - \bar{x}_k\}$ is given by

$$\begin{aligned} &\frac{\sum_i n_i\left(x_i - \bar{x}_k\right)\left(\bar{\theta}_i - \bar{\theta}\right)}{\sqrt{\sum_i n_i\left(\bar{\theta}_i - \bar{\theta}\right)^2\sum_i n_i\left(x_i - \bar{x}_k\right)^2}} \\ &= \hat{\beta}\sqrt{\frac{\sum_i n_i\left(x_i - x_k\right)^2}{\sum_i n_i\left(\bar{\theta}_i - \bar{\theta}\right)^2}}, \end{aligned}$$

we may view the test of monotonicity as equivalent to a test that the correlation is 0. Alternatively, the test of homogeneity may be conducted by treating the data as coming from a 2 x $k$ contingency table. That test rejects $H_0$ in favor of $H_2$ for large values of the statistic

$$\sum_i \frac{\left(n_i\bar{\theta}_i - n_i\bar{\theta}\right)^2}{n_i\bar{\theta}} + \sum_i \frac{\left(n_i\left(1-\bar{\theta}_i\right) - n_i\left(1-\bar{\theta}\right)\right)^2}{n_i\left(1-\bar{\theta}\right)} \tag{7.4}$$

$$= \frac{\sum_i n_i\left(\bar{\theta}_i - \bar{\theta}\right)^2}{\bar{\theta}\left(1-\bar{\theta}\right)} \tag{7.5}$$

which for large samples $n_i \to \infty$ has a chi-square distribution with $(k-1)$ degrees of freedom. In that case there is no need to define scores. We note that even though $H_1$ is included in $H_2$, one-sided tests which focus strictly on $H_1$ will in general be more powerful.

The regression model may be extended to apply to two or more groups of individuals. For example, to model the birth rates of men and women in the population, the functional regression model becomes

$$\theta_i = \alpha + \beta(x_i - \bar{x}_k) + \gamma\left(\alpha' + \beta'(x_i - \bar{x}_k)\right) \tag{7.6}$$

where $\gamma$ takes value 1 for the first group and 0 otherwise.

In the next section, we make use of the general theory of hypothesis testing to construct a nonparametric test statistic based on the ranks of the observations to test $H_0$ against $H_2$.

### 7.3.1 The Construction of the Test Statistics

We recall the following measures of similarities due to Spearman, Kendall, and Hamming, respectively, for rankings $\mu = (\mu(1), \ldots, \mu(n))'$, $\nu = (\nu(1), \ldots, \nu(n))'$:

$$\mathcal{A}_S(\mu, \nu) = \sum_{i=1}^{n}\left(\mu(i) - \frac{n+1}{2}\right)\left(\nu(i) - \frac{n+1}{2}\right)$$

$$\mathcal{A}_K(\mu, \nu) = \sum_{i<j} sgn\,(\mu(j) - \mu(i))\, sgn\,(\nu(j) - \nu(i))$$

$$\mathcal{A}_H(\mu, \nu) = \sum_{i=1}^{n}\sum_{j=1}^{n}\left(I\,[\mu(i) = j] - \frac{1}{n}\right)\left(I\,[\nu(i) = j] - \frac{1}{n}\right)$$

where $sgn(x)$ is either 1 or $-1$ depending on whether $x > 0$ or $x < 0$ and where $I\,[\cdot]$ is the indicator function which is 1 or 0 depending on whether the statement in brackets holds or not. In the model considered, the ranking which describes the time points may be viewed as a tied ranking with tie pattern

$$\boldsymbol{\delta}_1 = (n_1, n_2, \ldots, n_k) \tag{7.7}$$

and with ordering

$$\left\langle (1, \ldots, n_1)\,(n_1 + 1, \ldots, n_1 + n_2) \ldots \left(\sum_{i=1}^{k-1} n_i + 1, \ldots, \sum_{i=1}^{k} n_i\right)\right\rangle. \tag{7.8}$$

On the other hand, the binary variables $y_{ij}$ have with $e = 2$ the simple tie pattern

$$\boldsymbol{\delta}_2 = (y, n - y)\,. \tag{7.9}$$

### 7.3.2 *The Test Statistic Corresponding to Spearman and Kendall Similarity*

To compute the test statistic corresponding to Spearman similarity, note that

$$\mathcal{A}_S\left(\mu_{\delta_1}, \nu_{\delta_2}\right)$$
$$= \sum_{i=1}^{n} E\left[\left(\mu(i) - \frac{n+1}{2}\right)|C\left(\mu\right)\right] E\left[\left(\nu(i) - \frac{n+1}{2}\right)|C\left(\nu\right)\right].$$

The average of the compatible ranks at time $t_1$ is $g_1 = \left(\frac{n_1+1}{2}\right)$, at time $t_2$ it is $g_2 = n_1 + \left(\frac{n_2+1}{2}\right)$, and so on. In general at time $t_l$ the average rank is $g_l = \sum_{i=1}^{l-1} n_i + \left(\frac{n_l+1}{2}\right), l = 1, .., k$. Clearly, $\sum_{i=1}^{k} n_i g_i = \frac{n(n+1)}{2}$. Hence, the conditional expectation

$$E\left[\nu(l)|C\left(\nu\right)\right] = g_l, l = 1, \ldots, n_i.$$

Turning attention now to the binary observations, the average rank for the $n - y$ which take value 0 is $l_1 = \frac{n-y+1}{2}$, whereas the average rank for the $y$ observations which take value 1 is $l_2 = n - y + \left(\frac{y+1}{2}\right) = n - \frac{y}{2} + \frac{1}{2}$. It follows that at time $t_i$,

$$E\left[\mu(l)|C\left(\mu\right)\right] = \begin{cases} l_1 & l = 1, \ldots, n_i - y_i, \\ l_2 & l = n_i - y_i + 1, \ldots, n_i. \end{cases}$$

Since $l_2 - l_1 = \frac{n}{2}$ and $\sum_{i=1}^{k} n_i g_i = \frac{n(n+1)}{2}$, we have that

$$\mathcal{A}_S\left(\mu_{\delta_1}, \nu_{\delta_2}\right) = \sum_{i=1}^{k}\left[g_i - \frac{n+1}{2}\right]\left[l_2 y_i + (n_i - y_i)\, l_1\right]$$
$$= \frac{n}{2}\sum_{i=1}^{k}\left[g_i - \frac{n+1}{2}\right] y_i.$$

Hence, we may define the Spearman statistic to be

$$S = \sum_{i=1}^{k}\left[g_i - \frac{n+1}{2}\right] y_i \tag{7.10}$$

$$= \sum_{i=1}^{k} n_i c_i \left(\overline{\theta_i} - \overline{\theta}\right) \tag{7.11}$$

where $c_i = g_i - \frac{n+1}{2}$ and $\sum_{i=1}^{k} n_i c_i = 0$.

We now consider the test statistic corresponding to Kendall's similarity measure. Note that within a given time period, the difference between ties is zero and hence there is no contribution to the distance. Between different time periods we have

$$\sum_{i<j}^{n} E\left[sgn\left(\mu(j)-\mu(i)\right)|C\left(\mu\right)\right] E\left[sgn\left(\nu(j)-\nu(i)\right)|C\left(\nu\right)\right]$$

$$=\sum_{i<j}^{k}\left[\left(n_i-y_i\right)y_j-\left(n_j-y_j\right)y_i\right]$$

$$=\sum_{i<j}^{k}\left[n_i y_j-n_j y_i\right].$$

Now

$$\sum_{j=i+1}^{k} n_j=\left[n-n_i-\sum_{j=1}^{i-1} n_j\right]$$
$$=\left[n-g_i-\frac{n_i-1}{2}\right].$$

Hence,

$$\mathcal{A}_K\left(\mu_{\delta_1},\nu_{\delta_2}\right)=\sum_{i=1}^{k} y_i\left[g_i-\frac{n_i+1}{2}\right]-\sum_{i=1}^{k} y_i\left[n-g_i-\frac{n_i-1}{2}\right]$$
$$=2\sum_{i=1}^{k}\left[g_i-\frac{n+1}{2}\right]y_i$$

and it is seen that the Kendall and Spearman statistics are equivalent.

### 7.3.3 The Test Statistic Corresponding to Hamming Similarity

For the Hamming similarity, we have for time $t_i, l=\sum_{q=1}^{i-1} n_q+1,\ldots,\sum_{q=1}^{i} n_q$ and consequently

$$E\left[I\left(\nu(l)=j\right)|C\left(\nu\right)\right]=\begin{Bmatrix} \frac{1}{n_i}, \sum_{q=1}^{i-1} n_q+1\leq j\leq\sum_{q=1}^{i} n_q, \\ \\ 0, \text{ otherwise}, \end{Bmatrix}$$

For the data, it is seen that at time $t_i$ and $l = \sum_{q=1}^{i-1} n_q + 1, \ldots, \sum_{q=1}^{i} n_q$,

$$E\left[I\left(\mu(l) = j\right) | C\left(\mu\right)\right] = P\left(\mu\left(l\right) = j | C\left(\mu\right)\right) = \begin{cases} \frac{y_i}{n_i y} & j = 1, \\ \frac{(n_i - y_i)}{n_i (n-y)} & j = 0. \end{cases}$$

Consequently, the Hamming similarity measure for tied data becomes

$$\begin{aligned} &\mathcal{A}_H\left(\mu_{\delta_1}, \nu_{\delta_2}\right) \\ &= \sum_{i,j}^{n} \left(E\left[I\left(\mu(i) = j\right) | C\left(\mu\right)\right] E\left[I\left(\nu(i) = j\right) | C\left(\nu\right)\right]\right) - 1 \\ &= \sum_{i=1}^{k} \left[ y_i \frac{y_i}{n_i y} + \left(n_i - y_i\right) \frac{\left(n_i - y_i\right)}{n_i \left(n - y\right)} \right] - 1 \\ &= \sum_{i=1}^{k} n_i \left[ \frac{\bar{\theta}_i^2}{y} + \frac{\left(1 - \bar{\theta}_i\right)^2}{\left(n - y\right)} \right] - 1 \\ &= \frac{n}{\bar{\theta}\left(1 - \bar{\theta}\right)} \sum_{i=1}^{k} \left[ n_i \left(\bar{\theta}_i - \bar{\theta}\right)^2 \right]. \end{aligned} \tag{7.12}$$

This is the usual goodness-of-fit statistic.

### *7.3.4 The Asymptotic Distribution of the Test Statistics Under the Null Hypothesis*

The asymptotic distribution of the Spearman test statistic under the null hypothesis may be determined under two different assumptions. Let $\{y_i\}$ be $k$ independent binomials $(n_i, \theta_i)$ and suppose we would like to test $H_0$ vs $H_1$. In most applications the asymptotic situation of interest occurs when

$$n_i \to \infty, \text{ with } \frac{n_i}{n} \to \lambda_i > 0, \; i = 1, \ldots, k. \tag{7.13}$$

We now show that the Spearman statistic has an asymptotic normal distribution under either (7.13) or under the condition that the $\{n_i\}$ are bounded while $k \to \infty$.

**Theorem 7.5.** *Suppose that both $y \to \infty$ and $n - y \to \infty$ as $n \to \infty$. Under $H_o$ the Spearman test statistic has asymptotically a standard normal distribution, i.e.,*

$$\frac{\sum_{i=1}^{k} n_i c_i \left(\bar{\theta}_i - \bar{\theta}\right)}{\sqrt{\sum_{i=1}^{k} c_i^2 n_i \theta_i \left(1 - \theta_i\right)}} \to_d N\left(0, 1\right)$$

*under either (i) (7.13) or (ii) the $\{n_i\}$ are bounded and $k \to \infty$.*

*Proof.* (i) The Spearman statistic (7.11) is expressible as a linear combination of independent binomials, $\sum_{i=1}^{k} c_i y_i$ where $c_i = g_i - \frac{n+1}{2}$ and $g_i = \sum_{i=1}^{i-1} n_i + \left(\frac{n_i+1}{2}\right)$. Hence, under (7.13), we have approximately $y_i \approx_d N\left(n_i\theta_i, n_i\theta_i\left(1 - \theta_i\right)\right)$. In view of the independence of the $\{y_i\}$,

$$\sum c_i y_i \approx_d N\left(\sum c_i n_i \theta_i, \sum c_i^2 n_i \theta_i \left(1 - \theta_i\right)\right)$$

and hence

$$\frac{\left(\sum c_i y_i - \sum c_i n_i \theta_i\right)}{\sqrt{\sum c_i^2 n_i \theta_i \left(1 - \theta_i\right)}} \to_d N\left(0, 1\right).$$

Under $H_0, \sum c_i n_i \theta_i = 0$ and hence

$$\frac{\sum c_i y_i}{\sqrt{\sum c_i^2 n_i \theta_i \left(1 - \theta_i\right)}} \to_d N\left(0, 1\right).$$

(ii) For this part, note that the Spearman statistic is expressible as

$$\sum_{i=1}^{k} \sum_{j=1}^{n_i} c_i y_{ij}$$

where the $\{y_{ij}\}$ are Bernoulli $(\theta_i)$. We need to show that

$$\frac{\max_{1 \le i \le k} c_i^2}{\sum_{i=1}^{k} \sum_{j=1}^{n_i} c_i^2} \to 0, as\ k \to \infty.$$

Note that if $n_i \le M$, for all $i$,

$$\begin{aligned} c_i^2 &\le g_i^2 + \frac{(n+1)^2}{4} \\ &\le M^2 \left(i - \frac{1}{2}\right)^2 + \frac{1}{4} + \frac{(n+1)^2}{4} = O\left(k^2\right). \end{aligned}$$

Moreover, since $n = \sum_{i=1}^{k} n_i \leq Mk$

$$\begin{aligned}
\sum_{i=1}^{k}\sum_{j=1}^{n_i} c_i^2 &= \sum_{i=1}^{k} n_i g_i^2 - \frac{n(n+1)^2}{4} \\
&\geq \sum_{i=1}^{k} n_i g_i^2 - \frac{M^3k^3}{4} \\
&= \sum_{i=1}^{k} n_i \left(\sum_{j=1}^{i-1} n_j\right)^2 + \sum_{i=1}^{k}(n_i+1)\left(\sum_{j=1}^{i-1} n_j\right) + \sum_{i=1}^{k} n_i \left(\frac{n_i+1}{2}\right)^2 - \frac{M^3k^3}{4} \\
&\geq \sum_{i=1}^{k} n_i (i-1)^2 + 2\sum_{i=1}^{k}(i-1) = O\left(k^3\right).
\end{aligned}$$

It follows that

$$\frac{\max_{1\leq i\leq k} c_i^2}{\sum_{i=1}^{k}\sum_{j=1}^{n_i} c_i^2} = O(\frac{1}{k}) \to 0, as\ k \to \infty$$

and the theorem is proved. □

We may estimate $\theta_i$ either by $\bar{\theta}_i$ or by $\bar{\theta}$. In the first case, the test rejects whenever

$$\frac{\sum_{i=1}^{k} c_i y_i}{\sqrt{\sum_{i=1}^{k} c_i^2 n_i \bar{\theta}_i \left(1-\bar{\theta}_i\right)}} \geq z_\alpha$$

where $z_\alpha$ is the upper $100(1-\alpha)$ percentage point from a standard normal distribution. The expression for the estimate of the asymptotic power becomes

$$1-\Phi\left(z_\alpha - \frac{\sum_{i=1}^{k} n_i c_i \theta_i}{\sqrt{\sum c_i^2 n_i \theta_i (1-\theta_i)}}\right) = \Phi\left(\frac{\sum_{i=1}^{k} n_i c_i \theta_i}{\sqrt{\sum_{i=1}^{k} c_i^2 n_i \theta_i (1-\theta_i)}} - z_\alpha\right).$$

It is seen that the power converges to 1 with increasing $n$.

Alternatively, we may use the statistic

$$\frac{\sum_{i=1}^{k} n_i c_i \left(\bar{\theta}_i - \bar{\theta}\right)}{\sqrt{\bar{\theta}\left(1-\bar{\theta}\right)}\sqrt{\sum_{i=1}^{k} n_i c_i^2}} \tag{7.14}$$

**Table 7.5** Mortality statistics for South Africa 2000–2008

| Year | Number of deaths | Population size |
|---|---|---|
| 2000 | 416, 155 | 43, 789, 115 |
| 2001 | 454, 882 | 43, 997, 828 |
| 2002 | 502, 050 | 44, 187, 637 |
| 2003 | 556, 779 | 44, 344, 136 |
| 2004 | 576, 709 | 42, 718, 530 |
| 2005 | 598, 131 | 42, 768, 678 |
| 2006 | 612, 778 | 43, 647, 658 |
| 2007 | 603, 094 | 43, 586, 097 |
| 2008 | 592, 073 | 43, 421, 021 |

which in the simulation studies reported appears to more closely attain the prescribed significance level. In the case of equal sample sizes, $n_i = n_0$, say, the test statistic (7.14) takes the simpler form

$$\frac{\sum_{i=1}^{k}\left(i-\frac{k+1}{2}\right)\left(\bar{\theta}_i-\bar{\theta}\right)}{\sqrt{n_0}\sqrt{\bar{\theta}\left(1-\bar{\theta}\right)}\sqrt{\frac{k\left(k^2-1\right)}{12}}}.$$

*Example 7.3.* Returning to the data on mortality rates in South Africa in Table 7.5, we calculated values of 260.1 and 70.3 for the Spearman and Hamming similarity measures, respectively. These yielded p-values $< 10^{-4}$.

We consider another application for 6 beaches in Hong Kong.

*Example 7.4.* Table 7.6 shows the geometric *E. Coli* count for each of 6 beaches in the Sai Kung District of Hong Kong during the period 1986–2009. A beach is classified as good if the count is at most 24. The Spearman test statistic yielded a value of 22.98 which points to strong evidence of an upward trend in the annual proportion of good beaches.

## Chapter Notes

It was seen that the Spearman distance induces the Page statistic (Page 1963) whereas the Kendall distance does not induce the Jonckheere statistic (Jonckheere 1954).

There are a number of different approaches to the problem of testing for trend in proportions which are reported in Alvo and Berthelot (2012). For the well-known Cochran-Armitage test (Cochran 1954; Armitage 1955), it is known that when the expected values $n_i\theta_i$ or $n_i(1-\theta_i)$ are small, the normal approximation

**Table 7.6** Annual geometric mean *E. Coli* level (per 100 ml) in the Sai Kung District. Beaches: Clear Water Bay First (1), Clear Water Bay Second (2), Hap Mun Bay (3), Kiu Tsui (4), Silverstrand (5), Trio (6), and number of good beaches (7)

| Year | (1) | (2) | (3) | (4) | (5) | (6) | (7) |
|---|---|---|---|---|---|---|---|
| 1986 | 102 | 69 | 9 | 18 | 255 | 49 | 2 |
| 1987 | 133 | 52 | 6 | 9 | 62 | 32 | 2 |
| 1988 | 39 | 35 | 4 | 3 | 129 | 35 | 2 |
| 1989 | 80 | 38 | 3 | 5 | 192 | 23 | 3 |
| 1990 | 51 | 42 | 4 | 5 | 89 | 31 | 2 |
| 1991 | 30 | 14 | 2 | 4 | 106 | 14 | 4 |
| 1992 | 52 | 42 | 2 | 5 | 94 | 32 | 2 |
| 1993 | 31 | 16 | 3 | 4 | 56 | 20 | 4 |
| 1994 | 30 | 35 | 3 | 3 | 72 | 14 | 3 |
| 1995 | 55 | 39 | 6 | 3 | 226 | 16 | 3 |
| 1996 | 34 | 43 | 5 | 5 | 126 | 29 | 2 |
| 1997 | 62 | 66 | 3 | 5 | 148 | 30 | 2 |
| 1998 | 41 | 44 | 2 | 4 | 99 | 21 | 3 |
| 1999 | 11 | 12 | 2 | 4 | 32 | 17 | 5 |
| 2000 | 16 | 26 | 2 | 5 | 61 | 10 | 4 |
| 2001 | 28 | 22 | 1 | 5 | 100 | 12 | 4 |
| 2002 | 28 | 14 | 2 | 4 | 133 | 6 | 4 |
| 2003 | 17 | 21 | 4 | 5 | 97 | 10 | 5 |
| 2004 | 9 | 10 | 3 | 17 | 74 | 2 | 5 |
| 2005 | 16 | 19 | 4 | 14 | 67 | 6 | 5 |
| 2006 | 20 | 13 | 4 | 11 | 30 | 5 | 5 |
| 2007 | 14 | 9 | 3 | 6 | 33 | 2 | 5 |
| 2008 | 11 | 19 | 5 | 12 | 35 | 12 | 5 |
| 2009 | 15 | 27 | 3 | 19 | 31 | 5 | 4 |

may become unreliable. As a consequence the Cochran-Armitage test becomes conservative and may lead to a type I error rate greater than to the prescribed significance level. As well, the Cochran-Armitage test is very sensitive to the choice of scores.

A small simulation study was conducted in Alvo and Berthelot (2012). For k=5, three cases were considered: proportions which are strictly increasing, proportions which are nondecreasing, and some which have no particular pattern. It was seen that the Spearman measure is quite often superior in the first two cases. Predictably the power is smaller when the $\{\theta_i\}$ are closer together than when they are further apart.

# Chapter 8
# Probability Models for Ranking Data

Probability modeling for ranking data is an efficient way to understand people's perception and preference on different objects. Various probability models for ranking data have been developed, particularly in the last decade where many new problems involving a large number of objects emerged. In their review paper on probability models for ranking data, Critchlow et al. (1991) broadly categorized these models into four classes: (1) order statistics models, (2) paired comparison models, (3) distance-based models, and (4) multistage models. Since their publication in 1991, variants of these models and new models have been developed. In this chapter, we will introduce these four classes of models and describe their properties.

Before introducing these models, we would like to describe several distinctive features of these models, which may affect the choice of models to be considered in our study:

(a) *Some models allow for the presence of covariates*
In collecting data on rankings of a set of objects from a sample of judges, we may also obtain information on some covariates from the judges (judge-specific covariates) and covariates of the objects (object-specific covariates). Some covariates may even be judge-object-specific. For example, in collecting customers' preferences on a list of mobile phones, the judge-specific covariates could be age, gender, and income, and the object-specific covariates could be prices, weights, and brands, and the judge-object-specific covariates could be some personal experience on using each phone or brand. Most models except for the distance-based models and multistage models can allow for the presence of covariates.

(b) *Some models are predictive*
If we want to build a model to predict a ranking assigned by an individual, we need to have a predictive model for ranking data. In this case, the presence of covariates is a must and it is expected that the population is heterogeneous and different covariates may lead to different ranking of objects predicted from the



M. Alvo, P.L.H. Yu, *Statistical Methods for Ranking Data*, Frontiers in Probability and the Statistical Sciences, DOI 10.1007/978-1-4939-1471-5_8

fitted model. However when the population is homogeneous, the rankings given by judges can be assumed to be generated from a probability model on rankings. A distance-based model is a typical example.

(c) *Some models can handle big ranking data with a large number of objects* Most ranking models should work well for a small number of objects, say less than 10 or 15. Some may become computationally demanding or even infeasible to use for a large number of objects, examples of which will be the Thurstone order statistics models as its likelihood requires the computation of a high-dimensional integration. Recently, the development of social networks and the competitive pressure to provide customized services motivated many new ranking problems on hundreds or thousands of objects. Recommendations on products such as movies, books, and songs are typical examples in which the number of objects is extraordinarily large. In recent years, many researchers in statistics and computer science have developed models to handle such big data.

## 8.1 Order Statistics Models

Among the above four classes of probability models for ranking data, the class of order statistics models has the longest history in the statistical and psychological literature. Dating back to 1927, Thurstone published his/her famous paper *A law of comparative judgment* in *Psychological Review* in which the ranking of two objects was considered. The basic idea behind this approach is that a judge may have tastes that fluctuate from one instant to another according to the perception of each object which is not perfectly predictable and hence is a random variable. The ordering of these random variables then determines the judge's ranking of the objects. Thurstone (1927) proposed a ranking process where the ranking $\boldsymbol{\pi}_j$ of $t$ objects given by a random sample of judge $j$ ($j = 1, 2, \ldots, n$) is determined by the relative ordering of $t$ random utilities $y_{1j}, y_{2j}, \cdots, y_{tj}$, where $\boldsymbol{y}_j = (y_{1j}, \cdots, y_{tj})', j = 1, \cdots, n$ are independent.

The probability of observing a ranking $\boldsymbol{\pi}_j$ under the class of order statistics models is

$$P(\boldsymbol{\pi}_j) = P(y_{[1]_j j} > y_{[2]_j j} > \ldots > y_{[t]_j j}), \qquad \boldsymbol{\pi}_n \in \mathcal{P} \tag{8.1}$$

where $< [1]_j, [2]_j, \cdots, [t]_j >$ is the ordering of objects corresponding to ranking $\boldsymbol{\pi}_j$ such that judge $j$ assigns rank $i$ to object $[i]_j$ (i.e., $\pi_j([i]_j) = i$ or $\pi_j^{-1}(i) = [i]_j$) and $\mathcal{P}$ is the set of all $t!$ possible rankings. It should be noted that the order statistics model (8.1) is invariant under any strictly increasing transformation of the $y_{ij}$'s for which the ordering of the $y_{ij}$'s is preserved.

Critchlow et al. (1991) observed that if the utilities $y_{1j}, \cdots, y_{tj}$ are allowed to have arbitrary dependencies, any probability distribution on rankings can be expressed as in (8.1). To simplify the model, some probabilistic structures on $y$'s are assumed. The most common one is to assume that the $y_{ij}$'s are independent with cumulative distribution function. $F_i(y) = F(y - \alpha_{ij})$ or equivalently

$$y_{ij} = \alpha_{ij} + \varepsilon_{ij}, \tag{8.2}$$

where $\alpha_{ij}$ is the expected utility determined by judge $j$ to object $i$ and $\boldsymbol{\varepsilon}_j = (\varepsilon_{1j}, \cdots, \varepsilon_{tj})'$, $j = 1, \cdots, n$ are i.i.d. random vectors with cumulative distribution function $F$. Such models are referred to as *Thurstone order statistics models* (see Yellot 1977; Critchlow et al. 1991). Two famous Thurstone models studied extensively in the literature are

- Thurstone model (Thurstone 1927; Daniels 1950; Mosteller 1951):

  $F$ is the standard normal.

- Luce model (Bradley and Terry 1952; Luce 1959):

  $F$ is Gumbel (type I extreme value)[1], i.e., $F(\varepsilon) = \exp(-\exp(-\varepsilon))$.

Since the Luce model leads to a closed form,

$$P(\boldsymbol{\pi}_j) = \prod_{i=1}^{t-1} \frac{\exp(\alpha_{[i]_j j})}{\sum_{l=i}^{t} \exp(\alpha_{[l]_j j})}, \tag{8.3}$$

most applications and extensions are based on the Luce model. As the exponential distribution satisfies the memoryless property, it may not be appropriate in modeling the running times in many track competitions. Henery (1983) and Stern (1990a,b) thus extended the Luce model to the Thurstone model with error $\varepsilon_{ij} = \ln(u_{ij})$, where $u_{ij}$ follows a Gamma distribution with shape $r$ and scale 1. Properties of the Thurstone order statistics model can be found in Henery (1981) and Critchlow et al. (1991).

### 8.1.1 Luce Model

The Luce model can be viewed as an extension of the *multinomial (conditional) logit model* for top choice (McFadden 1974). For example, in examining 3 objects by judge $j$, object 2 is selected as the top-choice, i.e., the ordering is $< 2, \_, \_ >$, with the following top choice probability:

$$P(< 2, \_, \_ >) = P(y_{2j} > y_{1j}, y_{3j}) = \frac{\exp(\alpha_{2j})}{\exp(\alpha_{1j}) + \exp(\alpha_{2j}) + \exp(\alpha_{3j})}.$$

[1]Note $e^{-\varepsilon}$ follows an exponential distribution with mean 1.

Also, the probability of observing the ranking of 3 objects (3, 1, 2) (i.e., ordering: $< 2, 3, 1 >$) under the Luce model is

$$P(y_{2j} > y_{3j} > y_{1j}) = \frac{e^{\alpha_{2j}}}{e^{\alpha_{1j}} + e^{\alpha_{2j}} + e^{\alpha_{3j}}} \cdot \frac{e^{\alpha_{3j}}}{e^{\alpha_{1j}} + e^{\alpha_{3j}}}.$$

It is not difficult to see that the ranking probability under the Luce model can be expressed as a function of top-choice probabilities only.

**Theorem 8.1.** *Let $p_{ij}$ be the probability that object $i$ is ranked first by judge $j$ among the full list of $t$ objects. That is, $p_{ij} = P(y_{ij} > y_{kj}\ \forall k \neq i)$. Then the probability of ranking $\boldsymbol{\pi}_j$ with ordering $< [1]_j, [2]_j, \cdots, [t]_j >$ under the Luce model is given by*

$$P(\boldsymbol{\pi}_j) = p_{[1]_j j} \frac{p_{[2]_j j}}{1 - p_{[1]_j j}} \frac{p_{[3]_j j}}{1 - p_{[1]_j j} - p_{[2]_j j}} \cdots \frac{p_{[t-1]_j j}}{1 - p_{[1]_j j} - p_{[2]_j j} - \cdots - p_{[t-2]_j j}}.$$

*Proof.* The proof follows by observing that under the Luce model,

$$p_{ij} = \frac{\exp(\alpha_{ij})}{\exp(\alpha_{1j}) + \cdots + \exp(\alpha_{tj})}.$$ □

**Definition 8.1 (Independence of Irrelevant Alternatives (IIA) Tversky 1972).** Let $P(a|S)$ be the probability of choosing an object $a$ from a choice set $S \subseteq \{1, 2, \cdots, t\}$. The independence of irrelevant alternatives asserts that object $a$ is preferred to object $b$, by the (top) choice probability, is independent of the choice set $S$.

From Definition 8.1, we have,

$$P(a|S) > P(b|S) \Longleftrightarrow P(a|\{a,b\}) > P(b|\{a,b\}) \Longleftrightarrow P(a|\{a,b\}) > \frac{1}{2}.$$

If object $a$ is preferred to object $b$ out of the choice set $\{a, b\}$, then introducing a third alternative object $c$, thus expanding the choice set to $\{a, b, c\}$, must not make object $b$ preferable to object $a$. In other words, the choices between $a$ and $b$ depend on the preferences between $a$ and $b$ only, i.e., it is irrelevant to $c$.

**Theorem 8.2 (Luce 1959).** *The Luce model satisfies the IIA.*

*Proof.* Under the Luce model, it is easy to show that

$$P(a|S) = \frac{\exp(\alpha_a)}{\sum_{i \in S} \exp(\alpha_i)}$$

and thus $\frac{P(a|S)}{P(b|S)} > 1 \Longleftrightarrow \frac{\exp(\alpha_a)}{\exp(\alpha_b)} > 1 \Longleftrightarrow \frac{\exp(\alpha_a)/(\exp(\alpha_a)+\exp(\alpha_b))}{\exp(\alpha_b)/(\exp(\alpha_a)+\exp(\alpha_b))} > 1 \Longleftrightarrow \frac{P(a|\{a,b\})}{P(b|\{a,b\})} > 1.$ □

*Example 8.1.* Is IIA a good property? No. Let us consider the problem of selecting a travel mode to work among a car (C), a blue bus (B), or a red bus (R). Initially a traveler has a choice of going to work by car or taking a blue bus with $P(C) = P(B) = \frac{1}{2}$ . Now a red bus is introduced and the traveler considers the red bus to be exactly like the blue bus (i.e., $P(R) = P(B)$). However, in the Luce model, the odds $P(C)/P(B)$ is the same whether or not another alternative exists. The only probabilities for which $P(C)/P(B)=1$ and $P(R)/P(B)=1$ are $P(C)=P(B)=P(R)=\frac{1}{3}$. In real life, we would expect $P(C)=\frac{1}{2}$ and $P(B)=P(R)=\frac{1}{4}$.

It is natural that our choice on an object (such as the blue bus) will depend on our preference on similar objects or even its substitutes (like the red bus). By ignoring such dependency, the estimation of choice/ranking probabilities of course will be biased. In other words, if the list of all travel modes contains many irrelevant objects such as walking, bicycling, and skateboarding, it might be acceptable to estimate the probability for choosing car/bus based on the subset {car, bus} instead of the full list {car, bus, walking, bicycling, skateboarding}. However the estimation in this case will be relatively less efficient.

### 8.1.2 Rank-Ordered Logit Models

The Luce model can be extended to incorporate covariates as well. For example, we may include $M$ covariates of judge $j$, $x_{mj}$, $m = 1, 2, \ldots, M$, into the mean utility, i.e.,

$$\alpha_{ij} = \beta_{i0} + \sum_{m=1}^{M} \beta_{im} x_{mj}, \tag{8.4}$$

where $\beta_{im}$, $m = 0, 1, \ldots, M$ are the parameters specific to object $i$, and $P$ covariates of object $i$, $z_{pi}$, $p = 1, 2, \cdots, P$, into the mean utility, i.e.,

$$\alpha_{ij} = \beta_{i0} + \sum_{p=1}^{P} \gamma_p z_{pi}, \tag{8.5}$$

where $\gamma_p$, $p = 1, 2, \cdots, P$ are the parameters specific to all judges.

A further extension of the Luce model (specified in Allison and Christakis (1994)) includes judge-specific covariates, object-specific covariates, and their interactions or judge-object-specific covariates ($w_{qij}, q = 1, 2, \cdots, Q$) into the mean utility:

$$\alpha_{ij} = \beta_{i0} + \sum_{p=1}^{P} \gamma_p z_{pi} + \sum_{m=1}^{M} \beta_{im} x_{mj} + \sum_{q=1}^{Q} \theta_q w_{qij}, \tag{8.6}$$

where $\theta_q, q = 1, 2, \cdots, Q$ are the parameters specific to all judges and objects. These extensions of the Luce models are known as rank-ordered logit (ROL) model in the field of econometrics (see for example Chapman and Staelin 1982; Beggs et al. 1981; Hausman and Ruud 1987).

In the Luce and ROL models described above, the log-likelihood function is globally concave, and hence a global maximum exists (Beggs et al. 1981). The maximum likelihood estimates (MLE) of the model parameters can thus be obtained using standard methods, e.g., Newton-Raphson algorithm. Besides MLE, Koop and Poirier (1994) used a Bayesian method to estimate the parameters.

Both the Luce and ROL models can be built using the R package `mlogit`. Here, we use an example to demonstrate these two models.

*Example 8.2.* Consider a ranking data set for gaming platforms in which 91 Dutch students were asked to rank 6 gaming platforms: Xbox, PlayStation, GameCube, PlayStation Portable, Gameboy, or a personal computer (PC). The data set also contains information on whether the student currently owns each platform (own), the age of the student (age), and the number of hours spent on gaming per week (time). This data set was first studied in Fok et al. (2012) and can be accessed in the R package, `mlogit`.

First, we fit a Luce model (9.3) with PC as the reference level and the parameter estimates are shown in Table 8.1. It is noticed that students prefer Xbox for playing games the most and then PC, PlayStation, PlayStation Portable, GameCube, and Gameboy. However, playing games on Xbox and PlayStation are not significantly different from that on PC.

Now, we extend the Luce model by including object-specific covariate (own) and judge-specific covariate (time) into the model. This leads to the rank-ordered logit (ROL) model and the parameter estimates are shown in Table 8.1. It can be observed that owning a platform has a positive effect on the preference for the same platform and that students who spend more time playing games prefer a PC more than other gaming platforms. Applying the likelihood ratio test to compare the two models, it is clearly that the ROL model is substantially better. Notice that including age as another judge-specific covariate does not significantly improve the likelihood of the ROL (from −517.37 to −516.55), and hence the results are omitted here.

ROL models are popular for ranking data, and many extensions have been developed by different scholars. Koop and Poirier (1994) extended the use of ROL models to more general cases of ranking data. The number of objects ranked by $n$ judges can be different. The rank given by each judge is not necessarily complete. The objects that each judge is assigned to rank can be different as well. Fok et al. (2012) studied the mixtures of ROL models and found them to be useful in analyzing ranking capabilities.

**Table 8.1** Parameter estimates of the fitted Luce and ROL models for the gaming platform data

| Variable | Luce | ROL |
|---|---|---|
| *Intercept* | | |
| Xbox | 0.13 (0.18) | 1.40 (0.29) |
| PlayStation | −0.00 (0.18) | 0.94 (0.27) |
| PlayStation Portable | −0.65 (0.18) | 0.80 (0.28) |
| GameCube | −1.22 (0.20) | 0.05 (0.30) |
| Gameboy | −1.28 (0.19) | 0.09 (0.28) |
| *Platform ownership* | | 0.96 (0.19) |
| *Hours spent on gaming* | | |
| Xbox | | −0.17 (0.05) |
| PlayStation | | −0.13 (0.04) |
| PlayStation Portable | | −0.23 (0.05) |
| GameCube | | −0.19 (0.05) |
| Gameboy | | −0.24 (0.05) |
| Log-likelihood | −547.00 | −517.37 |

### *8.1.3 Some Non-IIA Order Statistics Models*

In spite of the fact that the ranking probability (8.1) under both the Luce and ROL models has a closed form, the unrealistic IIA property makes them fit some data not so well (see for example Brook and Upton 1974; Tallis and Dansie 1983; Bockenholt 1993). The main reason is that no correlation is assumed among the errors over the objects. This lack of correlation translates into an unrealistic substitution pattern among objects in some situations (see Example 8.1). Therefore, to overcome these problems, dependency structures other than those in the Thurstone order statistics model are required.

#### 8.1.3.1 Multivariate (Generalized) Extreme Value (GEV) Models

McFadden (1978) introduced the multivariate (or generalized) extreme value model which provides closed-form top-choice probabilities without the IIA restriction. The GEV assumes that the error terms in (8.2) follows a generalized extreme value distribution with cumulative distribution function

$$F(\varepsilon_1, \cdots, \varepsilon_t) = \exp[-H(e^{-\varepsilon_1}, \cdots, e^{-\varepsilon_t})],$$

where $G = \exp(-H)$ is a $t$-dimensional copula and all the univariate marginals are Gumbel distributed. Of course, when $H(x_1, \cdots, x_I) = \sum_{i=1}^{I} x_i$, the model degenerates to the Luce model. The GEV model is very flexible and Joe (2001) showed that the GEV model can fit various types of ranking data. Note that this result does not provide a way to construct the function $H$. In fact, the popular GEV model used in the literature is the nested logit model in which the function $H$ is expressed in a hierarchical form:

$$H(x_1, \cdots, x_I) = \sum_{k=1}^{K} (\sum_{j \in B_k} x_j^{1/\lambda_k})^{\lambda_k},$$

where $B_1, \cdots, B_K$ are $K$ nonoverlapping subsets (called nests) formed from a partition of all objects.

Under the nested logit model, the $\varepsilon_i$'s are correlated within nests but uncorrelated between nests. For example, suppose $K = 2$, $B_1 = \{\text{car}\}$, and $B_2 = \{\text{red bus, blue bus}\}$, it is reasonable that one who prefers traveling with the red bus may also prefer traveling with the blue bus and vice versa, but one's preference on car may not depend on his/her preference on the two buses. Such dependency structure can be represented by the following hierarchical form:

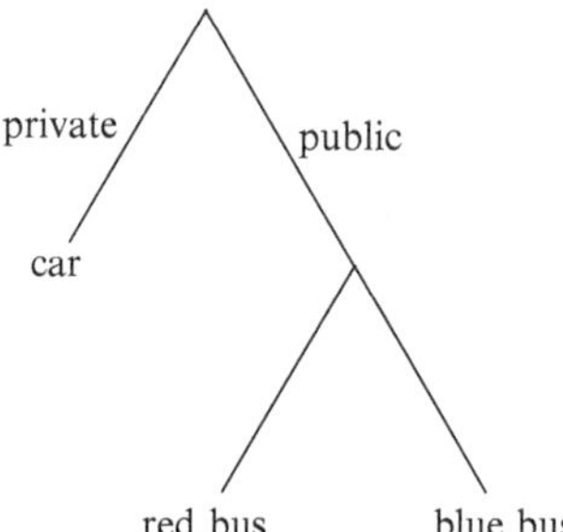


#### 8.1.3.2 Mixed Logit Models

Note that the rank-ordered logit model assumes that the utility for each object follows the linear model:

$$y_{ij} = \boldsymbol{x}_{ij}' \boldsymbol{\beta} + \varepsilon_{ij}$$

where the error terms $\varepsilon_{ij}$'s are independent and identically (type I) extreme value distributed. To allow dependency among the utilities, mixed logit models assume that the beta coefficients are judge-specific:

$$y_{ij} = \boldsymbol{x}_{ij}' \boldsymbol{\beta}_j + \varepsilon_{ij}$$

and further assume that $\boldsymbol{\beta}_j$'s are random and independent identically distributed with density $f(\boldsymbol{\beta}|\theta)$, where $\theta$ are some unknown parameters. A typical choice of $f$ is the normal density with mean $\boldsymbol{\beta}_0$ and covariance matrix $\boldsymbol{\Omega}$. Such randomness in

$\boldsymbol{\beta}_j$ allows unexplainable variation of covariates' impacts over judges and correlation of utilities across objects. McFadden and Train (2000) showed that any discrete choice model can be well approximated by a mixed logit model with appropriate specification of the distribution of $\boldsymbol{\beta}_j$ and the covariates $\boldsymbol{x}$.

Conditional on $\boldsymbol{\beta}_j$, the probability of observing $\boldsymbol{\pi}_j$ by judge $j$ is given in (8.3) with $\alpha_{ij} = \boldsymbol{x}'_{ij}\boldsymbol{\beta}_j$. Integrating it over the density of $\boldsymbol{\beta}_j$ then gives the unconditional probability under the mixed logit model:

$$P(\boldsymbol{\pi}_j) = \int \prod_{i=1}^{t-1} \left( \frac{e^{\boldsymbol{x}'_{[i]_j j}\boldsymbol{\beta}}}{\sum_{l=i}^{t} e^{\boldsymbol{x}'_{[l]_j j}\boldsymbol{\beta}}} \right) f(\boldsymbol{\beta})\, d\boldsymbol{\beta}.$$

If the mixing distribution $f(\boldsymbol{\beta})$ is discrete, with $\boldsymbol{\beta}$ taking a finite set of distinct values, the mixed logit model becomes the **latent class model** and sometimes called the **finite mixture model**.

Both nested logit and mixed logit models can be built using the R package `mlogit` which provides maximum likelihood estimation and the numerical integration (if any) in the likelihood is estimated using simulation techniques such as quasi-Monte Carlo method.

*Example 8.3.* The R package `mlogit` contains a top-choice data set named `Electricity` in which 361 individuals were asked in a series of at most 12 choice experiments. In each experiment, each individual was asked to choose the best out of four hypothetical electricity suppliers with different combination of characteristics including electricity price ($pf$) (in cents per kWh) and length of contract ($cl$, in years) offered, whether a time-of-day rate ($tod$) is included, whether a seasonal rate ($seas$) is included, and whether the supplier is local ($loc$) or is well known ($wk$).

We first fit a multinomial logit (MNL) model (i.e., the Luce model with the top choice only) using `mlogit` and its parameter estimates are shown in Table 8.2. The significant negative coefficients for $pf$, $tod$, $seas$, and $cl$ and the significant positive coefficients for $loc$ and $wk$ indicate that individuals tend to prefer a local and well-known supplier which offers a shorter length of contract with a lower fee.

Note that `Electricity` is a clustered data set as each individual was involved in a number of choice experiments. The independence assumption of the choice responses used in the multinomial logit model is therefore invalid. To incorporate such clustered effect, we use the mixed logit model with utility $y_{ijk}$ for supplier $i$ given by individual $j$ in the $k$th experiment as follows:

$$y_{ijk} = \boldsymbol{x}'_{ijk}\boldsymbol{\beta}_j + \varepsilon_{ijk},$$

where $\boldsymbol{\beta}_j$'s are independent identically distributed. As the utilities made by individual $j$ share the same random $\boldsymbol{\beta}_j$, the utilities given by the same individual are correlated whereas the utilities given by different individuals are uncorrelated. This helps describe the clustered effect.

**Table 8.2** Parameter estimates of the fitted MNL and mixed logit models for the electricity supplier data

| Variable | MNL | Mixed logit |
|---|---|---|
| *Fixed effect* | | |
| Electricity price $(pf)$ | −0.625 (0.023) | −0.933 (0.034) |
| Length of contract $(cl)$ | −0.108 (0.008) | −0.196 (0.013) |
| Time-of-day rate $(tod)$? | −5.463 (0.184) | −8.838 (0.286) |
| Seasonal rate $(seas)$? | −5.840 (0.187) | −8.860 (0.287) |
| Local $(loc)$? | 1.442 (0.051) | 2.105 (0.080) |
| Well known $(wk)$? | 0.996 (0.045) | 1.493 (0.065) |
| *Random effect (standard deviation)* | | |
| Electricity price $(pf)$ | | 0.200 (0.011) |
| Length of contract $(cl)$ | | 0.357 (0.018) |
| Time-of-day rate $(tod)$? | | 2.489 (0.120) |
| Seasonal rate $(seas)$? | | 1.274 (0.107) |
| Local $(loc)$? | | 1.503 (0.089) |
| Well known $(wk)$? | | 0.885 (0.075) |
| Log-likelihood | −4958.6 | −3970.3 |

Using the independent normal assumption for the $\boldsymbol{\beta}_j$'s, the mixed logit model is fitted using `mlogit` and the parameter estimates are shown in Table 8.2. It can be seen from the log-likelihood that the mixed logit model significantly performs better than the multinomial logit model and, in fact, all random effects in the mixed logit model are highly significant. Based on the fitted model, it is easy to see that an individual with mean coefficients for $pf$ and $cl$ is willing to pay $0.196/0.933 = 0.21$ cent per kWh extra in order to shorten the contract length by one year.

#### 8.1.3.3 Multilevel Logit Models

Notice that the above mixed logit model is basically a mixed model with both fixed and random effects. If more sampling information and dependency structures are available, more structured mixed models can be considered. For instance, Skrondal and Rabe-Hesketh (2003) applied a three-level logit model to analyze ranking data collected from the 1987–1992 panel of the British Election Study for rankings on three political parties: Conservative, Labour, and Liberal (Alliance) (indexed by $a$), given by a sample of voters (indexed by $j$) casting votes at different elections (indexed by $i$) over different constituencies (indexed by $k$). Note that in this three-level model, elections are nested within voters and voters nested within constituencies. One model considered is the random intercepts model at voter and constituency levels:

$$y_{aijk} = \alpha_{aijk} + \gamma_{ajk} + \gamma_{ak} + \varepsilon_{aijk}$$

where $\alpha_{aijk} = z_{aijk}b + \mathbf{x}'_{aijk}\boldsymbol{\beta}_a$ represents the fixed effects while $\gamma_{ajk}$ and $\gamma_{ak}$ represent, respectively, the random intercepts at both voter and constituency levels, and constituency level only. This special kind of multilevel logit models for ranking data can be built using the Stata program `gllamm` (http://gllamm.org/examples.html) which provides maximum likelihood estimation with integration approximated by quadrature methods.

## 8.2 Paired Comparison Models

Motivated by the connection between a ranking of objects and all pairwise comparisons of objects, paired comparison models aim at combining models for paired comparisons to generate a probabilistic model for ranking data. Note that a ranking of $t$ objects can be indexed by $t(t-1)/2$ pairwise preferences $I_{ab}$, $a < b$, where $I_{ab} = 1$ means object $a$ is preferred to object $b$. Smith (1950) assumed that the ranking is deduced from a set of $t(t-1)/2$ arbitrary paired comparison probabilities $p_{ab}$, $a < b$, where $p_{ab}$ is the probability of object $a$ being preferred to object $b$. The model does not allow ties, so that $p_{ab} = 1 - p_{ba}$. Assuming mutual independence of these $t(t-1)/2$ paired comparisons under the Smith model, the probability of observing a ranking $\boldsymbol{\pi}_j$ is thus given by

$$P(\boldsymbol{\pi}_j) = C \prod_{\{(a,b):\pi_j(a)<\pi_j(b)\}} p_{ab}, \tag{8.7}$$

where the constant $C$ is chosen to make the probabilities sum to 1. Note that the Smith model is indexed by $t(t-1)/2$ parameters $\{p_{ab}\}$. Imposing additional constraints on the $\{p_{ab}\}$ proposed by Mallows (1957) leads to two important subclasses of the Smith model: the Mallows-Bradley-Terry model and the Mallows model.

*The Class of Mallows-Bradley-Terry (MBT) Models.* To reduce the number of parameters in (8.7), Bradley and Terry (1952) proposed to re-parametrize $p_{ab}$ as

$$p_{ab} = \frac{v_a}{v_a + v_b}$$

where $v_i$ is a positive value associated with object $i$ and the sum of all $v_{i's}$ is equal to 1. Mallows (1957) substituted this form into the Smith model, which leads to the following ranking model. For any ranking $\boldsymbol{\pi}_j$ with associated ordering $< [1]_j, [2]_j, \cdots, [t]_j >$,

$$P(\boldsymbol{\pi}_j) = C(\mathbf{v}) \prod_{s=1}^{t-1} (v_{[s]_j})^{t-s}$$

where $C(\boldsymbol{v})$ is the proportionality constant. Since the Bradley-Terry paired comparison probabilities are invariant when multiplying the $v_i$'s by a positive constant, the number of free parameters is reduced to $t - 1$. Larger values of $v_i$ correspond to more preferred objects, just as the Thurstone order statistics model.

*The Class of Mallows Models.* Before discussing details of the model, we first give the definition of *modal ranking*.

**Definition 8.2.** A probability model is said to be strongly unimodal with modal ranking $\boldsymbol{\pi}_0$, if its ranking probability has the unique maximum at $\boldsymbol{\pi} = \boldsymbol{\pi}_0$.

Mallows (1957) further simplified the MBT model by re-expressing $p_{ab}$ as

$$p_{ab} = \frac{1}{2} + \frac{1}{2}\tanh[(\pi(a) - \pi(b))\ln(\theta) + \ln(\phi)],$$

where $\theta, \phi \in (0, 1)$. Thus, the Mallows model is given by

$$P(\boldsymbol{\pi}_j) = c(\theta, \phi)\theta^{d_S(\pi,\pi_0)}\phi^{d_K(\pi,\pi_0)},$$

where $c(\theta, \phi)$ is chosen to make the probabilities sum to 1 and $d_S(\boldsymbol{\pi}, \boldsymbol{\pi}_0)$ and $d_K(\boldsymbol{\pi}, \boldsymbol{\pi}_0)$ are the Spearman and Kendall distances between $\boldsymbol{\pi}$ and $\boldsymbol{\pi}_0$ (c.f. Sect. 3.1). The Mallows model has the interpretation that the ranking probability decreases geometrically according to increasing distance from $\boldsymbol{\pi}$ to the modal ranking $\boldsymbol{\pi}_0$.

For a detailed review on paired comparison models, readers can refer to David (1988). Pendergrass and Bradley (1960) further extended the paired comparison models to triple comparison models.

## 8.3 Distance-Based Models

A distance function is useful in measuring the discrepancy between two rankings. The usual properties of a distance function between two rankings $\boldsymbol{\mu}$ and $\boldsymbol{v}$ are: (1) reflexivity, $d(\boldsymbol{\mu}, \boldsymbol{\mu}) = 0$; (2) positivity, $d(\boldsymbol{\mu}, \boldsymbol{v}) > 0$ if $\boldsymbol{\mu} \neq \boldsymbol{v}$; and (3) symmetry, $d(\boldsymbol{\mu}, \boldsymbol{v}) = d(\boldsymbol{v}, \boldsymbol{\mu})$. For ranking data, we require that the distance, apart from having these usual properties, must be right invariant,

$$d(\boldsymbol{\mu}, \boldsymbol{v}) = d(\boldsymbol{\mu} \circ \boldsymbol{\tau}, \boldsymbol{v} \circ \boldsymbol{\tau}), \text{ where } \boldsymbol{\mu} \circ \boldsymbol{\tau}(i) = \boldsymbol{\mu}(\boldsymbol{\tau}(i)).$$

This requirement ensures that a relabeling of the objects has no effect on the distance. If a distance function satisfies the triangle inequality $d(\boldsymbol{\mu}, \boldsymbol{v}) \leq d(\boldsymbol{\mu}, \boldsymbol{\sigma}) + d(\boldsymbol{\sigma}, \boldsymbol{v})$, the distance is said to be a *metric*.

Some popular right-invariant distances have been given in Chap. 3. Note that the Spearman Footrule and Kendall distance are metrics, but the Spearman distance is not, just as the squared Euclidean distance is not. To produce a metric version of the Spearman distance, we may take the square root of the Spearman distance, as given by

$$\left(\sum_{i=1}^{t}[\mu(i)-\nu(i)]^2\right)^{0.5}. \tag{8.8}$$

Readers can refer to Critchlow et al. (1991) for further examples of distance functions.

It is reasonable to assume that there is a modal ranking $\boldsymbol{\pi}_0$, and we expect most of the judges to have rankings close to $\boldsymbol{\pi}_0$. According to this framework, Diaconis (1988) developed a class of distance-based models,

$$P(\boldsymbol{\pi}|\lambda,\boldsymbol{\pi}_0)=\frac{e^{-\lambda d(\boldsymbol{\pi},\boldsymbol{\pi}_0)}}{C(\lambda)}, \tag{8.9}$$

where $\lambda \geq 0$ is the dispersion parameter and $d(\boldsymbol{\pi},\boldsymbol{\sigma})$ is an arbitrary right-invariant distance. In the particular case where we use Kendall as the distance function, the model is called the Mallows' $\phi$-model (Mallows 1957). Note that Mallows' $\phi$-models also belong to the class of paired comparison models (Critchlow et al. 1991). Critchlow and Verducci (1992) and Feigin (1993) provided more details about the relationship between distance-based models and paired comparison models.

In distance-based models, the ranking probability is the greatest at the modal ranking $\boldsymbol{\pi}_0$ and the probability of a ranking will decay the further it is away from the modal ranking $\boldsymbol{\pi}_0$. The rate of the decay is governed by the parameter $\lambda$. For a small value of $\lambda$, the distribution of rankings will be more concentrated around $\boldsymbol{\pi}_0$. When $\lambda$ becomes very large, the distribution of rankings will look more uniform. The closed form for the proportionality constant $C(\lambda)$ only exists for some distances. In principle, it can be solved numerically by summing the value $e^{-\lambda d(\boldsymbol{\pi},\boldsymbol{\pi}_0)}$ over all possible $\boldsymbol{\pi}$ in $\mathcal{P}$. This numerical calculation could be time-consuming, as the computational time increases exponentially with the number of objects.

Given a ranking data set $\{\boldsymbol{\pi}_k, k=1,\ldots,n\}$ and a known modal ranking $\boldsymbol{\pi}_0$, the maximum likelihood estimator (MLE) $\hat{\lambda}$ of the distance-based model can be found by solving the following equation:

$$\frac{1}{n}\sum_{k=1}^{n}d(\boldsymbol{\pi}_k,\boldsymbol{\pi}_0)=E_{\hat{\lambda},\sigma}[d(\boldsymbol{\pi},\boldsymbol{\pi}_0)], \tag{8.10}$$

which equates the observed mean distance with the expected distance under the distance-based model.

The MLE can be found numerically because the observed mean distance is a constant and the expected distance is a strictly decreasing function of $\hat{\lambda}$. For the ease of solving, we re-parametrize $\lambda$ with $\phi$ where $\phi = e^{-\lambda}$. The range of $\phi$ lies in

(0, 1] and the value of $\hat{\phi}$ can be obtained using the method of bisection. Critchlow (1985) suggested applying the method with 15 iterations, which yields an error of less than $2^{-15}$. Also, the central limit theorem holds for the MLE $\hat{\lambda}$, which is shown in Marden (1995).

If the modal ranking $\boldsymbol{\pi}_0$ is unknown, it can be estimated by the MLE $\hat{\boldsymbol{\pi}}_0$ which minimizes the sum of distance over $\mathcal{P}$, that is,

$$\hat{\boldsymbol{\pi}}_0 = \operatorname*{argmin}_{\boldsymbol{\pi}_0 \in \mathcal{P}} \sum_{k=1}^{n} d(\boldsymbol{\pi}_k, \boldsymbol{\pi}_0). \tag{8.11}$$

For a large $t$, a global search algorithm for MLE $\hat{\boldsymbol{\pi}}_0$ is not practical because the number of possible rankings is too large. Instead, as suggested in Busse et al. (2007), a local search algorithm should be used. They suggested iteratively searching for the optimal model ranking with the smallest sum of distances $\sum_{k=1}^{n} d(\boldsymbol{\pi}_k, \boldsymbol{\pi}_0)$ over $\boldsymbol{\pi}_0 \in \Pi^{(m)}$, where $\Pi^{(m)}$ is the set of all rankings having a Cayley distance (Sect. 8.3.2) of 0 or 1 to the optimal modal ranking found in the $m$th iteration:

$$\hat{\boldsymbol{\pi}}_0^{(m+1)} = \operatorname*{argmin}_{\boldsymbol{\pi}_0 \in \Pi^{(m)}} \sum_{k=1}^{n} d(\boldsymbol{\pi}_k, \boldsymbol{\pi}_0).$$

Cayley's distance $d_C(\boldsymbol{\pi}, \boldsymbol{\sigma})$ is defined to be the minimal number of transpositions needed to transform $\boldsymbol{\pi}$ to $\boldsymbol{\sigma}$. A reasonable choice of the initial ranking $\hat{\boldsymbol{\pi}}_0^{(0)}$ can be formed by ordering the mean ranks.

Distance-based models can handle partial ranking, with some modifications in the distance measures. There are several ways to handle partially ranked data in distance-based models. Beckett (1993) estimated the model parameters using the EM algorithm. On the other hand, Adkins and Fligner (1998) offered a non-iterative maximum likelihood estimation procedure for Mallows' $\phi$-model without using the EM algorithm. Critchlow (1985) suggested replacing the distance metric $d$ by the Hausdorff metric $d^*$. The Hausdorff metric between two partial rankings $\boldsymbol{\pi}^*$ and $\boldsymbol{\sigma}^*$ equals

$$d^*(\boldsymbol{\pi}^*, \boldsymbol{\sigma}^*) = \max[\max_{\pi \in C(\pi^*)} \min_{\sigma \in C(\sigma^*)} d(\boldsymbol{\pi}, \boldsymbol{\sigma}), \max_{\sigma \in C(\sigma^*)} \min_{\pi \in C(\pi^*)} d(\boldsymbol{\pi}, \boldsymbol{\sigma})], \tag{8.12}$$

where $C(\mu^*)$ is the set of complete rankings compatible with $\mu^*$ (see Definition 3.1).

### 8.3.1 *ϕ-Component Models*

Fligner and Verducci (1986) extended the distance-based models by decomposing the distance metric $d(\boldsymbol{\pi}, \boldsymbol{\sigma})$ into $t-1$ distance metrics,

$$d(\boldsymbol{\pi}, \boldsymbol{\sigma}) = \sum_{i=1}^{t-1} d_i(\boldsymbol{\pi}, \boldsymbol{\sigma}), \tag{8.13}$$

where $d_i(\boldsymbol{\pi}, \boldsymbol{\sigma})$'s are statistically independent. Kendall's distance can be decomposed in this form. Fligner and Verducci (1986) developed two new classes of ranking models, called $\phi$-component models and cyclic structure models, for the decomposition.

Fligner and Verducci (1986) showed that Kendall distance satisfies (8.13):

$$d_K(\boldsymbol{\pi}, \boldsymbol{\pi}_0) = \sum_{i=1}^{t-1} V_i, \tag{8.14}$$

where

$$V_i = \sum_{j=i+1}^{t} I\{[\pi(\pi_0^{-1}(i)) - \pi(\pi_0^{-1}(j))] > 0\}. \tag{8.15}$$

Here, $V_1$ represents the number of adjacent transpositions required to place the best object in $\boldsymbol{\pi}_0$ in the first position and then remove this item in both $\boldsymbol{\pi}$ and $\boldsymbol{\pi}_0$, and $V_2$ is the number of adjacent transpositions required to place the best remaining object in $\boldsymbol{\pi}_0$ in the first position of the remaining items, and so on. Therefore, the ranking can be described as $t-1$ stages, $V_1$ to $V_{t-1}$, where $V_i = m$ can be interpreted as $m$ mistakes made in stage $i$.

By applying dispersion parameter $\lambda_i$ at stage $V_i$, the Mallow's $\phi$-model is extended to

$$P(\boldsymbol{\pi}|\boldsymbol{\lambda}, \boldsymbol{\pi}_0) = \frac{e^{-\sum_{i=1}^{t-1} \lambda_i V_i}}{C(\boldsymbol{\lambda})}, \tag{8.16}$$

where $\boldsymbol{\lambda} = \{\lambda_i, i = 1, \ldots, t-1\}$ and $C(\boldsymbol{\lambda})$ is the proportionality constant, which equals

$$\prod_{i=1}^{t-1} \frac{1 - e^{-(t-i+1)\lambda_i}}{1 - e^{-\lambda_i}}. \tag{8.17}$$

These models were named $t-1$ parameter models in Fligner and Verducci (1986), but were also named $\phi$-component models in other papers (e.g., Critchlow et al. 1991). Mallow's $\phi$-models are special cases of $\phi$-component models when $\lambda_1 = \ldots = \lambda_{t-1}$.

Based on a ranking data set $\{\boldsymbol{\pi}_k, k = 1, \ldots, n\}$ and a given modal ranking $\boldsymbol{\pi}_0$, the maximum likelihood estimates $\hat{\lambda}_i$, $i = 1, 2, \ldots, t-1$ can be found by solving the equation

$$\frac{1}{n}\sum_{k=1}^{n} V_{k,i} = \frac{e^{-\hat{\lambda}_i}}{1 - e^{-\hat{\lambda}_i}} - \frac{(t-i+1)e^{-(t-i+1)\hat{\lambda}_i}}{1 - e^{-(t-i+1)\hat{\lambda}_i}}, \tag{8.18}$$

where

$$V_{k,i} = \sum_{j=i+1}^{t} I\{[\pi_k(\pi_0^{-1}(i)) - \pi_k(\pi_0^{-1}(j))] > 0\}. \tag{8.19}$$

The left- and right hand sides of (8.18) can be interpreted as the observed mean and theoretical mean of $V_i$, respectively.

The extension of distance-based models to $t - 1$ parameters allows more flexibility in the model, but unfortunately, the symmetric property of distance is lost. Notice here that the so-called "distance" in $\phi$-component models can be expressed as

$$\sum_{i<j} \lambda_i I\{[\pi(\pi_0^{-1}(i)) - \pi(\pi_0^{-1}(j))] > 0\}, \tag{8.20}$$

which is obviously not symmetric, and hence it is not a proper distance measure. For example, in $\phi$-component model, let $\boldsymbol{\pi} = (2, 3, 4, 1), \boldsymbol{\pi}_0 = (4, 3, 1, 2)$:

$$\begin{aligned} d(\boldsymbol{\pi}, \boldsymbol{\pi}_0) &= \lambda_1 V_1 + \lambda_2 V_2 + \lambda_3 V_3 = 3\lambda_1 + 0\lambda_2 + 1\lambda_3 \\ &\neq 1\lambda_1 + 2\lambda_2 + 1\lambda_3 = d(\boldsymbol{\pi}_0, \boldsymbol{\pi}). \end{aligned}$$

The symmetric property of distance is thus not satisfied. Lee and Yu (2012) introduced new weighted distance measures which can retain the properties of a distance and also allow different weights for different ranks. For the details, read Chap. 11.

### 8.3.2 Cyclic Structure Models

Cayley's distance can also be decomposed into $t - 1$ statistical independent metrics. Fligner and Verducci (1986) showed that $d_C(\boldsymbol{\pi}, \boldsymbol{\pi}_0)$ can be decomposed as

$$d_C(\boldsymbol{\pi}, \boldsymbol{\pi}_0) = \sum_{i=1}^{t-1} X_i(\boldsymbol{\pi}, \boldsymbol{\pi}_0), \tag{8.21}$$

where $X_i(\boldsymbol{\pi}, \boldsymbol{\pi}_0) = I(i \neq \max\{\sigma(i), \sigma(\sigma(i)), \ldots\})$ and $\sigma(i) = \pi(\pi_0^{-1}(i))$.

This generalization can be illustrated by an example found in Fligner and Verducci (1986). Suppose there are $t$ lockers and each locker has one key that can open it. The key for locker $j$ is placed in locker $\sigma(j)$. Without loss of generality, let the cost of breaking a locker be one. The minimum cost of opening all lockers will then be $C(\boldsymbol{\pi}, \boldsymbol{\pi}_0)$, and it can be decomposed as the sum of costs of opening locker $\pi^{-1}(i), i = 1, 2, \ldots t - 1$, which equals $X_i(\boldsymbol{\pi}, \boldsymbol{\pi}_0)$.

If we relax the assumption that the costs of breaking every locker are equal, the total cost will become

$$\sum_{i=1}^{t-1} \theta_i X_i(\boldsymbol{\pi}, \boldsymbol{\pi}_0), \tag{8.22}$$

where $\theta_i$ is the cost of opening locker $i$. This "total cost" can be interpreted as a weighted version of Cayley's distance. Similar to the extension of Mallow's $\phi$-models to $\phi$-component models, Fligner and Verducci (1986) developed the cyclic structure models using the weighted Cayley distance. Under this model assumption, the probability of observing a ranking $\boldsymbol{\pi}$ is

$$P(\boldsymbol{\pi}|\boldsymbol{\theta}, \boldsymbol{\pi}_0) = \frac{e^{-\sum_{i=1}^{t-1} \theta_i X_i(\boldsymbol{\pi}, \boldsymbol{\pi}_0)}}{C(\boldsymbol{\theta})}, \tag{8.23}$$

where $\boldsymbol{\theta} = \{\theta_i, i = 1, \ldots, t-1\}$ and $C(\boldsymbol{\theta})$ is the proportionality constant, which equals

$$\prod_{i=1}^{t-1} \{1 + (t-i)e^{-\theta_i}\}. \tag{8.24}$$

For a ranking data set $\{\boldsymbol{\pi}_k, k = 1, \ldots, n\}$ with a given modal ranking $\boldsymbol{\pi}_0$, the MLEs $\hat{\theta}_i, i = 1, 2, \ldots, t-1$ can be found from the equation

$$\hat{\theta}_i = \log(t-i) - \log \frac{\bar{X}_i}{1 - \bar{X}_i}, \tag{8.25}$$

where

$$\bar{X}_i = \frac{\sum_{k=1}^{n} X_i(\boldsymbol{\pi}_k, \boldsymbol{\pi}_0)}{n}. \tag{8.26}$$

## 8.4 Multistage Models

The class of multistage models includes ranking data models that postulate the ranking process can be decomposed into a sequence of independent stages. For a ranking of $t$ objects, the ranking process can be decomposed into $t-1$ stages, where at stage $i$, the $i$th object is selected. In this respect, the Luce models and $\phi$-component models described above clearly belong to the class of multistage models.

Fligner and Verducci (1988) proposed the general multistage models with $\frac{t(t-1)}{2}$ parameters. They are

$$p(m,r) = \text{Prob}(V_r = m), \tag{8.27}$$

where

$$\sum_{m=0}^{t-r} p(m,r) = 1 \tag{8.28}$$

and $V$'s are defined as in the previous section.

A total of three multistage models are proposed in Fligner and Verducci (1988), namely the free model, the strongly unimodal model, and the exponential factor model. Under the free model, which is the most general (least constraints) multistage models, the probability of observing a ranking $\boldsymbol{\pi}$ is

$$\prod_{r=1}^{t-1} p(m,r). \tag{8.29}$$

Under the strongly unimodal model, the parameters will have additional constraints, which are

$$p(0,r) > p(1,r) \tag{8.30}$$

and

$$p(m,r) \text{ is a nonincreasing function of } m, \tag{8.31}$$

for both $m$ and $r = 1, 2, \ldots t$.

Under the exponential factor model, the parameters will be in the form of

$$p(m,r) = C(r)e^{-\lambda_r f(m)}, \tag{8.32}$$

where $f(\cdot)$ is a nonnegative and strictly increasing arbitrary function, and $C(r)$ is the proportionality constant. To avoid the identification problem, the convention that $f(0) = 0$ and $f(1) = 1$ is suggested. Note that if $f(x) = x$, the model will become the $\phi$-component model.

Besides the multistage model proposed by Fligner and Verducci (1988), Xu (2000) also proposed a multistage model with $(t-1)^2$ parameters $c_{ij}$, both $i$ and $j = 1, 2, \ldots t-1$. The parameters $c_{rj}$, $j = 1, 2, \ldots, t-1$ determine which object will be selected in stage $r$.

## 8.5 Properties of Ranking Models

As defined in Critchlow et al. (1991), some properties for ranking models are as follows:

(1) Label invariance
The relabeling of objects has no effect on the probability models.

(2) Reversibility
A reverse function $\gamma(\boldsymbol{\pi})$ for a ranking of $t$ objects is defined as

$$\gamma(i) = t + 1 - i. \tag{8.33}$$

Reversing the ranking $\boldsymbol{\pi}$ has no effect on the probability models.

(3) $L$-decomposability
The ranking of $t$ objects can be decomposed into $t - 1$ stages. At stage $i$, where $i = 1, 2, \ldots, t-1$, the best among the objects remaining at that stage is selected, and then this object will be removed in the following stages.

(4) Strong unimodality (weak transposition property)
A transposition function $\tau_{ij}$ is defined to mean that $i$ and $j$ are interchanged as

$$\tau(i) = j, \tau(j) = i, \tau(m) = m \text{ for all } m \neq i, j. \tag{8.34}$$

With modal ranking $\boldsymbol{\pi}_0$, for every pair of objects $i$ and $j$ such that $\pi_0(i) < \pi_0(j)$ and every $\boldsymbol{\pi}$ such that $\pi(i) = \pi(j) - 1$,

$$P(\boldsymbol{\pi}) \geq P(\boldsymbol{\pi} \circ \tau_{ij}), \tag{8.35}$$

with equality attained at $\boldsymbol{\pi} = \boldsymbol{\pi}_0$. It guarantees the probability is nonincreasing as $\boldsymbol{\pi}$ moves one step away from $\boldsymbol{\pi}_0$, for objects having adjacent ranks.

(5) Complete consensus (transposition property)
As compared with the strong unimodality, complete consensus is an even stronger property which guarantees for every pair of objects $(i, j)$ such that $\pi_0(i) < \pi_0(j)$ and every $\boldsymbol{\pi}$ such that $\pi(i) < \pi(j)$, $P(\boldsymbol{\pi}) \geq P(\boldsymbol{\pi} \circ \tau_{ij})$. From this definition, we can see that complete consensus implies strong unimodality.

All four classes of models satisfy property (1). However, not all of them satisfy properties (2) to (5). We will discuss them in the following.

### *8.5.1 Properties of Order Statistics Models*

Critchlow et al. (1991) showed that, for order statistics models, if the random error distribution is symmetric, then the models will satisfy property (2).

Property (3) is difficult to verify for the order statistics model, because it involves a multiple integral which may not have a closed form, except for the special case of the Luce (1959) model, which can satisfy property (3).

Savage (1956, 1957), and Henery (1981) showed that, if for all $i$, $\mu_{ij}$ is distinct for $j = 1, 2, \ldots, t$ and

$$\frac{F'(y - \mu_{iu})}{F'(y - \mu_{iv})} \tag{8.36}$$

is a nonincreasing function of $x$ for $\mu_{iu} < \mu_{iv}$, where $F(\cdot)$ is the cumulative distribution function of the random error, the order statistics models will satisfy properties (4) and (5).

### 8.5.2 Properties of Paired Comparison Models

Marley (1968) showed that the class of paired comparison models satisfy properties (2) and (3), which can be easily verified from the definition of paired comparison models.

Critchlow et al. (1991) showed that paired comparison models will satisfy property (4) under the following conditions:

- $p_{ij} > 0.5$ and $p_{jm} > 0.5$ imply $p_{im} > 0.5$,
- $p_{ij} \neq 0.5$,

for all $i, j, m = 1, 2, \ldots, t$.

Property (5) will be satisfied under the following conditions:

- $p_{ij} > 0.5$ and $p_{jm} > 0.5$ imply $p_{im} > \max(p_{ij}, p_{jm})$,
- $p_{ij} \neq 0.5$,

for all $i, j, m = 1, 2, \ldots, t$.

### 8.5.3 Properties of Distance-Based and Multistage Models

Critchlow et al. (1991) showed that all distance-based models satisfy properties (1) and (2) and models with the four distances in Sect. 8.3 satisfy properties (3) to (5). The Hausdorff metric extension of Critchlow (1985) with the four distances in Sect. 8.3 also satisfies properties (1) to (5).

It is obvious that multistage models satisfy property (3) but not (2). Fligner and Verducci (1988) showed that the strongly unimodal model, but not the free model, satisfies property (3). Furthermore, the exponential factor model satisfies property (4), and hence the $\phi$-component model also satisfies property (4) as it is a special case of the exponential factor model.

## Chapter Notes

In this chapter, we have introduced several important probability models for ranking data. Extension of order statistics models and distance-based models will be discussed in Chaps. 9 and 11, respectively. Other models not considered here are a variety of exponential family models based on marginals (spectral decomposition of Diaconis (1988, 1989)) or pairwise and higher-way comparisons (inversion models of McCullagh (1993b)), nested orthogonal contrast models (Marden 1992), and models based on insertion sorting (Doignon et al. 2004; Biernacki and Jacques 2013).

# Chapter 9
# Probit Models for Ranking Data

In 1980, the American Psychological Association (APA) conducted an election in which five candidates $(A, B, C, D,$ and $E)$ were running for president and voters were asked to rank order all of the candidates. Candidates $A$ and $B$ are research psychologists, $C$ is a community psychologist, and $D$ and $E$ are clinical psychologists. Among those voters, 5738 gave complete rankings. These complete rankings are considered here (Diaconis (1988)). Note that lower rank implies more favorable. Then the average ranks received by candidates $A, B, C, D,$ and $E$ are 2.84, 3.16, 2.92, 3.09, and 2.99, respectively. This means that voters generally prefer candidate $A$ the most, candidate $C$ the second, etc. However, in order to make inferences on the preferences of the candidates, modeling of the ranking data is needed. In Sect. 9.1 we consider a model for this data which takes into account covariates.

In Sect. 9.2 we consider the following example for which factor analysis would be appropriate. In 1997, a mainland marketing research firm conducted a survey on people's attitude toward career and living style in three major cities in Mainland China – Beijing, Shanghai, and Guangzhou. Five hundred responses from each city were obtained. A question regarding the behavior, conditions, and criteria for job selection of the 500 respondents in Guangzhou will be discussed here. In the survey, respondents were asked to rank the three most important criteria on choosing a job among 13 criteria: (1) favorable company reputation, (2) large company scale, (3) more promotion opportunities, (4) more training opportunities, (5) comfortable working environment, (6) high income, (7) stable working hours, (8) fringe benefits, (9) well matched with employees' profession or talent, (10) short distance between working place and home, (11) challenging, (12) corporate structure of the company, and (13) low working pressure.

 
M. Alvo, P.L.H. Yu, *Statistical Methods for Ranking Data*, Frontiers in Probability and the Statistical Sciences, DOI 10.1007/978-1-4939-1471-5_9

## 9.1 Multivariate Normal Order Statistics Models

In light of the Thurstone order statistics model mentioned in Sect. 8.1, the multivariate normal order statistics (MVNOS) model assumes that the ranking of $t$ objects given by a judge is determined by ordering $t$ latent utilities for the objects assigned by the judge. However, unlike the Thurstone order statistics model that assumes independent utilities, the MVNOS model assumes that the utilities are possibly correlated and the ranking $\boldsymbol{\pi}_j$ given by judge $j$ has the following probability:

$$P(\boldsymbol{\pi}_j) = P(y_{[1]_j,j} > y_{[2]_j,j} > \cdots > y_{[t]_j,j}), \quad i = 1, \cdots, t, \tag{9.1}$$

where $< [1]_j, \cdots, [t]_j >$ is the ordering of the $t$ objects corresponding to the ranking $\boldsymbol{\pi}_j$ and the latent utility vector $\boldsymbol{y}_j = (y_{1j}, \cdots, y_{tj})'$ of judge $j$ is assumed to follow multivariate normal distribution with mean utility vector $\boldsymbol{\mu}_j = (\mu_{1j}, \cdots, \mu_{tj})'$ and a general covariance matrix $\boldsymbol{V}$, i.e.,

$$\boldsymbol{y}_j = \boldsymbol{\mu}_j + \boldsymbol{e}_j \tag{9.2}$$

$$\boldsymbol{e}_j \overset{\text{iid}}{\sim} N(\boldsymbol{0}, \boldsymbol{V}). \tag{9.3}$$

The MVNOS model is sometimes termed the *multinomial probit model* for ranking data.

### 9.1.1 The MVNOS Model with Covariates

When there are some covariates associated with the judges and objects, it is natural to impose the following linear model for $\boldsymbol{\mu}_j$:

$$\boldsymbol{\mu}_j = \boldsymbol{Z}_j \boldsymbol{\beta}, \tag{9.4}$$

where $\boldsymbol{Z}_j$ is a $t \times p$ matrix of covariates associated with judge $j$ and $\boldsymbol{\beta}$ is a $p \times 1$ vector of unknown parameters. For example, in a marketing survey, respondents are asked to rank products according to their preference. Usually, apart from the ranking given by the respondents, some socioeconomic variables ($\boldsymbol{s}_j$) about the respondents and the attributes ($\boldsymbol{a}_i$) of the products are also available. Then one may study the heterogeneity of the preference due to these variables by assuming the following model:

$$\mu_{ij} = \boldsymbol{a}_i'\boldsymbol{\gamma} + \boldsymbol{s}_j'\boldsymbol{\delta}_i, \quad i = 1, \cdots, t. \tag{9.5}$$

The parameter vector $\boldsymbol{\gamma}$ represents the attribute effect common to all the respondents while the vector $\boldsymbol{\delta}_i$ represents the respondents' socioeconomic background which may affect their preference of product $i$. It is easily seen that equation (9.5) is a particular case of the model in (9.4) when

$$\boldsymbol{Z}_j = \begin{pmatrix} \boldsymbol{a}_1' & \boldsymbol{s}_j' & \boldsymbol{0} & \cdots & \boldsymbol{0} \\ \boldsymbol{a}_2' & \boldsymbol{0} & \boldsymbol{s}_j' & & \boldsymbol{0} \\ \vdots & \vdots & & \ddots & \\ \boldsymbol{a}_t' & \boldsymbol{0} & \boldsymbol{0} & & \boldsymbol{s}_j' \end{pmatrix} \quad \text{and} \quad \boldsymbol{\beta} = \begin{pmatrix} \boldsymbol{\gamma} \\ \boldsymbol{\delta}_1 \\ \boldsymbol{\delta}_2 \\ \vdots \\ \boldsymbol{\delta}_t \end{pmatrix}.$$

In what follows, we shall consider the MVNOS model with the mean given in equation (9.4).

## 9.1.2 Parameter Identifiability of the MVNOS Model

Note that one can add an arbitrary constant (location shift) or multiply a positive constant (scale shift) to both sides of (9.2) while leaving the ranking probability unchanged. The location-shift problem is commonly dealt with by subtracting the first $t-1$ rows by the last row leading to the model

$$\boldsymbol{w}_j = \boldsymbol{X}_j\boldsymbol{\beta} + \boldsymbol{\varepsilon}_j \tag{9.6}$$

$$\boldsymbol{\varepsilon}_j \overset{\text{iid}}{\sim} N(\boldsymbol{0}, \boldsymbol{\Sigma}), \tag{9.7}$$

where $w_{ij} = y_{ij} - y_{tj}$, $\boldsymbol{X}_j = [\boldsymbol{I}_{t-1}, -\boldsymbol{1}_{t-1}]\boldsymbol{Z}_j$, $\varepsilon_{ij} = e_{ij} - e_{tj}$, and

$$\boldsymbol{\Sigma} = [\boldsymbol{I}_{t-1}, -\boldsymbol{1}_{t-1}]\boldsymbol{V}[\boldsymbol{I}_{t-1}, -\boldsymbol{1}_{t-1}]'.$$

Here, $\boldsymbol{I}$ denotes an identity matrix and $\boldsymbol{1}$ denotes a vector of 1's. Then the ranking $\boldsymbol{\pi}_j$ with respective ordering $< [1]_j, \cdots, [t]_j >$ corresponds to the event

$$E_j = \{\boldsymbol{w}_j : w_{[1]_j,j} > \cdots > w_{[r-1]_j,j} > 0 > w_{[r+1]_j,j} > \cdots > w_{[t],j}\}$$
$$\text{whenever } [r]_j = t. \tag{9.8}$$

For the sake of simplicity, we use the convention $w_{[0]_j,j} = +\infty$ and $w_{[t+1]_j,j} = -\infty$. Notice that the scale-shift problem still exists in the model given by (9.6) and it can be easily resolved by adding a constraint on $\boldsymbol{\Sigma}$ such as $\sigma_{11} = 1$.

Since rankings of objects only depend on utility differences, $\boldsymbol{\beta}$ and $\boldsymbol{\Sigma}$ (with $\sigma_{11}$ fixed) are estimable, but the original parameters $\mu_j$ and $\boldsymbol{V}$ still cannot be fully identified. For example, suppose $t = 3$ and $\mu_j = \mu$. Then the following three sets of parameters under the MVNOS model lead to the same ranking probabilities:

Set A: $\mu_A = (1, 0, -1)'$ $V_A = \begin{pmatrix} 1 & 0 & 0.2 \\ 0 & 1 & 0.8 \\ 0.2 & 0.8 & 1 \end{pmatrix}$

Set B: $\mu_B = (-1, -2, -3)'$ $V_B = \begin{pmatrix} 0.4 & 0 & 0.4 \\ 0 & 1.6 & 1.6 \\ 0.4 & 1.6 & 2 \end{pmatrix}$

Set C: $\mu_C = (1, 0, -1)'$ $V_C = \begin{pmatrix} 0.756 & -0.444 & -0.311 \\ -0.444 & 0.356 & 0.089 \\ -0.311 & 0.089 & 0.222 \end{pmatrix}$

This is because they all have the same utility differences $y_1 - y_3$ and $y_2 - y_3$ whose joint distribution is

$$N\left(\begin{bmatrix} 2 \\ 1 \end{bmatrix}, \begin{bmatrix} 1.6 & 0 \\ 0 & 0.4 \end{bmatrix}\right).$$

Generally speaking, the parameter $\boldsymbol{\beta}$ can be identified in the presence of covariates $\boldsymbol{X}_j$. However, when there are no covariates, i.e., $\mu_j = \mu$, the values of the $\mu_i$'s will be determined only within a location shift. This indeterminacy can be eliminated by imposing one constraint on the $\mu_i$'s, say, $\mu_t = 0$.

The major identification problem is due to indeterminacy of the covariance matrix $\boldsymbol{V}$ of the utilities. Owing to the fact that the utilities $y_{ij}, i = 1, \cdots, t$, are invariant under any scale shift of $\boldsymbol{V}$ and any transformation of $\boldsymbol{V}$ of the form:

$$\boldsymbol{V} \longrightarrow \boldsymbol{V} + \boldsymbol{c}\boldsymbol{1}_t' + \boldsymbol{1}_t\boldsymbol{c}', \tag{9.9}$$

for any constant vector $\boldsymbol{c}$ (Arbuckle and Nugent 1973), $\boldsymbol{V}$ can never be identified unless it is structured. In the previous example, it can be seen that $\boldsymbol{V}_A$ can be transformed to $\boldsymbol{V}_B$ and $\boldsymbol{V}_C$ by setting $\boldsymbol{c} = (-0.3, 0.3, 0.5)'$ and $\boldsymbol{c} = (-0.122, -0.322, -0.389)'$, respectively. This identification problem is well known in the context of Thurstone order statistics models and multinomial probit models (Arbuckle and Nugent 1973; Dansie 1985; Bunch 1991; Yai et al. 1997; Train 2003).

Various solutions which impose constraints on the covariance matrix $\boldsymbol{V}$ have been proposed in the literature. Among them, the methods proposed by Chintagunta (1992) and Yu (2000) provide the most flexible form for $\boldsymbol{V}$ which does not require fixing any cell. Chintagunta's method restricts each column sum of $\boldsymbol{V}$ to zero (and $\sigma_{11} = 1$), resulting in $\boldsymbol{V} = \boldsymbol{B}^{-}\boldsymbol{\Sigma}(\boldsymbol{B}')^{-}$, with $\boldsymbol{B} = [\boldsymbol{I}_{t-1}, -\boldsymbol{1}_{t-1}]$ while Yu's method restricts each column sum of $\boldsymbol{V}$ to 1 (and $\sigma_{11} = 1$), leading to

$$\boldsymbol{V} = \boldsymbol{A}^{-1}\begin{bmatrix} \boldsymbol{\Sigma} & \boldsymbol{0} \\ \boldsymbol{0} & t \end{bmatrix}(\boldsymbol{A}')^{-1} \text{ with } \boldsymbol{A} = \begin{bmatrix} \boldsymbol{I}_{t-1} & -\boldsymbol{1}_{t-1} \\ \boldsymbol{1}_{t-1}' & 1 \end{bmatrix}.$$

Note that the $\boldsymbol{V}$ identified by Chintagunta's method is singular and the associated utilities must be correlated whereas Yu's method always produces a non-singular matrix $\boldsymbol{V}$ and includes the identity matrix (or its scale shift) as a special case. In addition, it is easy to show that this non-singular matrix is an invariant transformation of the matrix used by Chintagunta (1992) under the transformation (9.9) with $\boldsymbol{c} = \frac{1}{2t}\boldsymbol{1}_t$.

### 9.1.3 Bayesian Analysis of the MVNOS Model

Given a sample of $n$ judges, the likelihood function of $(\boldsymbol{\beta}, \boldsymbol{\Sigma})$ is given by

$$L(\boldsymbol{\beta}, \boldsymbol{\Sigma}) = \prod_{j=1}^{n} P(E_j), \tag{9.10}$$

where the event $E_j$ is given in (9.8). Note that the evaluation of the above likelihood function requires the numerical approximation of the $(t-1)$-dimensional integral (e.g., Genz 1992) which can be done relatively accurately provided that the number of objects $(t)$ is small, say less than 15. To avoid a high-dimensional numerical integration, limited information methods using the induced paired/triple-wise comparisons from the ranking data (e.g., structural equation models by Maydeu-Olivares and Bockenholt (2005) fitted using Mplus) have been proposed. Another approach is to use a Monte Carlo Expectation-Maximization (MCEM) algorithm (e.g., Yu et al. 2005; see also Sect. 9.2) which can avoid the direct maximization of the above likelihood function.

In this section we will consider a simulation-based Bayesian approach which can also avoid the evaluation and maximization of the above likelihood function. Recently, a number of R packages have become available for the Bayesian estimation of the MVNOS models for ranking data, including `MNP` (Imai and van Dyk 2005), `rJAGS` (Johnson and Kuhn 2013) as well as our own package `StatMethRank`.

#### 9.1.3.1 Bayesian Estimation and Prior Distribution

In a Bayesian approach, the first step is to specify the prior distribution of the identified parameters. As mentioned previously one constraint on $\boldsymbol{\Sigma}$ could be added in order to fix the scale and hence to identify all the parameters. Under this condition, the usual Wishart prior distribution for the constrained $\boldsymbol{\Sigma}$ could not be used. In the context of multinomial probit model studied by McCulloch and Rossi (1994), instead of imposing the scale constraint on $\boldsymbol{\Sigma}$, we may compute the full posterior distribution of $\boldsymbol{\beta}$ and $\boldsymbol{\Sigma}$ and obtain the marginal posterior distribution of the identified parameters such as $\boldsymbol{\beta}/\sqrt{\sigma_{11}}, \sigma_{ii}/\sigma_{11}$ and $\rho_{ij} = \sigma_{ij}/\sqrt{\sigma_{ii}\sigma_{jj}}$.

Let $f(\boldsymbol{\beta}, \boldsymbol{\Sigma})$ denote the joint prior distribution of $(\boldsymbol{\beta}, \boldsymbol{\Sigma})$. Then the posterior density of $(\boldsymbol{\beta}, \boldsymbol{\Sigma})$ is

$$f(\boldsymbol{\beta}, \boldsymbol{\Sigma}|\boldsymbol{\Pi}) \propto L(\boldsymbol{\beta}, \boldsymbol{\Sigma}) f(\boldsymbol{\beta}, \boldsymbol{\Sigma}), \tag{9.11}$$

where $\boldsymbol{\Pi} = \{\boldsymbol{\pi}_1, \cdots, \boldsymbol{\pi}_n\}$ is the data set of all $n$ observed rankings. It is convenient to use a normal prior on $\boldsymbol{\beta}$,

$$\boldsymbol{\beta} \sim N(\boldsymbol{\beta}_0, \boldsymbol{A}_0^{-1}),$$

and an independent Wishart prior on $\boldsymbol{G} \equiv \boldsymbol{\Sigma}^{-1}$,

$$\boldsymbol{G} \equiv \boldsymbol{\Sigma}^{-1} \sim W_{t-1}(\alpha, \boldsymbol{P}).$$

Note that our parametrization of the Wishart distribution is such that $E(\boldsymbol{\Sigma}^{-1}) = \alpha \boldsymbol{P}^{-1}$.

Although (9.11) is intractable for Bayesian calculations, we may use the method of Gibbs sampling with data augmentation. We augment the parameter $(\boldsymbol{\beta}, \boldsymbol{\Sigma})$ by the latent variable $\boldsymbol{W} = (\boldsymbol{w}_1, \cdots, \boldsymbol{w}_n)$. Now, the joint posterior density of $(\boldsymbol{\beta}, \boldsymbol{\Sigma}, \boldsymbol{W})$ is

$$f(\boldsymbol{\beta}, \boldsymbol{\Sigma}, \boldsymbol{W}|\boldsymbol{\pi}) \propto f(\boldsymbol{\Pi}|\boldsymbol{W}) f(\boldsymbol{W}|\boldsymbol{\beta}, \boldsymbol{\Sigma}) f(\boldsymbol{\beta}, \boldsymbol{\Sigma}), \tag{9.12}$$

which allows us to sample from the full conditional posterior distributions. The details are provided in the next section.

#### 9.1.3.2 Gibbs Sampling Algorithm for the MVNOS Model

The Gibbs sampling algorithm for the MVNOS model consists of drawing samples consecutively from the full conditional posterior distributions, as follows:

1. Draw $\boldsymbol{w}_j$ from $f(\boldsymbol{w}_j|\boldsymbol{\beta}, \boldsymbol{\Sigma}, \boldsymbol{\Pi})$, for $j = 1, \cdots, n$.
2. Draw $\boldsymbol{\beta}$ from $f(\boldsymbol{\beta}|\boldsymbol{\Sigma}, \boldsymbol{W}, \boldsymbol{\Pi}) \propto f(\boldsymbol{\beta}|\boldsymbol{\Sigma}, \boldsymbol{W})$.
3. Draw $\boldsymbol{\Sigma}$ from $f(\boldsymbol{\Sigma}|\boldsymbol{\beta}, \boldsymbol{W}, \boldsymbol{\Pi}) \propto f(\boldsymbol{\Sigma}|\boldsymbol{\beta}, \boldsymbol{W})$.

In step (1), it can be shown that given $\boldsymbol{\beta}$, $\boldsymbol{\Sigma}$, and $\boldsymbol{\Pi}$, the $\boldsymbol{w}_j$'s are independent and $\boldsymbol{w}_j$ follows a truncated multivariate normal distribution, $N(\boldsymbol{X}_j\boldsymbol{\beta}, \boldsymbol{\Sigma})I(\boldsymbol{w}_j \in E_j)$. One may simulate $\boldsymbol{w}_j$ by using the acceptance-rejection technique, but this may lead to a high rejection rate when the number of objects is fairly large. Instead of drawing the whole vector $\boldsymbol{w}_j$ at one time, we successively simulate each entry of $\boldsymbol{w}_j$ by conditioning on the other $t-2$ entries. More specifically, we replace step (1) by

1. draw $w_{ij}$ from $f(w_{ij}|w_{-i,j}, \boldsymbol{\beta}, \boldsymbol{\Sigma}, \boldsymbol{\Pi})$, for $i = 1, \cdots, t-1$, $j = 1, \cdots, n$, where $w_{-i,j}$ is $\boldsymbol{w}_j$ with $w_{ij}$ deleted.

Let $\boldsymbol{x}'_{ij}$ be the $i$th row of $\boldsymbol{X}_j$, $\boldsymbol{X}_{-i,j}$ be $\boldsymbol{X}_j$ with the $i$th row deleted, and $\boldsymbol{g}_{-i,i}$ be the $i$th column of $\boldsymbol{G}$ with $g_{ii}$ deleted. Suppose $< [1]_j, \cdots, [t]_j >$ is the ordering of objects corresponding to their ranks $\boldsymbol{\pi}_j = (\pi_{1j}, \cdots, \pi_{tj})$. Then $\pi_{ij} = r$ if and only if $[r]_j = i$. Now we have

$$\begin{aligned} &w_{ij}|w_{-i,j}, \boldsymbol{\beta}, \boldsymbol{\Sigma}, \boldsymbol{\Pi} \sim N(m_{ij}, \tau_{ij}^2) \\ &\text{subject to } w_{[r+1]_j j} < w_{ij} < w_{[r-1]_j j} \text{ whenever } \pi_{ij} = r, \end{aligned} \tag{9.13}$$

where

$$m_{ij} = \boldsymbol{x}'_{ij}\boldsymbol{\beta} - g_{ii}^{-1}\boldsymbol{g}'_{-i,i}(w_{-i,j} - \boldsymbol{X}_{-i,j}\boldsymbol{\beta})$$

and $\tau_{ij}^2 = g_{ii}^{-1}$.

Although it still involves simulation from a truncated univariate normal distribution, we can adopt the inverse method to sample from this distribution without using the acceptance-rejection technique which may not be efficient (Devroye 1986).

Returning to steps (2) and (3), since we are conditioning on $\boldsymbol{W}$, the MVNOS model is simply a standard Bayesian linear model setup. Therefore, the full conditional distribution of $\boldsymbol{\beta}$ is

$$\boldsymbol{\beta}|\boldsymbol{\Sigma}, \boldsymbol{W} \sim N_p(\boldsymbol{\beta}_1, \boldsymbol{A}_1^{-1}), \tag{9.14}$$

where

$$\boldsymbol{A}_1 = \boldsymbol{A}_0 + \sum_{j=1}^{n} \boldsymbol{X}'_j \boldsymbol{\Sigma}^{-1} \boldsymbol{X}_j \quad \text{and} \quad \boldsymbol{\beta}_1 = \boldsymbol{A}_1^{-1}(\boldsymbol{A}_0\boldsymbol{\beta}_0 + \sum_{j=1}^{n} \boldsymbol{X}'_j \boldsymbol{\Sigma}^{-1} \boldsymbol{w}_j).$$

Finally, the full conditional distribution of $\boldsymbol{\Sigma}$ is such that $\boldsymbol{\Sigma} = \boldsymbol{G}^{-1}$ with

$$\boldsymbol{G}|\boldsymbol{\beta}, \boldsymbol{W} \sim W_{t-1}\left(\alpha + n, \boldsymbol{P} + \sum_{j=1}^{n}(\boldsymbol{w}_j - \boldsymbol{X}'_j\boldsymbol{\beta})(\boldsymbol{w}_j - \boldsymbol{X}'_j\boldsymbol{\beta})'\right). \tag{9.15}$$

With a starting value for $(\boldsymbol{\beta}, \boldsymbol{\Sigma}, \boldsymbol{W})$, we draw in turn from each of the full conditional distributions given by (9.13), (9.14), and (9.15). When this process is repeated many times, the draws obtained will converge to a single draw from the full joint posterior distribution of $\boldsymbol{\beta}$, $\boldsymbol{\Sigma}$, and $\boldsymbol{W}$ In practice, we iterate the process $M + N$ times. The first $M$ burn-in iterations are discarded. Because the iterates in the Gibbs sample are autocorrelated, we keep every $s$th draw in the last $N$ iterates so that the resulting sample contains approximately independent draws from the joint posterior distribution. The value $s$ here can be determined based on the graph of the sample autocorrelation of the Gibbs iterates.

A natural choice for a starting value for $(\boldsymbol{\beta}, \boldsymbol{\Sigma})$ is to use $(0, \boldsymbol{I})$. However, it is nonstandard to find a starting value for $\boldsymbol{W}$. We adopt an approach motivated by the fact that the ranking of $\{w_{1j}, \cdots, w_{t-1,j}, 0\}$ must be consistent with the observed ranking $\{\pi_{1j}, \cdots, \pi_{tj}\}$. Using this fact, a simple choice for the starting value of the $w$'s is to use $w_{ij} = (\pi_{ij} - \pi_{tj})/\sqrt{(t^2 - 1)/12}$, a type of standardized rank score.

It should be remarked that since Thurstone's normal order statistics model is a MVNOS model with $\boldsymbol{V} = \boldsymbol{I}_t$, its parameters can be estimated by fixing $\boldsymbol{V}$ to $\boldsymbol{I}_t$, or equivalently, fixing $\boldsymbol{\Sigma}$ to $\boldsymbol{I}_{t-1} + \mathbf{1}_{t-1}\mathbf{1}'_{t-1}$ and skipping the step of generating $\boldsymbol{\Sigma}$ in the above Gibbs sampling algorithm.

*Remark.* Although the MVNOS model discussed here considered the case of the complete ranking of $t$ objects, it is not difficult to extend it to incorporate incomplete or partial ranking by modifying the event $E_j$ in (9.8) and the corresponding truncation rule in (9.13) used to sample $w_{ij}$ in the Gibbs sampling. For instance a partial ordering of 4 objects $A, B, C$, and $D$ given by judge $j$ is $B \succ C \succ A, D$.

The event $E_j$ will then be modified to $\{\boldsymbol{w}_j : \max\{w_{Aj}, 0\} < w_{Cj} < w_{Bj}\}$, and hence $w_{Aj}, w_{Bj}$, and $w_{Cj}$ will be separately simulated from truncated normal over intervals $(-\infty, w_{Cj})$, $(w_{Cj}, +\infty)$, and $(\max\{w_{Aj}, 0\}, w_{Bj})$, respectively. So far, we assume that the data does not contain tied ranks or, equivalently, the observed ordering of the objects that are tied is unknown. For example we will treat the tied ranking $B \succ C \succ A = D$ as if the partial ranking $B \succ C \succ A, D$. See Sect. 9.2.1 for similar treatments of incomplete rankings in the context of factor analysis.

### 9.1.4 Adequacy of the Model Fit

To test for the adequacy of the model, we may group the $t!$ rankings into a small number of meaningful subgroups and examine the fit for each subgroup. In particular, let $n_i$ be the observed frequency that object $i$ is ranked as the top object. Also let

$$\hat{p}_i = P(Y_i > Y_1, \cdots, Y_{i-1}, Y_{i+1}, \cdots, Y_t | \hat{\boldsymbol{\beta}}, \hat{\boldsymbol{\Sigma}})$$

be the estimated partial probability of ranking object $i$ as first under the fitted MVNOS model with posterior mean estimates $\hat{\boldsymbol{\beta}}$ and $\hat{\boldsymbol{\Sigma}}$. The fit can be examined by comparing the observed frequency $n_i$ with the expected frequency $n\hat{p}_i$, $i = 1, 2, \cdots, t$, or by calculating the standardized residuals:

$$r_i = \frac{n_i - n\hat{p}_i}{\sqrt{n\hat{p}_i(1-\hat{p}_i)}}, \qquad j = 1, 2, \cdots, t.$$

If the expected frequencies match the observed frequencies well or the absolute values of the residuals are small enough, say, $< 2$, the MVNOS model adequately fits the data. The same argument can be applied to other ranking models.

In performing these calculations, it is necessary to evaluate numerically the estimated probability $\hat{p}_i$ which may be expressed as

$$\int_{-\infty}^{0} \cdots \int_{-\infty}^{0} \phi(\boldsymbol{v}|\boldsymbol{\beta}^*, \boldsymbol{\Sigma}^*) d\boldsymbol{v},$$

where $\boldsymbol{v} = (Y_1 - Y_i, \cdots, Y_{i-1} - Y_i, Y_{i+1} - Y_i, \cdots, Y_t - Y_i)' \sim N(\boldsymbol{\beta}^*, \boldsymbol{\Sigma}^*)$ and $\boldsymbol{\beta}^*$ and $\boldsymbol{\Sigma}^*$ can be obtained from $\hat{\boldsymbol{\beta}}$ and $\hat{\boldsymbol{\Sigma}}$, respectively. We employ the Geweke-Hajivassiliou-Keane (GHK) method (see Geweke 1991; Hajivassiliou 1993; Keane 1994). Let $\boldsymbol{L} = (\ell_{ij})$ be the unique lower triangular matrix obtained from the Cholesky decomposition of $\boldsymbol{\Sigma}^*$ (i.e., $\boldsymbol{\Sigma}^* = \boldsymbol{L}\boldsymbol{L}'$). The GHK simulator for the estimated partial probability $\hat{p}_i$ is constructed via the following steps:

1. Compute

$$P(v_1 < 0|\boldsymbol{\beta}^*, \boldsymbol{\Sigma}^*) = \Phi(-\frac{\beta_1^*}{\ell_{11}}),$$

and draw a $\eta_1 \sim N(0,1)$ with $\eta_1 < -\frac{\beta_1^*}{\ell_{11}}$.
2. For $s = 2, \cdots, t-1$, compute $P(v_s < 0|\eta_1, \eta_2, \cdots, \eta_{s-1}, \boldsymbol{\beta}^*, \boldsymbol{\Sigma}^*) = \Phi(-\frac{\beta_s^* + \sum_{j=1}^{s-1} \ell_{sj}\eta_j}{\ell_{ss}})$, and draw a $\eta_s \sim N(0,1)$ with $\eta_s < -\frac{\beta_s^* + \sum_{j=1}^{s-1} \ell_{sj}\eta_j}{\ell_{ss}}$.
3. Estimate $\hat{p}_i$ by $P(v_1 < 0|\boldsymbol{\beta}^*, \boldsymbol{\Sigma}^*)\Pi_{s=2}^{t-1} P(v_s < 0|\eta_1, \eta_2, \cdots, \eta_{s-1}, \boldsymbol{\beta}^*, \boldsymbol{\Sigma}^*)$.
4. Repeat steps 1–3 a large number of times to obtain independent estimates of $\hat{p}_i$, and finally by taking the average of these estimates, the GHK simulator for $\hat{p}_i$ is obtained. In a later application, we will use 10,000 replications.

### 9.1.5 Analysis of the APA Election Data

We now consider the APA election data. Let $Y_{ij}$ be the $j$th voter's utility of selecting candidate $i$, $i = A, B, C, D, E$. We apply the MVNOS model in which (i) the $j$th voter's ranking is assumed to be formed by the relative ordering of $Y_{Aj}, Y_{Bj}, Y_{Cj}, Y_{Dj}, Y_{Ej}$; and (ii) the $Y$'s satisfy the following model:

$$Y_{ij} = \mu_i + e_{ij}, \quad i = A, B, C, D, E, \ j = 1, \cdots, 5738,$$
$$(e_{Aj}, e_{Bj}, e_{Cj}, e_{Dj}, e_{Ej})' \overset{\text{iid}}{\sim} N(\mathbf{0}, \boldsymbol{V}),$$

or equivalently, the model could be formed by the relative ordering of $w_{Aj}, w_{Bj}, w_{Cj}, w_{Dj}, 0$, and the $w$'s satisfy

$$w_{ij} = \beta_i + \varepsilon_{ij}, \quad i = A, B, C, D, \ j = 1, \cdots, 5738,$$
$$(\varepsilon_{Aj}, \varepsilon_{Bj}, \varepsilon_{Cj}, \varepsilon_{Dj})' \overset{\text{iid}}{\sim} N(\mathbf{0}, \boldsymbol{\Sigma}),$$

where $\beta_i = \mu_i - \mu_E$ and $\boldsymbol{\Sigma} = (\sigma_{ij})$ with $\sigma_{ij} = v_{ij} + v_{EE} - v_{iE} - v_{jE}$.

Using the proper priors, $\beta \sim N(\beta_0 = 0, A_0^{-1} = 100)$ and $\boldsymbol{\Sigma}^{-1} \sim W_{t-1}(\alpha = t+1, \boldsymbol{P} = (t+1\boldsymbol{I})$, 11,000 Gibbs iterations are generated. The first 1000 burn-in iterations were discarded. As evidenced from the sample autocorrelation of the Gibbs samples (not shown here), keeping every 20th draw in the last 10,000 Gibbs iterations gives approximately independent draws from the joint posterior distribution of the parameters $\boldsymbol{\beta}$ and $\boldsymbol{\Sigma}$ of the MVNOS model. By imposing the constraint $\mu_E = 0$ and our constraint for $\boldsymbol{V}$ to the Gibbs sequences, we obtain estimates for $\mu_i (i = A, B, C, D, E)$ and $v_{ij}, (i \leq j)$.

**Table 9.1** Observed proportions and estimated probabilities that a candidate is ranked as first under various models for the APA election data (*the value in bracket is the residual, $r_i$*)

| Candidate | Observed proportion, $n_i/n$ | Estimated probabilities, $\hat{p}_i$, under the following models | | | |
|---|---|---|---|---|---|
| | | MVNOS model | Thurstone's model | Stern's mixture model | |
| | | | | 2 components | 3 components |
| A | 0.184 | 0.193 (−1.87) | 0.170 (2.78) | 0.199 (−2.89) | 0.189 (−1.16) |
| B | 0.135 | 0.130 (1.17) | 0.231 (−17.24) | 0.153 (−3.84) | 0.155 (−4.27) |
| C | 0.280 | 0.276 (0.70) | 0.179 (20.12) | 0.276 (0.80) | 0.272 (1.19) |
| D | 0.204 | 0.198 (1.25) | 0.220 (−2.93) | 0.186 (3.47) | 0.192 (2.28) |
| E | 0.197 | 0.200 (−0.58) | 0.200 (−0.65) | 0.186 (2.12) | 0.189 (1.41) |

#### 9.1.5.1 Adequacy of Model Fit and Model Comparison

To examine the goodness of fit of the MVNOS model, Table 9.1 shows the observed proportions and estimated partial probabilities under the MVNOS model. The two statistics for Thurstone's normal order statistics model and Stern's mixture of Luce (called BTL in his/her paper) models are also listed in Table 9.1 as alternatives to the MVNOS model. Thurstone's model is fitted by repeating the Gibbs sampling with $\boldsymbol{V}$ fixed at $\boldsymbol{I}_t$, while Stern's mixture models were fitted by Stern (1993). Stern found that the data seem to be a mixture of 2 or 3 groups of voters. This feature is also supported by Diaconis's (1989) spectral analysis and McCullagh's (1993b) model of inversions.

As seen from Table 9.1, the estimated partial probabilities for the MVNOS model match the observed proportions very well. Also the magnitudes of the standardized residuals $r_i$ for the MVNOS model only are all very small ($< 2$), indicating that among the four models considered in Table 9.1, the MVNOS model gives the best fit to the APA election data.

#### 9.1.5.2 Interpretation of the Fitted MVNOS Model

Table 9.2 shows the posterior means, standard deviations, and 90 % posterior intervals for the parameters of the MVNOS model. It is not surprising to see that the ordering of the posterior means of the $\mu_i$'s is the same as that of the average ranks. Apart from the posterior means, the Gibbs samples can also provide estimates of the probability that candidate $i$ is more favorable than candidate $j$. For instance, the probability that candidate $A$ is more favorable than candidate $C$ is estimated by the sample mean of $\Phi(\frac{\mu_A - \mu_C}{\sqrt{v_{AA} + v_{CC} - 2v_{AC}}})$ in the Gibbs samples, which is found to be 0.509 (posterior standard deviation $= 0.006$).

According to the boxplots of $\mu_i$, $v_{ii}$, and $r_{ij} = v_{ij}/\sqrt{v_{ii}v_{jj}} (i \neq j)$ shown in Fig. 9.1, distributions of some parameters are fairly symmetric. In addition, a large estimate of $v_{CC}$ indicates that voters have fairly large variation of the preference

**Table 9.2** Parameter estimates of the MVNOS model for the APA election data

| | Posterior moments | | |
|---|---|---|---|
| Parameter | Mean | SD | 90% interval |
| $\mu_A$ | 0.086 | 0.015 | (0.062, 0.111) |
| $\mu_B$ | −0.071 | 0.014 | (−0.097, −0.048) |
| $\mu_C$ | 0.067 | 0.018 | (0.037, 0.095) |
| $\mu_D$ | −0.048 | 0.014 | (−0.071, −0.026) |
| $v_{AA}$ | 0.524 | 0.008 | (0.511, 0.537) |
| $v_{AB}$ | 0.116 | 0.006 | (0.106, 0.126) |
| $v_{AC}$ | 0.246 | 0.008 | (0.233, 0.257) |
| $v_{AD}$ | 0.041 | 0.008 | (0.027, 0.053) |
| $v_{AE}$ | 0.074 | 0.004 | (0.067, 0.081) |
| $v_{BB}$ | 0.498 | 0.011 | (0.479, 0.516) |
| $v_{BC}$ | 0.087 | 0.009 | (0.072, 0.100) |
| $v_{BD}$ | 0.178 | 0.007 | (0.166, 0.191) |
| $v_{BE}$ | 0.121 | 0.007 | (0.109, 0.132) |
| $v_{CC}$ | 0.833 | 0.024 | (0.795, 0.870) |
| $v_{CD}$ | −0.123 | 0.014 | (−0.146, −0.101) |
| $v_{CE}$ | −0.043 | 0.010 | (−0.060, −0.026) |
| $v_{DD}$ | 0.679 | 0.018 | (0.651, 0.708) |
| $v_{DE}$ | 0.224 | 0.008 | (0.212, 0.239) |
| $v_{EE}$ | 0.624 | 0.008 | (0.610, 0.638) |

on candidate C. To further investigate the structure of the covariance matrix $\boldsymbol{V}$, a principal components analysis of the posterior mean estimate for $\boldsymbol{V}$ is performed and the result is presented in Table 9.3.

A principal components analysis of the posterior mean estimate for $\boldsymbol{V}$ produces the utilities of the five candidates $\{A, B, C, D, E\}$ as

$$\begin{bmatrix} y_A \\ y_B \\ y_C \\ y_D \\ y_E \end{bmatrix} = \begin{bmatrix} 0.086 \\ \text{-}0.071 \\ 0.067 \\ \text{-}0.048 \\ 0 \end{bmatrix} + \sqrt{1.015}\boldsymbol{a}_1 z_1 + \sqrt{0.21}_5 z_2 + \sqrt{0.440}\boldsymbol{a}_3 z_3$$
$$+\sqrt{0.357}\boldsymbol{a}_4 z_4 + \sqrt{0.346}\boldsymbol{a}_5 z_5 \quad (9.16)$$

where the $z$'s are independently and identically distributed as $N(0, 1)$ and the principal components $\boldsymbol{a}$'s are given in Table 9.3. Since rankings of objects only depend on utility differences, the term $\sqrt{0.21}_5 z_2$ does not affect the rankings and hence, interpretation is based on the remaining four components.

Component 1 separates two groups of candidates, {A, C} and {D, E}, implying that there are two groups of voters: voters who prefer candidates A and C more and those who prefer candidates D and E more. Component 3 contrasts candidate E with

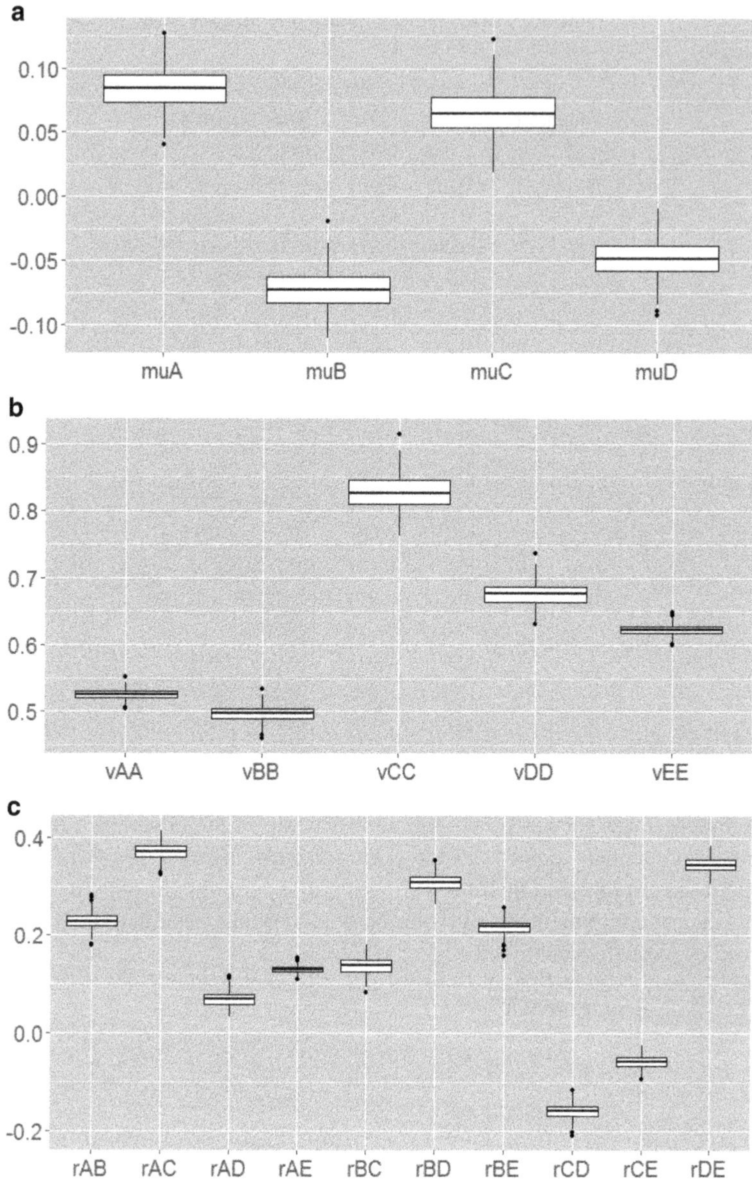


**Fig. 9.1** Boxplots of $\mu_i$, $v_{ii}$, and $r_{ij} = v_{ij}/\sqrt{v_{ii}v_{jj}}(i \neq j)$ for the APA election data

candidates B and D, indicating that voters either prefer B and D to E or prefer E to B and D. For instance, if voters like B, they prefer D to E. Finally, components 4 and 5 indicate a contrast between A and C and a contrast between B and D, respectively. Based on the variances of the components, we can see that component 1 dominates and hence it plays a major role on ranking the five candidates.

**Table 9.3** Principal components analysis of the posterior mean estimate for $\boldsymbol{V}$

| | Principal component, $\boldsymbol{a}_i$ | | | | |
|---|---|---|---|---|---|
| Candidate | 1 | 2 | 3 | 4 | 5 |
| A | 0.245 | 0.447 | 0.046 | 0.789 | −0.340 |
| B | −0.087 | 0.447 | −0.412 | 0.133 | 0.778 |
| C | 0.726 | 0.447 | 0.012 | −0.515 | −0.081 |
| D | −0.524 | 0.447 | −0.442 | −0.281 | −0.502 |
| E | −0.361 | 0.447 | 0.796 | −0.125 | 0.145 |
| Variance | 1.015 | 1.000 | 0.440 | 0.357 | 0.346 |

## 9.2 Factor Analysis

It was mentioned in Sect. 9.1.2 that the parameters of a MVNOS model cannot be fully identified unless the covariance matrix $\boldsymbol{V}$ is structured. One possibility to resolve this problem is to impose a factor covariance structure used in factor analysis onto $\boldsymbol{V}$.

Factor analysis is widely used in social sciences and marketing research to identify the common characteristics among a set of variables. The classical $d$-factor model for a set of continuous variables $y_1, y_2, \cdots, y_t$ is defined as

$$y_{ij} = \boldsymbol{z}_j' \boldsymbol{a}_i + \varepsilon_{ij}, \quad i = 1, \ldots, t; \sim j = 1, \ldots, n \tag{9.17}$$

where $\boldsymbol{y}_j = (y_{1j}, \ldots, y_{tj})'$ is a $t$-dimensional vector of response variables from individual $j$, $\boldsymbol{z}_j = (z_{1j}, \ldots, z_{dj})'$ is a vector of unobserved common factors associated with individual $j$, $\boldsymbol{a}_i = (a_{i1}, \ldots, a_{id})'$ is a vector of factor loadings associated with object $i$ on the $d$ factors, and $\varepsilon_{ij}$ represents the error of the factor model. By adopting the MVNOS framework with the latent utilities satisfying the above factor model, we can generalize the classical factor model to analyze ranking data. In what follows, we shall assume that the reader has a basic familiarity with the statistical concepts of factor scores, factor loadings, and varimax rotation as can be found in most textbooks on multivariate analysis.

### 9.2.1 The Factor Model

Suppose we have a random sample of $n$ individuals from the population and each individual is asked to rank $t$ objects under study according to their own preferences. Within the framework of the MVNOS model, the ranking of the $t$ objects given by individual $j$ in the factor model is determined by the ordering of $t$ latent utilities $y_{1j}, \ldots, y_{tj}$ which satisfies a more general $d$-factor model:

$$y_{ij} = \boldsymbol{z}_j' \boldsymbol{a}_i + b_i + \varepsilon_{ij} \quad j = 1, \ldots, n; \; i = 1, \ldots, t (> d) \tag{9.18}$$

where $\boldsymbol{b} = (b_1, \dots, b_t)'$ is the mean utility vector reflecting the relative importance of the $t$ objects and $\boldsymbol{a}_i = (a_{i1}, \dots, a_{id})'$ represents the factor loadings. It is assumed that the latent common factors $z_1, \dots, z_n$ are independent and identically distributed according to the standard $d$-variate normal distribution, $N_d(\boldsymbol{0}, \boldsymbol{I})$. The error term, $\varepsilon_{ij}$, is the unique factor which is assumed to follow a $N(0, \sigma_i^2)$ distribution, independent of the $z_i$'s.

Denote a complete ranking by $\boldsymbol{\pi}_j = (\pi_{1j}, \dots, \pi_{tj})'$ where $\pi_{ij}$ is the rank of object $i$ from individual $j$. Smaller ranks refer to the more preferred objects and hence higher utilities. For example, if $\boldsymbol{\pi}_j = (2, 3, 1)'$ is recorded, it corresponds to the unobservable utilities $\boldsymbol{y}_j = (y_{1j}, y_{2j}, y_{3j})'$ with $y_{2j} < y_{1j} < y_{3j}$. Note that the only observable quantities are the $\pi_{ij}$'s but not the $y_{ij}$'s.

*Remark.* Extension of the above factor model to incorporate incomplete ranking data is quite straightforward. In the case of the top $q$ partial rankings with the top $q$ objects being $[1]_j, \dots, [q]_j$ for individual $j$, it is natural to assign objects $[1]_j, \dots, [q]_j$ with ranks $1, \dots, q$, respectively, and the rest of objects with midrank, i.e., $[(q+1) + \cdots + t]/(t-q)$. The factor model can be extended to restrict the utilities $y_{1j}, \cdots, y_{tj}$ to satisfy $y_{[1]_j j} > y_{[2]_j j} > \cdots > y_{[q]_j j} > y_{[q+1\}_j j}, \cdots, y_{[t]_j j}$. For subset rankings, individuals are asked to rank a subset of the $t$ objects only. Ranking of the set of remaining objects is unknown and we can simply restrict the ordering of the utilities of objects in the subset consistent to the ranking of these objects. Generally speaking, a ranking $\boldsymbol{\pi}$, complete or incomplete, corresponds to an event $\{\boldsymbol{y} : C\boldsymbol{y} < \boldsymbol{0}\}$, for some contrast matrix $C$. For instance in the case of ranking $t = 4$ objects, the complete ranking $\boldsymbol{\pi}_1 = (2, 3, 1, 4)'$, top 2 partial ranking $\boldsymbol{\pi}_2 = (2, 3.5, 1, 3.5)'$, and the subset ranking $\boldsymbol{\pi}_3 = (2, \_, 1, \_)'$ refer to the events with their respective matrices $C$ being

$$\begin{pmatrix} 1 & 0 & -1 & 0 \\ -1 & 1 & 0 & 0 \\ 0 & -1 & 0 & 1 \end{pmatrix}, \quad \begin{pmatrix} 1 & 0 & -1 & 0 \\ -1 & 1 & 0 & 0 \\ -1 & 0 & 0 & 1 \end{pmatrix}, \quad \text{and} \quad \begin{pmatrix} 1 & 0 & -1 & 0 \end{pmatrix}.$$

Notationally, let

$$\boldsymbol{A}_{d \times t} = [\boldsymbol{a}_1 \cdots \boldsymbol{a}_t],$$

$\boldsymbol{\Psi}_{t \times t}$ be the diagonal matrix with $diag(\boldsymbol{\Psi}) = (\sigma_1^2, \dots, \sigma_t^2)$, and all other entries equal to zero, and

$$\boldsymbol{\theta} = \{\boldsymbol{A}, \boldsymbol{b}, \boldsymbol{\Psi}\}$$

the set of parameters of interest. We shall discuss the maximum likelihood estimation of $\boldsymbol{\theta}$ based on various types of ranking data via the Monte Carlo Expectation-Maximization (MCEM) algorithm in the next section.

### 9.2.2 *Monte Carlo Expectation-Maximization Algorithm*

In order to deal with missing data, the EM algorithm is a broadly applicable approach for the computation of maximum likelihood estimates having the advantages of simplicity and stability. It requires one to compute the conditional expectation of the complete-data log-likelihood function given the observed data (E-step) and then to maximize the likelihood function with respect to the parameters of interest (M-step).

Let $\boldsymbol{Y}_{n\times t}$, $\boldsymbol{Z}_{n\times d}$ be the matrices of the unobservable response utilities and latent common factors, respectively, with their $j$th rows corresponding to individual $j$. Denote by $\boldsymbol{\Pi}_{n\times t} = [\boldsymbol{\pi}_1, \ldots, \boldsymbol{\pi}_n]'$ the matrix of the observed ranked data. Under an EM setting, we denote by $\{\boldsymbol{Y}, \boldsymbol{Z}\}$ the missing data and by $\boldsymbol{\Pi}$ the observed data.

#### 9.2.2.1 Implementing the E-step via the Gibbs Sampler

Since the complete-data log-likelihood function, apart from a constant, is given by

$$\ell(\boldsymbol{\theta}|\boldsymbol{Y}, \boldsymbol{Z}) = -\frac{n}{2}\sum_{i=1}^{t}\log\sigma_i^2 - \frac{1}{2}\sum_{i=1}^{t}\sum_{j=1}^{n}\frac{(y_{ij} - \boldsymbol{z}_j'\boldsymbol{a}_i - b_i)^2}{\sigma_i^2}, \tag{9.19}$$

the E-step here only involves computation of the conditional expectations of the complete-data sufficient statistics $\{\boldsymbol{Y}'\boldsymbol{Y}, \boldsymbol{Z}'\boldsymbol{Z}, \boldsymbol{Z}'\boldsymbol{Y}, \boldsymbol{Y}'\boldsymbol{1}, \boldsymbol{Z}'\boldsymbol{1}\}$ given $\boldsymbol{\Pi}$ and $\boldsymbol{\theta}$. This can be done by using the Gibbs sampling algorithm which consists of drawing samples consecutively from the full conditional posterior distributions, as shown below:

1. Draw $z_j$ from $f(z_j|\boldsymbol{y}_j, \boldsymbol{\pi}_j, \boldsymbol{\theta})$.
2. Draw $\boldsymbol{y}_j$ from $f(\boldsymbol{y}_j|z_j, \boldsymbol{\pi}_j, \boldsymbol{\theta})$ for $j = 1, \ldots, n$.

For step 1, making draws from $f\left(z_j|\boldsymbol{y}_j, \boldsymbol{\pi}_j, \boldsymbol{\theta}\right)$ is simple because

$$f(z_j|\boldsymbol{y}_j, \boldsymbol{\pi}_j, \boldsymbol{\theta}) = f(z_j|\boldsymbol{y}_j, \boldsymbol{\theta}),$$

which is independent of $\boldsymbol{\pi}_j$. Draws of $\boldsymbol{Z}$ can be made from the conditional distribution

$$z_j|\boldsymbol{y}_j, \boldsymbol{\theta} \sim N_d(\boldsymbol{A}(\boldsymbol{A}'\boldsymbol{A} + \boldsymbol{\Psi})^{-1}(\boldsymbol{y}_j - \boldsymbol{b}), \boldsymbol{I} - \boldsymbol{A}(\boldsymbol{A}'\boldsymbol{A} + \boldsymbol{\Psi})^{-1}\boldsymbol{A}'). \tag{9.20}$$

For step 2, $\boldsymbol{y}_j$ requires to have consistent orderings with the observed ranking $\boldsymbol{\pi}_j$. Suppose that $< [1]_j, \cdots, [t]_j >$ represents the ordering of the $t$ objects with respect to the complete ranking $\boldsymbol{\pi}_j$ such that $[1]_j$ is the most preferred object, $[2]_j$ is the second most preferred object, and so on. Define $y_{[0]_j j} = +\infty$ and $y_{[t+1]_j j} = -\infty$.

*Complete Rankings*

For the cases with complete rankings, we can draw $y_{ij}$ sequentially for $i = 1, \cdots, t$ from

$$y_{ij}|y_{1j}, \ldots, y_{i-1,j}, y_{i+1,j}, \ldots, y_{tj}, \boldsymbol{\pi}_j, \boldsymbol{z}_j, \boldsymbol{\theta} \sim N(\boldsymbol{z}_j'\boldsymbol{a}_i, \sigma_i^2) \tag{9.21}$$

with the constraint $y_{[r-1]_j j} > y_{ij} > y_{[r+1]_j j}$ for $\pi_{ij} = r$ (or $[r]_j = i$).

*Top q Partial Rankings*

For *top q partial* rankings, we draw the top $q$ objects (i.e., $\{x_{[1]_j j}, \ldots, x_{[q]_j j}\}$) as in the complete case and simulate the other objects by

$$y_{ij} \sim N(\boldsymbol{z}_j'\boldsymbol{a}_i, \sigma_i^2) \tag{9.22}$$

with the constraint $-\infty < y_{ij} < y_{[q]_j j}$ for $\pi_{ij} = r$ (or $[r]_j = i$).

*Subset Rankings*

For *subset* rankings, individuals are asked to rank a subset of the $t$ objects only. Rankings of the set of remaining objects, $\{y_{i'j}\}$, are unknown and we can simulate $\{y_{i'j}\}$ from

$$\{y_{i'j}|i' \notin \{\text{ranked objects}\}\} \sim N(\boldsymbol{z}_j'\boldsymbol{a}_i, \sigma_i^2). \tag{9.23}$$

The conditional expectation of $\boldsymbol{Y}'\boldsymbol{1}$ and $\boldsymbol{Y}'\boldsymbol{Y}$ can be approximated by taking the average of the random draws of $\sum_j \boldsymbol{y}_j$ and the average of their product sum $\sum_j \boldsymbol{y}_j\boldsymbol{y}_j'$, respectively. Finally, conditional expectations of $\boldsymbol{Z}'\boldsymbol{1}$, $\boldsymbol{Z}'\boldsymbol{Z}$, and $\boldsymbol{Z}'\boldsymbol{Y}$ can be obtained by

$$\begin{aligned}
E[\boldsymbol{Z}'\boldsymbol{1}|\boldsymbol{\Pi}, \boldsymbol{\theta}] &= \boldsymbol{A}(\boldsymbol{A}'\boldsymbol{A} + \boldsymbol{\Psi})^{-1}(E[\boldsymbol{Y}'\boldsymbol{1}|\boldsymbol{\Pi}, \boldsymbol{\theta}] - n\boldsymbol{b}),\\
E[\boldsymbol{Z}'\boldsymbol{Z}|\boldsymbol{\Pi}, \boldsymbol{\theta}] &= n[\boldsymbol{I} - \boldsymbol{A}(\boldsymbol{A}'\boldsymbol{A} + \boldsymbol{\Psi})^{-1}\boldsymbol{A}']\\
&\quad + \boldsymbol{A}(\boldsymbol{A}'\boldsymbol{A} + \boldsymbol{\Psi})^{-1}E[(\boldsymbol{Y} - \boldsymbol{1}\boldsymbol{b}')'(\boldsymbol{Y} - \boldsymbol{1}\boldsymbol{b}')|\boldsymbol{\Pi}, \boldsymbol{\theta}](\boldsymbol{A}'\boldsymbol{A} + \boldsymbol{\Psi})^{-1}\boldsymbol{A}'\\
E[\boldsymbol{Z}'\boldsymbol{Y}|\boldsymbol{\Pi}, \boldsymbol{\theta}] &= \boldsymbol{A}(\boldsymbol{A}'\boldsymbol{A} + \boldsymbol{\Psi})^{-1}E[(\boldsymbol{Y}' - \boldsymbol{b}\boldsymbol{1}')\boldsymbol{Y}|\boldsymbol{\Pi}, \boldsymbol{\theta}].
\end{aligned}$$

#### 9.2.2.2 M-Step

By replacing the complete-data sufficient statistics $\{\boldsymbol{Y}'\boldsymbol{Y}, \boldsymbol{Z}'\boldsymbol{Z}, \boldsymbol{Z}'\boldsymbol{Y}, \boldsymbol{Y}'\boldsymbol{1}, \boldsymbol{Z}'\boldsymbol{1}\}$ with their corresponding conditional expectations obtained in E-step, we can compute the maximum likelihood estimate of $\boldsymbol{\theta}$ by

$$\begin{pmatrix} \hat{A} \\ \hat{b}' \end{pmatrix} = \left[ (Z\ 1)'(Z\ 1) \right]^{-1} (Z\ 1)'Y = \begin{bmatrix} Z'Z & Z'1 \\ 1'Z & 1'1 \end{bmatrix}^{-1} \begin{pmatrix} Z'Y \\ 1'Y \end{pmatrix}$$

and

$$\begin{aligned} \hat{\Psi} &= \frac{1}{n}\, diag\left( (Y - Z\hat{A} - 1\hat{b}')'(Y - Z\hat{A} - 1\hat{b}') \right) \\ &= \frac{1}{n}\, diag\left( Y'Y - 2\hat{A}'Z'Y - 2\hat{b}1'Y + \hat{A}'Z'Z\hat{A} + 2\hat{b}1'Z\hat{A} + n\hat{b}\hat{b}' \right). \end{aligned}$$

The new set of $\theta$ is then used for calculation of the conditional expectation of the sufficient statistics in the E-step and the algorithm is iterated until convergence is attained.

#### 9.2.2.3 Determining Convergence of MCEM via Bridge Sampling

To determine convergence of the EM algorithm we propose to use the bridge sampling criterion discussed by Meng and Wong (1996). The bridge sampling estimate for the likelihood ratio associated with the individual $j$ is given by

$$\frac{L(\theta^{(s+1)}|y_j, z_j)}{L(\theta^{(s)}|y_j, z_j)} = \frac{\sum_{m=1}^{M} \left[ \frac{L(\theta^{(s+1)}|y_j^{(s,m)}, z_j^{(s,m)})}{L(\theta^{(s)}|y_j^{(s,m)}, z_j^{(s,m)})} \right]^{1/2}}{\sum_{m=1}^{M} \left[ \frac{L(\theta^{(s)}|y_j^{(s+1,m)}, z_j^{(s+1,m)})}{L(\theta^{(s+1)}|y_j^{(s+1,m)}, z_j^{(s+1,m)})} \right]^{1/2}},$$

where $\{y_j^{(s,m)}, z_j^{(s,m)}, m = 1, \ldots, M\}$ denote the $M$ Gibbs samples from $f(y_j|z_j, \pi_j, \theta^{(s)})$ and $f(z_j|y_j, \pi_j, \theta^{(s)})$ with $\theta^{(s)}$ being the $s$th iterate of $\theta$. The estimate for the log-likelihood ratio of two consecutive iterates is then given by

$$\hat{h}(\theta^{(s+1)}, \theta^{(s)}) = \sum_{j=1}^{n} \log \frac{L(\theta^{(s+1)}|y_j, z_j)}{L(\theta^{(s)}|y_j, z_j)}.$$

We plot $\hat{h}(\theta^{(s+1)}, \theta^{(s)})$ against $s$ to determine the convergence of the MCEM algorithm. A curve converging to zero indicates a convergence because the EM algorithm should increase the likelihood at each step.

### *9.2.3 Simulation*

We adopt the parameter values listed in Table 9.4 used by Brady (1989) to study the MCEM algorithm for complete and incomplete rankings. Using the factor model and these parameter values, thirty sets of data with $n = 1,000$ and utility vectors

**Table 9.4** The Parameter values of a 2-factor model for seven objects

| Object | $a_1$ | $a_2$ | $b$ | $\sigma_j$ |
|---|---|---|---|---|
| 1 | 6.0 | 7.0 | 30.0 | 8.49 |
| 2 | −3.0 | 5.0 | 32.0 | 6.00 |
| 3 | 9.0 | −8.0 | 34.0 | 4.24 |
| 4 | −7.2 | −4.0 | 36.0 | 4.24 |
| 5 | 12.0 | 10.0 | 38.0 | 8.49 |
| 6 | −2.0 | −9.0 | 40.0 | 5.00 |
| 7 | −8.0 | 6.0 | 42.0 | 8.00 |

**Table 9.5** Incomplete block design for subset rankings

| | | |
|---|---|---|
| 1 | 2 | 4 |
| 2 | 3 | 5 |
| 3 | 4 | 6 |
| 4 | 5 | 7 |
| 1 | 5 | 6 |
| 2 | 6 | 7 |
| 1 | 3 | 7 |

of $t = 7$ objects were simulated. Three types of ranked data were *observed* from each data set. The first type corresponds to the complete rankings for seven objects by ranking the utilities of the 7 objects. The second type corresponds to the top 3 partial rankings constructed from the rankings of the three largest utilities while the third type corresponds to the subset rankings of 3 out the 7 objects chosen according to the incomplete block design as shown in Table 9.5.

In our simulation studies, the Gibbs sampler and the MCEM algorithm both converge fairly fast. Computation time required for each MC E-step in the case of subset rankings is shorter than that in complete rankings because the number of truncated normal variates to be drawn is smaller. For each E step, we discarded the first 100 burn-in cycles and selected one $\boldsymbol{x}_i$ systematically from every fifth cycle afterward until a total of 40 draws was reached. The MCEM algorithm converged within 10 iterations for all simulation data sets. The means of the 30 sets of estimates for the complete rankings, *top* 3 partial rankings and 3 *out of* 7 subset rankings, together with their biases and standard errors are shown in Table 9.6. Small values of biases and standard errors show that the estimation method for incomplete rankings is extremely efficient and reliable with high accuracy.

Intuitively, since more information is provided when complete rankings are observed, the estimation of the factor model should perform better than with partial or subset rankings. This is indeed the case as can be seen from Table 9.6. Larger biases and standard errors are obtained for the case of 3 out of 7 subset rankings.

**Table 9.6** Simulation results of MCEM algorithm

| | $a_1$ | | | $a_2$ | | | $b$ | | | $\sigma_j$ | | |
|---|---|---|---|---|---|---|---|---|---|---|---|---|
| Object | Mean | Bias | SE* | Mean | Bias | SE* | Mean | Bias | SE* | Mean | Bias | SE* |
| *Complete rankings* | | | | | | | | | | | | |
| 1 | 6.03 | 0.03 | 0.08 | 6.81 | −0.19 | 0.06 | 30.09 | 0.09 | 0.07 | 8.38 | −0.11 | 0.07 |
| 2 | −3.12 | −0.12 | 0.06 | 4.99 | −0.01 | 0.06 | 32.16 | 0.16 | 0.05 | 6.00 | 0.00 | 0.06 |
| 3 | 9.14 | 0.14 | 0.04 | −7.95 | 0.05 | 0.05 | 33.87 | −0.13 | 0.06 | 4.18 | −0.06 | 0.06 |
| 4 | −7.21 | −0.01 | 0.05 | −3.89 | 0.11 | 0.04 | 35.88 | −0.12 | 0.06 | 4.17 | −0.07 | 0.04 |
| 5 | 12.02 | 0.02 | 0.10 | 9.90 | −0.10 | 0.08 | 38.08 | 0.08 | 0.14 | 8.53 | 0.04 | 0.09 |
| 6 | −2.01 | −0.01 | 0.05 | −8.96 | 0.04 | 0.05 | 39.76 | −0.24 | 0.05 | 4.94 | −0.06 | 0.05 |
| 7 | −8.07 | −0.07 | 0.09 | 6.11 | 0.11 | 0.08 | 42.16 | 0.16 | 0.10 | 8.02 | 0.02 | 0.07 |
| *Top 3 partial rankings* | | | | | | | | | | | | |
| 1 | 5.96 | −0.04 | 0.12 | 6.87 | −0.13 | 0.12 | 29.98 | −0.02 | 0.13 | 8.62 | 0.13 | 0.12 |
| 2 | −2.92 | 0.08 | 0.10 | 4.96 | −0.04 | 0.09 | 32.06 | 0.06 | 0.08 | 5.97 | −0.03 | 0.11 |
| 3 | 8.97 | −0.03 | 0.07 | −8.01 | −0.01 | 0.08 | 33.94 | −0.06 | 0.09 | 4.29 | 0.05 | 0.04 |
| 4 | −7.19 | 0.01 | 0.06 | −4.03 | −0.03 | 0.05 | 36.08 | 0.08 | 0.06 | 4.20 | −0.04 | 0.07 |
| 5 | 11.81 | −0.19 | 0.16 | 10.05 | 0.05 | 0.15 | 37.88 | −0.12 | 0.11 | 8.53 | 0.04 | 0.15 |
| 6 | −1.97 | 0.03 | 0.07 | −8.89 | 0.11 | 0.08 | 40.03 | 0.03 | 0.07 | 5.07 | 0.07 | 0.09 |
| 7 | −7.88 | 0.12 | 0.10 | 6.07 | 0.07 | 0.10 | 42.05 | 0.05 | 0.10 | 7.84 | −0.16 | 0.09 |
| *3 out of 7 subset rankings* | | | | | | | | | | | | |
| 1 | 6.16 | 0.16 | 0.17 | 7.09 | 0.09 | 0.20 | 30.06 | 0.06 | 0.16 | 8.52 | 0.03 | 0.16 |
| 2 | −3.21 | −0.21 | 0.11 | 4.93 | −0.07 | 0.14 | 32.03 | 0.03 | 0.10 | 6.09 | 0.09 | 0.10 |
| 3 | 8.98 | −0.02 | 0.08 | −7.86 | 0.14 | 0.06 | 34.03 | 0.03 | 0.09 | 4.23 | −0.01 | 0.01 |
| 4 | −7.09 | 0.11 | 0.07 | −3.98 | 0.02 | 0.06 | 35.96 | −0.04 | 0.09 | 4.21 | −0.03 | 0.02 |
| 5 | 12.26 | 0.26 | 0.18 | 9.81 | −0.19 | 0.19 | 37.84 | −0.16 | 0.15 | 8.46 | −0.03 | 0.08 |
| 6 | −2.18 | −0.18 | 0.08 | −8.84 | 0.16 | 0.11 | 40.12 | 0.12 | 0.11 | 4.94 | −0.06 | 0.05 |
| 7 | −8.15 | −0.15 | 0.14 | 5.83 | −0.17 | 0.12 | 41.94 | 0.06 | 0.13 | 7.98 | −0.02 | 0.08 |

* The standard errors are obtained empirically based on the 30 estimates

### 9.2.4 Factor Score Estimation

So far we have been interested mainly in problems concerning the parameters in factor models and their estimation. Indeed, this frequently represents the main objective of factor analysis since the loading coefficients, to a large extent, determine the reduction of observed variables into a small number of common factors in terms of meaningful phenomena. While these problems constitute the primary interest of factor analysis, it is sometimes desirable to go one step further and to estimate the scores of an individual on the common factors in terms of the realizations of the variates for that individual. Factor scores provide information concerning the relative position of each individual corresponding to each factor whereas the loadings generally remain constant for all individuals. We therefore turn our attention to the problem of factor score estimation.

With the normality assumption, estimates of the factor score can be obtained via the regression approach and the generalized least squares approach that, respectively, minimize the variation of the estimator and the sum of squared standardized residuals (see Lawley and Maxwell 1971). However, these two approaches can only be used when the utility $\boldsymbol{Y}$ can be observed. Recently, Shi and Lee (1997a) developed a Bayesian approach for estimating the factor scores in factor models with polytomous data. By constructing appropriate posterior distribution, they proposed using the posterior mean as a factor score estimate. Their method involves computation of some multiple integrals which is handled by some Monte Carlo methods. To avoid tedious computation, Shi and Lee (1997b) applied the EM algorithm to obtain a Bayesian estimate of the factor score with polytomous variables. In this section, we will estimate the factor scores with ranked data via the MCEM algorithm discussed in Sect. 9.2.2.

#### 9.2.4.1 Factor Score Estimation Using the MCEM Algorithm

The factor score $\boldsymbol{z}_j$ can be estimated by the posterior mode of the posterior distribution $\boldsymbol{z}_j|\boldsymbol{\pi}_j,\boldsymbol{\theta}$. Hence, the MCEM algorithm can be used to find the estimate by viewing the $\boldsymbol{z}_j$'s as parameters in the complete-data log-likelihood function $\ell$ in (9.19) and the resulting maximum likelihood estimate of $\boldsymbol{z}_j$ will then be the posterior mode estimate. The MCEM iteration can be simplified as follows: given an initial value $\boldsymbol{z}_j^{(0)}$ and the estimate $\boldsymbol{\theta}$, at the $(s+1)^{th}$ MCEM iteration,

**E-step:** Find $E(\boldsymbol{y}_j|\boldsymbol{\pi}_j,\boldsymbol{z}_j^{(s)},\boldsymbol{\theta})$ via Gibbs sampler.

**M-step:** Update $\boldsymbol{z}_j^{(s)}$ to $\boldsymbol{z}_j^{(s+1)}$ by

$$\boldsymbol{z}_j^{(s+1)} = \boldsymbol{A}(\boldsymbol{A}'\boldsymbol{A}+\boldsymbol{\Psi})^{-1}[E(\boldsymbol{y}_j|\boldsymbol{\pi}_j,\boldsymbol{z}_j^{(s)},\boldsymbol{\theta})-\boldsymbol{b})]. \tag{9.24}$$

The Monte Carlo E-step is exactly the same as finding the conditional expectation of $\boldsymbol{y}_j$ while the M-step improves the estimate of $\boldsymbol{z}_j$ in a single step only. This iterative procedure will converge to the appropriate posterior mode which will be taken as an estimate of $\boldsymbol{z}_j$. We propose to stop the MCEM iteration when the likelihood function of $\boldsymbol{z}_j^{(s)}$ and $\boldsymbol{z}_j^{(s+1)}$ is very close to each other. A simple stopping criterion is to consider the following expression:

$$l(\boldsymbol{z}^{(s)},\boldsymbol{z}^{(s+1)}) = \log\frac{\exp\frac{\sum_i \boldsymbol{z}_j^{(s)'}\boldsymbol{z}_j^{(s)}}{2}}{\exp\frac{\sum_i \boldsymbol{z}_j^{(s+1)'}\boldsymbol{z}_j^{(s+1)}}{2}} = \frac{1}{2}\sum_i\left(\boldsymbol{z}_j^{(s)'}\boldsymbol{z}_j^{(s)}-\boldsymbol{z}_j^{(s+1)'}\boldsymbol{z}_j^{(s+1)}\right). \tag{9.25}$$

Convergence of the MCEM iteration is attained when $l(\boldsymbol{z}^{(s)},\boldsymbol{z}^{(s+1)})$ becomes stationary at zero level.

Note that it is possible to estimate the factor scores using the posterior mean based on the samples generated from the Gibbs sampler. We note that the posterior mode and the posterior mean are usually very close and, moreover, the covariance matrix of the posterior mode can be obtained as a by-product of the MCEM factor score estimation.

#### 9.2.4.2 The Covariance Matrix of the Factor Score Estimates

To provide more insight about the estimates and the impact of lost information from continuous to ranking measurements, it is desirable to derive the covariance matrix of the posterior distribution $f(\boldsymbol{z}_j|\boldsymbol{\pi}_j,\boldsymbol{\theta})$, which is given by the negative inverse of the Hessian matrix of $\log[f(\boldsymbol{z}_j|\boldsymbol{\pi}_j,\boldsymbol{\theta})]$. A convenient way to evaluate the Hessian matrix is via the following expression:

$$
\begin{aligned}
-\frac{\partial^2 \log[f(\boldsymbol{z}_j|\boldsymbol{\pi}_j,\boldsymbol{\theta})]}{\partial \boldsymbol{z}_j \partial \boldsymbol{z}_j'} &= -\int \frac{\partial^2 \log[f(\boldsymbol{z}_j|\boldsymbol{y}_j,\boldsymbol{\theta})]}{\partial \boldsymbol{z}_j \partial \boldsymbol{z}_j'} f(\boldsymbol{y}_j|\boldsymbol{z}_j,\boldsymbol{\pi}_j,\boldsymbol{\theta}) d\,\boldsymbol{y}_j \\
&\quad -\mathrm{Var}\left\{-\frac{\partial \log[f(\boldsymbol{z}_j|\boldsymbol{y}_j,\boldsymbol{\theta})]}{\partial \boldsymbol{z}_j}\right\}
\end{aligned} \tag{9.26}
$$

where the variance is with respect to $f(\boldsymbol{y}_j|\boldsymbol{z}_j,\boldsymbol{\pi}_j,\boldsymbol{\theta})$ (Tanner (1997)).

It can be shown that the covariance matrix of the factor score estimate $\hat{\boldsymbol{z}}_j$ is equal to

$$
\left[(\boldsymbol{I}-\boldsymbol{A}(\boldsymbol{A}'\boldsymbol{A}+\boldsymbol{\Psi})^{-1}\boldsymbol{A}')^{-1} - \boldsymbol{W}\ \mathrm{Var}(\boldsymbol{y}_j|\boldsymbol{z}_j,\boldsymbol{\pi}_j,\boldsymbol{\theta})\boldsymbol{W}'\right]^{-1} |_{\boldsymbol{z}_j=\hat{\boldsymbol{z}}_j}, \tag{9.27}
$$

where $\boldsymbol{W} = (\boldsymbol{I}-\boldsymbol{A}(\boldsymbol{A}'\boldsymbol{A}+\boldsymbol{\Psi})^{-1}\boldsymbol{A}')^{-1}\boldsymbol{A}(\boldsymbol{A}'\boldsymbol{A}+\boldsymbol{\Psi})^{-1}$ and $\mathrm{Var}(\boldsymbol{y}_j|\boldsymbol{z}_j,\boldsymbol{\pi}_j,\boldsymbol{\theta})$ can be approximated by the Gibbs sample variance, a by-product of the MCEM factor score estimation.

### *9.2.5 Application to the Job Selection Ranking Data*

We now consider the marketing survey on people's attitude toward career and living style in three main cities in Mainland China – Beijing, Shanghai, and Guangzhou. Five hundred responses from each city were obtained. A question regarding the behavior, conditions, and criteria for job selection of the 500 respondents in Guangzhou will be discussed here. Respondents were asked to rank the three most important criteria on choosing a job among the following 13 criteria: 1. favorable company reputation; 2. large company scale; 3. more promotion opportunities; 4. more training opportunities; 5. comfortable working environment; 6. high income; 7. stable working hours; 8. fringe benefits; 9. well matched with employees'

**Table 9.7** Summary statistics of job selection ranking data

| Criteria | Sample mean | Sample variance |
|---|---|---|
| 1. Favorable company reputation | 8.16 | 2.18 |
| 2. Large company scale | 7.01 | 6.27 |
| 3. More promotion opportunities | 7.43 | 5.69 |
| 4. More training opportunities | 8.12 | 2.43 |
| 5. Comfortable working environment | 6.89 | 7.68 |
| 6. High income | 6.03 | 11.14 |
| 7. Stable working hours | 7.64 | 4.72 |
| 8. Fringe benefits | 6.68 | 8.59 |
| 9. Well matched with employees' profession or talent | 8.14 | 2.18 |
| 10. Short distance between working place and home | 8.30 | 1.13 |
| 11. Challenging | 8.19 | 1.92 |
| 12. Corporate structure of the company | 8.40 | 0.63 |
| 13. Low working pressure | 8.22 | 1.43 |

profession or talent; 10. short distance between working place and home; 11. challenging; 12. corporate structure of the company; and 13. low working pressure.

This is a typical *top 3 out of 13 objects* partial ranking problem. The values "1", "2," and "3" were assigned to the most, second, and third important criteria for job selection, respectively. Regarding the other less important items, it is common to define the midrank, i.e., $\frac{1}{t-q}[(q+1)+\cdots+t]$, as their rank. In this case the midrank is $\frac{1}{10}[4+\cdots+13]=8.5$. Table 9.7 provides some preliminary statistics, including sample mean and sample variance for each of the 13 criteria based on these 500 incomplete rankings with the midrank imputations.

The factor model is assumed and the analysis is made possible by the MCEM algorithm. Initial values of $\boldsymbol{\theta}$ were obtained by principal factor analysis and a standardized rank score, $\frac{\pi_{ij}}{\sqrt{(t^2-1)/12}} \times \frac{1}{\sqrt{\sum_i 1/\sigma_i^2}}$, was used as starting value of $y_{ij}$ in the Gibbs sampler. The choice of standardized rank score was motivated by the fact that the rankings of $y_{ij}'s$ must be consistent with the observed ranking $\boldsymbol{\pi}_j$.

#### 9.2.5.1 Model Estimation

Factor models with the number of factors ranging from zero to five were estimated. The Gibbs sampler (in the MC E-step) converged quite rapidly. We discarded the first 100 burn-in cycles and selected one $\boldsymbol{y}_j$ systematically from every fifth cycle afterward until a total of 40 draws was reached.

We used the bridge sampling criterion discussed in Sect. 9.2.2.3 to detect the convergence of the MCEM algorithm. Figure 9.2 shows the plot of the log-likelihood ratio against the number of iterations of the 3-factor model. The MCEM algorithm converged after 20 iterations.

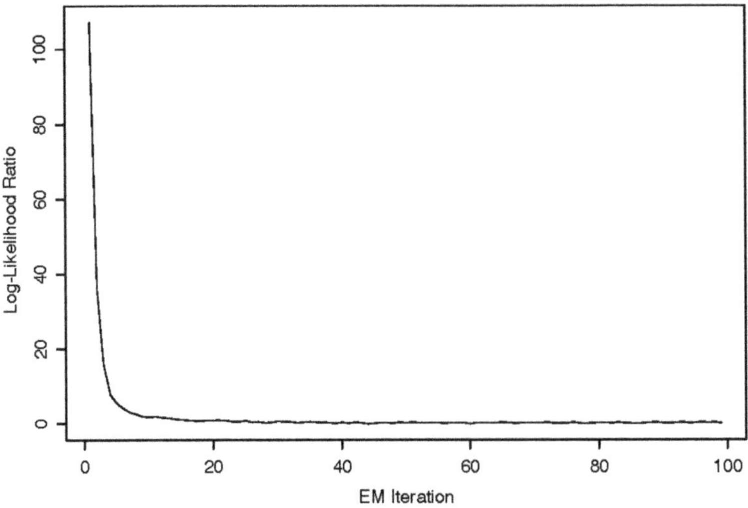


**Fig. 9.2** Bridge sampling criteria

**Table 9.8** AIC values and proportions of variance explained by the $d$-factor model with $d = 0, 1, 2, 3, 4, 5$

| Number of factors ($d$) | Number of free parameters | AIC value | Proportion of variance explained |
|---|---|---|---|
| 0 | 24 | 6086.2 | – |
| 1 | 36 | 5909.0 | 0.1933 |
| 2 | 47 | 5894.6 | 0.2745 |
| 3 | 57 | 5892.8 | 0.4125 |
| 4 | 66 | 5896.7 | 0.4861 |
| 5 | 74 | 5903.8 | 0.5385 |

The Akaike information criterion (AIC) was used to determine the appropriate number of factors. The observed likelihood function which can be written as a product of multivariate normal probabilities over the rectangular region was approximated by the Geweke-Hajivassiliou-Keane (GHK) method shown to be unbiased and most reliable. Table 9.8 exhibits the values of AIC approximated by GHK methods and the proportions of variation explained by the $d$-factor models with $d = 0, 1, 2, 3, 4, 5$. It can be seen that the "best" model according to AIC is the 3-factor model and the proportion of variation explained by the 3-factor model is 41 %.

To examine the goodness of fit of the 3-factor model, we compare the top-choice probability for each of the 13 objects based on the fitted model with its corresponding observed proportions. Here, the top-choice probabilities is estimated using the GHK method. Figure 9.3 provides a plot of the estimated top-choice probabilities vs the respective observed proportions. The points appear to lie on the straight line, indicating the 3-factor model fits the data reasonably well.

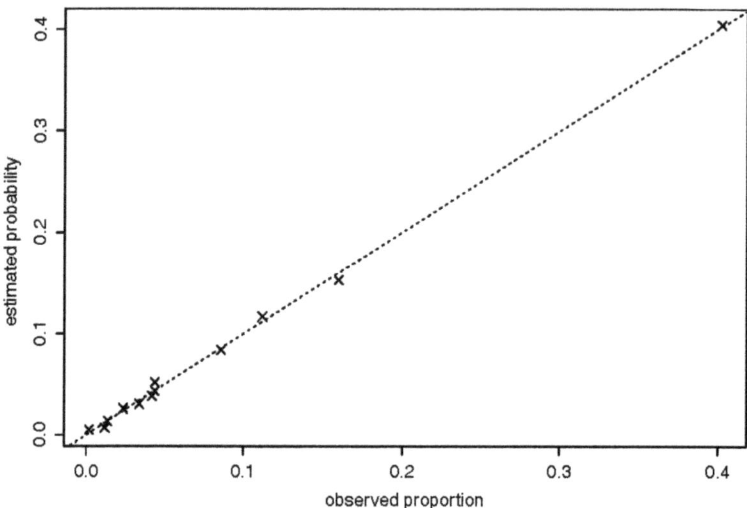


**Fig. 9.3** Estimated top-choice probabilities vs observed proportions

**Table 9.9** Parameter estimates of the 3-factor model

| Criteria | Factor 1 | Factor 2 | Factor 3 | $\boldsymbol{b}$ | $\sigma^2$ |
|---|---|---|---|---|---|
| 1 | −0.58 | −0.41 | 0.25 | −0.79 | 0.73 |
| 2 | −0.02 | −0.21 | 0.79 | −0.46 | 0.29 |
| 3 | 0.29 | −0.65 | −0.04 | 0.30 | 0.40 |
| 4 | 0.44 | −0.38 | 0.24 | −0.74 | 1.10 |
| 5 | −0.10 | 0.09 | −0.14 | 0.75 | 0.47 |
| 6 | −0.05 | 0.07 | −0.73 | 1.31 | 0.35 |
| 7 | −0.09 | 0.69 | −0.17 | 0.42 | 0.27 |
| 8 | −0.22 | 0.36 | −0.34 | 0.94 | 0.31 |
| 9 | 0.58 | 0.03 | 0.21 | −0.26 | 0.53 |
| 10 | 0.14 | 0.70 | −0.15 | −0.05 | 0.19 |
| 11 | 0.39 | −0.07 | 0.09 | −0.46 | 0.88 |
| 12 | −0.50 | 0.21 | 0.08 | −1.10 | 0.82 |
| 13 | −0.07 | 0.52 | −0.23 | −0.17 | 0.47 |
| Cumul. prop. exp. | 0.15 | 0.30 | 0.41 | | |

Estimated values of the factor loadings were obtained by varimax rotation. The values of factor loadings expressed as the correlation between factors and utilities together with the estimated values of $\boldsymbol{b}$ and $\boldsymbol{\sigma}^2$ are summarized in Table 9.9. The first factor can be regarded as a measure of career prospect. Utilization of one's talent and job aspiration are major concerns in this factor. The second dimension represents the undemanding job nature. Short distance between working place and home, stable working hours, and low working pressure all score high loadings in

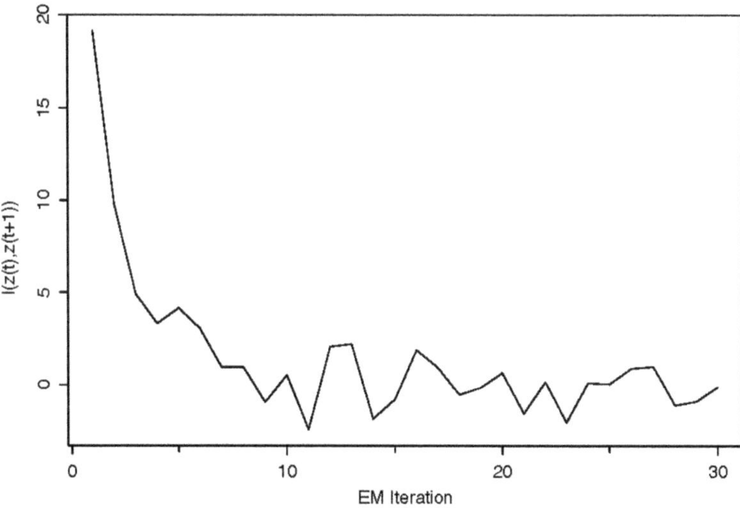


**Fig. 9.4** Stopping criterion

this factor. The third factor represents a contrast between the scale of the company and the salary. A large company offering lower income can be more attractive than a small company offering higher income.

Also, the mean vector $\boldsymbol{b}$ reflects the overall importance of the 13 criteria. Note that the ordering of $\hat{b}_1, \ldots, \hat{b}_t$ is consistent with the average of the 500 rankings. Criterion 6 has the largest mean value which implies salary is their major concern on choosing a job while factors regarding the company itself are least important because $\hat{b}_1$, $\hat{b}_4$, and $\hat{b}_{12}$ get large negative values.

#### 9.2.5.2 Factor Score Estimation

To estimate the factor scores of the fitted 3-factor model, we applied the MCEM method. It is found that the Gibbs sampler in the E-step converged quite rapidly. We discarded the first 100 burn-in cycles and selected one $\boldsymbol{y}_j$ systematically from every fifth cycle afterward until 40 draws were reached. We simulated a total of 300 cycles for each E-step. Also, we applied the stopping criterion to detect the convergence of the MCEM algorithm. Figure 9.4 gives the plot of $l(z^{(s)}, z^{(s+1)})$ against the number of iterations. According to the plot, the MCEM algorithm converged after 20 iterations.

It is often of interest to study the relationship between the factor scores and the covariates of each individual. In this survey, age group was collected in nine 5-year bands covering the ages from 15 to 59 ((1) 15–19, (2) 20–24, ….., (9) 55–59), while education level was recorded in five categories: primary (1), junior secondary (2), senior secondary (3), postsecondary (4), and university degree or above (5).

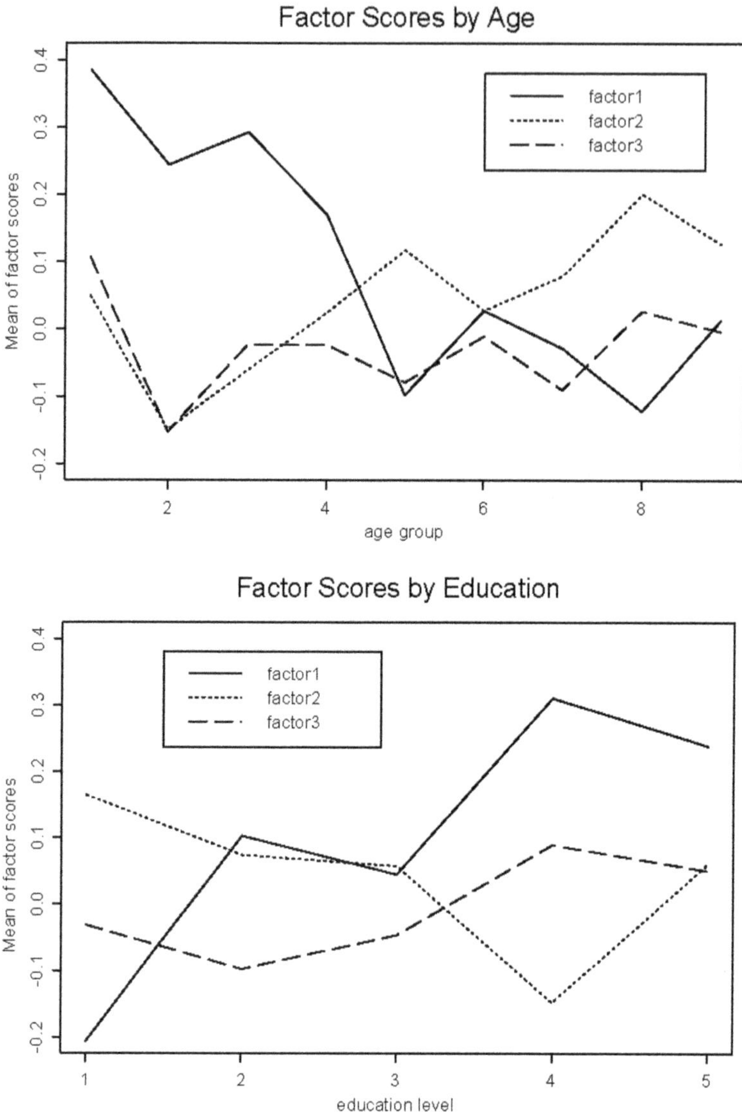


**Fig. 9.5** Factor scores vs age and education level

Figure 9.5 provides plots of the means of the factor score estimates of individuals of different age groups and education levels. From the plot of factor scores by age, a decreasing trend for factor 1 scores and an increasing trend for factor 2 scores are observed whereas from the plot of factor scores by education, an increasing trend for factor 1 scores and a decreasing trend for factor 2 scores are observed. For factor 3 scores, only a slightly increasing trend in education is observed. These observations

**Table 9.10** Standard error of factor scores

| | $S_{11}$ | $S_{22}$ | $S_{33}$ | $S_{12}$ | $S_{13}$ | $S_{23}$ |
|---|---|---|---|---|---|---|
| Mean | 0.1727 | 0.3279 | 0.3275 | 0.0119 | 0.0483 | 0.0174 |
| SD | 0.0052 | 0.0064 | 0.0140 | 0.0040 | 0.0064 | 0.0077 |
| Min. | 0.1606 | 0.3135 | 0.2993 | 0.0014 | 0.0354 | 0.0000 |
| Max. | 0.1903 | 0.3509 | 0.3825 | 0.0243 | 0.0699 | 0.0428 |

imply a young, well-educated person acquires more on career prospect while an old, less educated person may seek for a job with undemanding job nature. Finally, a better educated person is more willing to work in a large company offering lower salary.

To demonstrate the performance of our estimation on factor scores, Table 9.10 provides descriptive statistics on the covariance matrix $S$ of $\hat{z}_j, i = 1, \ldots, 500$. Small values of the standard error show that the estimation method is good and reliable. Also, it seems that the impact of unobservable information for this case is not serious.

## Chapter Notes

To address the robustness of the MVNOS model, Yu (2000) considered two approaches. The first one is to study the sensitivity of the parameter estimates if an outlying ranking is added to the data while the second one is to consider a more general distribution and look at the differences between the 2 sets of estimates.

In Sect. 9.1, we discussed that the parameters of a MVNOS model cannot be fully identified unless the variance-covariance matrix $V$ is structured. Of course, factor analysis mentioned in Sect. 9.2 provides a solution for the simplified but yet flexible dependency structure for $\boldsymbol{V}$. Other choices of dependency structure include wandering vector models (Yu and Chan 2001) and wandering ideal point models (Leung 2003).

In the factor analysis for ranking data, Yu et al. (2005) commented that apart from studying the relationship between the factors and individual's covariate via factor score estimation, we can incorporate the effect of covariates directly into the factor model: $y_{ij} = \boldsymbol{z}_j^{\prime}\boldsymbol{a}_i + b_i + \boldsymbol{w}_j^{\prime}\boldsymbol{c}_i + \varepsilon_{ij}$, where $\boldsymbol{w}_j$ is a vector of covariates of individual $j$ such as sex and age and $\boldsymbol{c}_i$ is a vector of regression parameters. The procedures of the MCEM algorithm can be implemented easily for this model but the details are omitted here. However, the number of parameters to be estimated would increase accordingly. Recently, Yu et al. (2013) further extended factor analysis to a data set of paired rankings such as rankings given by couples and identified the common factors between the individuals in each pair.

So far, we treated the ranking with ties as if the ordering of tied objects is unknown. Poon and Xu (2009) extended the MVNOS model to allow for tied objects by assuming that any two objects $a$ and $b$ have different ranks if and only if their utilities differ by more than a small value, i.e., $|y_a - y_b| \geq \delta$. For example, the ranking $B \succ A = C$ has utilities satisfying $y_B - y_A \geq \delta, y_B - y_C \geq \delta$, and $|y_A - y_C| < \delta$. However, the parameter $\delta$ in Poon and Xu (2009) must be fixed at a prespecified value because of the parameter identifiability problem.

# Chapter 10
# Decision Tree Models for Ranking Data

A number of models for ranking data were introduced in Chaps. 8 and 9. However, not all of these models are designed to incorporate individual/object-specific covariates. Distance-based models discussed in Sect. 8.3 are typical examples of ranking models that are not presently designed to incorporate covariates. As these models generally assume a homogeneous population of individuals, they always give the same predicted ranking. Order statistics models discussed in Sect. 8.3 and Chap. 9 are typical examples of models that are able to incorporate covariates in a "linear model" form. However, there are only a few diagnostic procedures available to determine whether a satisfactory model is found. For instance, is it necessary to transform some of the covariates? Which variables or interaction terms should be included into the model?

For those ranking models that are able to incorporate covariates, it will be difficult to interpret the coefficients of the fitted models if nonlinearity or higher-order interactions are present. For example, Holland and Wessells (1998) applied a rank-ordered logit model with more than 20 interaction terms to predict consumer preferences for fresh salmon. They reported that the model performance has greatly improved after including the interactions but at the same time they mentioned that interpretation of the coefficients is less clear.

The use of decision trees can provide a powerful nonparametric model capable of automatically detecting nonlinear and interaction effects. This could serve as a complement to existing parametric models for ranking data. Decision trees are so called because they can be constructed by a set of rules displayed in a treelike structure. Figure 10.1 exhibits a decision tree with 3 leaf (or terminal) nodes. Since the resulting trees are easy to interpret and provide insight into the data structure, they have been popularly used for classification and regression problems by statisticians, machine learning researchers, and many other data analysts.



M. Alvo, P.L.H. Yu, *Statistical Methods for Ranking Data*, Frontiers in Probability and the Statistical Sciences, DOI 10.1007/978-1-4939-1471-5_10

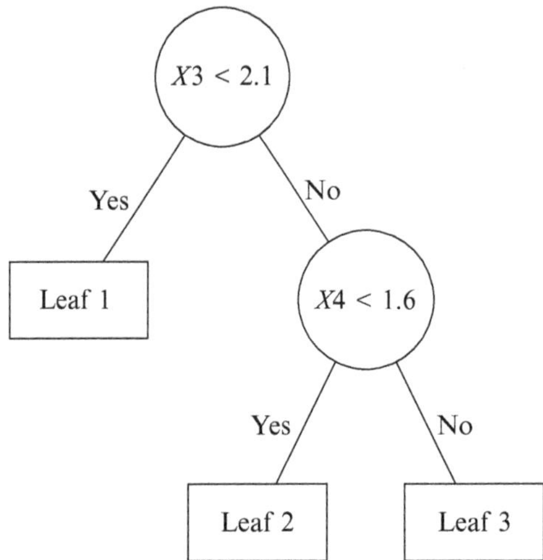


**Fig. 10.1** A hypothetical decision tree

In the literature, there are many variants of tree-construction methods such as CART (Breiman et al. 1984) and C4.5 (Quinlan 1992). Many of these decision tree models are constructed in a top-down manner: starting at the root node (the entire training data set) and recursively partitioning the data into two or more child nodes in such a way that each new generation of nodes has better performance than its parent node. The most important step in tree construction is to select the best split for each internal node according to a certain splitting criterion. One approach is to search for the best split based on an "impurity" function (*impurity function approach*). An impurity function defined for each node measures the degree of impurity of the node. The most frequently used impurity functions are the entropy and the Gini index. An alternative approach to do splitting is to apply a statistical test of homogeneity to test whether the split can make the child nodes with significant different distributions of the data (*statistical test approach*). Common statistical tests are the chi-square test and likelihood ratio test for independence in a two-way contingency table. Generally speaking, construction of a decision tree comprises two stages: *tree growing* and *tree pruning*. See Appendix C for a detailed review of decision trees.

Many popular statistical and data mining software such as SAS/Enterprise Miner, Salford Predictive Modeler, and R provide modeling tools for building decision trees for discrete choice data but not for ranking data. In this chapter, we will introduce recent works by Yu et al. (2010) and Wan (2011) which developed decision tree models based on impurity function and statistical test approaches, respectively.

## 10.1 Impurity Function Approach

In this section, we describe a methodology for constructing a decision tree for ranking data based on the impurity function approach. First of all, the ranking data is randomly partitioned into a training set and a testing set. Following the idea of the CART method, Yu et al. (2010) developed a decision tree algorithm which consists of two stages:

- Tree growing: starting from the root node (the training set), recursively partition each node to identify the best split according to the impurity function until some split-stopping criteria are met. Once tree growing is stopped, a tree is built.
- Tree pruning: using the tree found in the tree-growing stage, find the best subtree by removing branches that does not show any significant improvement in a cost-complexity measure based on a ten-fold cross-validation.

### *10.1.1 Building Decision Tree for Ranking Data*

#### 10.1.1.1 Tree Growing

In growing a tree, the most important step is to search for the best splitting rule in each node based on an impurity function. Here, we will introduce four impurity functions designed for ranking data. Suppose we are given an internal node $\tau$ which contains a data set of rankings of $t$ objects. We first define some notations.

**Definition 10.1.** Let $\boldsymbol{\pi}^{t,m}$ be a set of all possible rankings of $m$ objects from the $t$ objects. Let $\boldsymbol{\pi}^{t,m}_{\{a_1,a_2,\cdots,a_m\}}$ be a set of all possible rankings of the $m$ objects $\{a_1, a_2, \cdots, a_m\}$ from the $t$ objects. Let $\boldsymbol{\Omega}^t_m$ be a collection of all possible subsets of $m$ objects from the $t$ objects.

For example, we have $t = 4$ objects: $\{1, 2, 3, 4\}$. Then we have

$$\boldsymbol{\pi}^{4,2} = \{(1,2), (2,1), (1,3), (3,1), (1,4), (4,1), (2,3), (3,2), (2,4), (4,2), (3,4), (4,3)\}$$

$$\boldsymbol{\pi}^{4,2}_{\{1,2\}} = \{(1,2), (2,1)\}$$

$$\boldsymbol{\Omega}^4_2 = \{\{1,2\}, \{1,3\}, \{1,4\}, \{2,3\}, \{2,4\}, \{3,4\}\}.$$

**Definition 10.2 (Top $q$ notion).** For $\boldsymbol{r} = (a_1, a_2, \cdots, a_q) \in \boldsymbol{\pi}^{t,q}$, let $p_T(\boldsymbol{r} \mid \tau)$ be the proportion of individuals in node $\tau$ who rank object $a_1$ the first, object $a_2$ the second, and so on until object $a_q$ the $q$th. Ranks of the remaining $t - q$ objects are not considered.

**Definition 10.3** (**$m$-wise notion**). For $\boldsymbol{r} = (a_1, a_2, \cdots, a_m) \in \boldsymbol{\pi}^{t,m}_{B_m}$, where $B_m \in \boldsymbol{\Omega}^t_m$, let $p_W(\boldsymbol{r} \mid \tau)$ be the proportion of individuals in node $\tau$ who rank object $a_1$ higher than object $a_2$, object $a_2$ higher than object $a_3$, and so on until object $a_{m-1}$ higher than object $a_m$. Objects other than $a_1, a_2, \ldots, a_n$ are not taken into consideration.

In Appendix C.2.1 impurity functions for unordered categorical responses are described. We provide an extension of the Gini and entropy impurity functions to deal with ranking data.

Given a ranking data set of $t$ objects, the extended Gini and entropy developed using the top $q$ and $m$-wise notions are defined as follows:

$$\text{Top-}q\text{ Gini:} \qquad i_T^{(q)}(\tau) = 1 - \sum_{r \in \pi^{t,q}} [p_T(r \mid \tau)]^2 \tag{10.1}$$

$$\text{Top-}q\text{ entropy:} \qquad i_T^{(q)}(\tau) = -\sum_{r \in \pi^{t,q}} p_T(r \mid \tau) \log_2 p_T(r \mid \tau) \tag{10.2}$$

$$m\text{-wise Gini:} \qquad i_W^{(m)}(\tau) = \frac{1}{C^t_m} \sum_{\boldsymbol{B}_m \in \boldsymbol{\Omega}^t_m} \left( 1 - \sum_{r \in \boldsymbol{\pi}^{t,m}_{B_m}} [p_W(r \mid \tau)]^2 \right) \tag{10.3}$$

$$m\text{-wise entropy:} \qquad i_W^{(m)}(\tau) = \frac{-1}{C^t_m} \left( \sum_{B_m \in \boldsymbol{\Omega}^t_m} \sum_{r \in \pi^{t,m}_{B_m}} p_W(r \mid \tau) \log_2 p_W(r \mid \tau) \right) \tag{10.4}$$

The normalizing term $1/C^t_m$ is to bound $i_W^{(m)}(\tau)$ in the range of 0 and 1.

Given an impurity function $i(\tau)$ (one of the above measures), we define the goodness of (binary) split $s$ for node $\tau$, denoted by $\Delta i(s, \tau)$, as

$$\Delta i(s, \tau) = i(\tau) - p_L i(\tau_L) - p_R i(\tau_R).$$

$\Delta i(s, \tau)$ is the difference between the impurity measure for node $\tau$ and the weighted sum of the impurity measures for the left child and the right child nodes. The weights, $p_L$ and $p_R$, are the proportions of the samples in node $\tau$ that go to the left node $\tau_L$ and the right node $\tau_R$, respectively.

By going through all the possible splits, the best split of node $\tau$ is the one with the largest goodness of split $\Delta i(s, \tau)$. The node splitting will continue until the node size is less than a prespecified minimum node size value. In our application, the minimum node size is set to be one-tenth of the size of the training set.

#### 10.1.1.2 Tree Pruning

After the tree is fully constructed, we proceed to the tree pruning stage where the optimal subtree is determined to improve the prediction accuracy. The basic idea is to make use of the minimal cost-complexity algorithm described in Appendix C.2.2 with ten-fold cross-validation to obtain the final tree that minimizes the misclassification cost. See Yu et al. (2010) for details of the choice of misclassification cost and the implementation procedure of tree pruning.

### *10.1.2 Leaf Assignment*

Various approaches are proposed to make the assignment for every leaf node:

1. Calculate the mean rank of each object and deduce the predicted ranking by ordering the mean ranks.
2. Calculate the top-choice frequency of each object and decide the predicted ranking by ordering the frequency.
3. Use the most frequently observed ranking to represent the predicted ranking.
4. Look at the paired comparison probabilities of each object pair or the top-5 most frequently observed ranking responses.

The first three approaches reveal the predicted ranking of the objects. However, in some situations, the predicted rankings are not of primary concern. Instead, it is of interest to investigate the importance of the covariates to the rank order preference. For this kind of exploration, method 4 provides a more general idea of how the preference orders are distributed within a leaf node. In the following, we advocate a nonparametric procedure to examine the differences among the mean ranks of objects in each leaf node.

#### 10.1.2.1 Nonparametric Inference of the Leaf Nodes

Given a fitted tree, the training data is partitioned into a number of relatively more homogeneous leaf nodes. In this case, we would like to understand the preference of the objects in each leaf node. It is thus of interest to examine the rank order difference among the objects so that a clearer picture can be obtained in interpreting each leaf node.

The first thing we should consider is to test for randomness on ranking of $t$ objects in each leaf node. This is equivalent to apply the Friedman test to the ranking data in each leaf node. Under the null hypothesis $H_0$ of randomness in the Friedman test, the Friedman test statistic $Q_\chi$, corrected for the observed ties, in leaf node $\tau$ with $n_\tau$ individuals is given by

$$Q_\chi = \frac{(t-1)\sum_{i=1}^{t}[r_i - n_\tau(t+1)/2]^2}{\sum_{j=1}^{n_\tau}\sum_{i=1}^{t} r_{ij}^2 - n_\tau t(t+1)^2/4},$$

where $r_i$ is the sum of the ranks $r_{ij}$ for object $i$. When the data contains no ties, $Q_\chi$ becomes the test statistic used in Sect. 2.3.

Iman and Davenport (1980) argued that the chi-square approximation of $Q_\chi$ may be undesirably conservative and suggested using the following $F$ distribution approximation:

$$Q_F = \frac{(n_\tau - 1)Q_\chi}{n_\tau(t-1) - Q_\chi}.$$

Under $H_0$, $Q_F$ follows asymptotically an $F$ distribution with $(t-1)$ and $(n_\tau - 1)(t-1)$ degrees of freedom.

Upon rejection of $H_0$ in the Friedman test, it is possible to identify the difference between specific pairs of objects by the multiple comparison procedure (Conover 1999). Preferences on objects $a$ and $b$ are significantly different at the significance level $\alpha$ if the following inequality is satisfied:

$$|r_a - r_b| > t_{1-\alpha/2}\sqrt{\frac{2n_\tau\left(\sum_{i=1}^{n_\tau}\sum_{j=1}^{t} r_{ij}^2 - \sum_{j=1}^{t} r_{j/n_\tau}^2\right)}{(n_\tau - 1)(t-1)}}$$

where $t_{1-\alpha/2}$ is the value from the student-$t$ distribution with $(n_\tau - 1)(t-1)$ degrees of freedom.

### 10.1.3 Performance Assessment of Decision Tree for Ranking Data

Another important issue that should be addressed is the performance assessment of the fitted decision tree. The most frequently used performance measure is misclassification rate. However, this is not a good performance measure for assessing predictive accuracy of the fitted tree because a predicted ranking can only be classified either correctly or incorrectly, overlooking the fact that the predicted ranking can be partially agreed with the observed ranking. That means some objects in the rank permutation, but not all, are in the correct ordered position.

A widely used single measure for evaluating the overall performance of a binary classifier is the area under the receiver operating characteristic (ROC) curve. It is simple and attractive because it is not susceptible to the threshold choice and it is regardless of the costs of the different kinds of misclassification and class priors

(Bradley 1997; Hand and Till 2001) . The value of AUC always falls within [0.5, 1.0] – it equals 0.5 when the instances are predicted at random and equals 1.0 for perfect accuracy.

However, standard ROC curve can be used for binary data only. Hand and Till (2001) extended it to discrete choice data and Yu et al. (2010) further generalized it to ranking data. The general idea is to first convert the observed and predicted rankings in a testing set into many binary choices with each comparing preference between a pair of objects. Using the observed and predicted binary choice outcomes for each pair of objects, an ROC curve can be drawn and the corresponding AUC can be computed using the standard procedure. Finally, the AUC of the fitted decision tree for ranking data can be obtained by taking the average of the AUCs for all pairs of objects.

### *10.1.4 Analysis of 1993 European Value Priority Data*

The ranking data set was obtained from the International Social Service Programme (ISSP) in 1993 (Jowell et al. 1993), which is a continuing, annual program of cross-national collaboration on surveys covering a wide spectrum of topics for social science research. The survey was conducted using standardized questionnaire in 1993 at 20 countries around the world, such as Great Britain, Australia, the USA, Bulgaria, the Philippines, Israel, and Spain. It mainly focused on value orientations, attitudes, beliefs, and knowledge concerning nature and environmental issues and included the so called Inglehart Index, a collection of four indicators of materialism/post-materialism as well. Respondents were asked to pick the most important (rank "1") and the second most important (rank "2") goals for their government from the following four alternatives:

1. Maintain order in nation (ORDER).
2. Give people more to say in government decisions (SAY).
3. Fight rising prices (PRICES).
4. Protect freedom of speech (SPEECH).

After removing those invalid responses, the survey gave a ranked data set of 5,737 observations with top choice and top two rankings. In addition, the data provide some judge-specific characteristics and they are applied in tree partitioning. The candidate splitting variables are summarized in Table 10.1.

Respondents can be classified into value priority groups on the basis of their top two choices among the four goals. "Materialist" corresponds to an individual who gives priority to ORDER and PRICES regardless of the ordering, whereas those who choose SAY and SPEECH will be termed "post-materialist." The last category consists of judges giving all the other combinations of rankings, and they will be classified as holding "mixed" value orientations.

Inglehart's thesis of generational based values has been influential in political science since the early 1970s. He has argued that value priorities were shifting

**Table 10.1** Description of 1993 EVP data

| Covariate | Description/Code | Type |
|---|---|---|
| Country | West Germany = 1, East Germany = 2, Great Britain = 3, Italy = 4, Poland = 5 | Nominal |
| Gender | Male = 1, female = 2 | Binary |
| Education | 0–10 years = 1, 11–13 years = 2, 14 or more years = 3 | Ordinal |
| Age | Value ranges from 15 to 91 | Interval |
| Religion | Catholic and Greek Catholic=1, Protestant = 2, others = 3, none = 4 | Nominal |

**Table 10.2** Summary of the best pruned subtrees by four impurity measures

| Method | Avg. AUC | SE | AUC | No. of leaves | Depth |
|---|---|---|---|---|---|
| Top-2 entropy | 0.61947 | 0.0056 | 0.62951 | 12 | 5 |
| Pairwise Gini | 0.61896 | 0.0058 | 0.62902 | 12 | 5 |
| Pairwise entropy | 0.61857 | 0.0056 | 0.62709 | 11 | 5 |
| Top-2 Gini | 0.61425 | 0.0063 | 0.61931 | 9 | 4 |

profoundly in economically developed Western countries, from concern over sustenance and safety needs toward quality of life and freedom of self-expression, thus from a materialist orientation to a post-materialist orientation. In this analysis, we study the Inglehart hypothesis in five European countries by our decision tree approach, which helps to identify the attributes that affect Europeans' value priority.

The data are divided randomly into 2 sets, 70 % to the training set for growing the initial tree and finding the best pruned subtree for each of the four splitting criteria; and 30 % to the testing set for performance assessment and selection of the splitting criterion to build the final tree.

As a decision tree is an unstable classifier, small changes in the training set can cause major changes in the fitted tree structure; we therefore repeat this procedure 50 times and compare the four splitting criteria with their averaged AUC. The final tree model is created using the entire data set for interpretation. Notice that the testing set is not involved in the tree building process and pruned subtree selection. The four splitting criteria for rankings include top-2 and pairwise measures of Gini and entropy.

The second and third columns of Table 10.2 show the averaged AUC and their standard error of the best pruned subtrees for each splitting criterion based on 50 repetitions. The tree structure and performance of the final models are also presented. Figure 10.2 displays the six ROC curves of each object pairs that arise from the top-2 entropy tree. The tree did a better job of predicting the object pair "SAY vs PRICES", but poor for "SAY vs SPEECH." The performance of the four trees is comparable and it is hard to distinguish them in the graph.

The four tree models are found to have similar node partitions. The root node is split according to whether the judges came from Poland or not. At the second

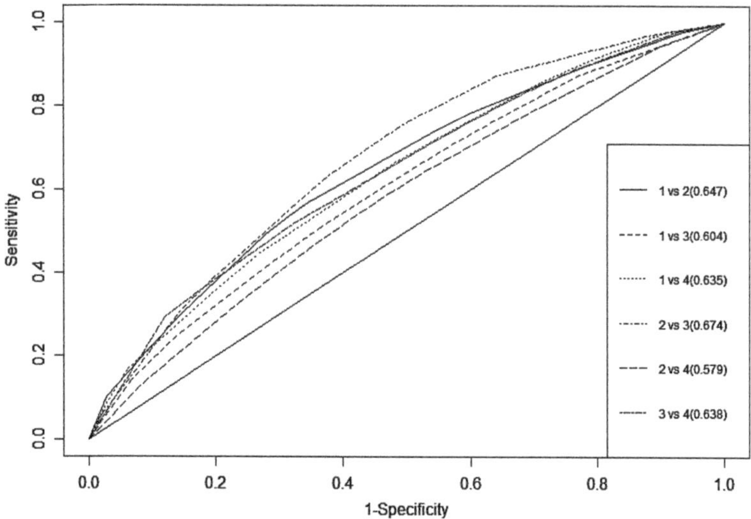


**Fig. 10.2** ROC curves of top-2 entropy tree
Remark: The four value objects are coded as follows 1 = [ORDER], 2 = [SAY], 3 = [PRICES] and 4 = [SPEECH]. The 45° diagonal line connecting (0,0) and (1,1) is the ROC curve corresponding to random chance. Given next to the legends are the areas under the corresponding dashed ROC curves

level, the splits are based on age. For Polish, the respondents are divided with the rule "age<59?", while the remaining judges are split according to age < 53 or not. Further partitions involve education level, country, and age. The factors religion and gender do not seem to be influential. It is observed that in the learning phase, top-2 Gini tends to give a smaller tree, while top-2 entropy gives a more complicated tree on average. Based on the assessment criterion, the top-2 entropy tree is chosen as the best model and it is applied for further analysis.

The tree with 5 levels of depth and 12 leaves is sketched in Figure 10.3. For the sake of brevity, we do not show the other three tree structures. A summary of the leaf nodes of the final tree is reported in two tables. Table 10.3 shows the individuals' value priority, the three most frequent top two rankings, together with the proportion of six pairs of political goals in each leaf node whereas Table 10.4 lists the mean rank of the four political goals. In order to determine if the observed rankings in the leaf node imply statistically significant differences across alternatives, the Friedman test is used, and highly significant results are obtained in all leaves, indicating that the respondents had different priority to at least one political goal. The post hoc multiple comparison procedures are thus further performed and based on the results, the rankings of the four goals are deduced (see Table 10.4).

We now turn to examine the covariate and interaction effects based on the final tree model. In Poland, individuals were more likely to favor materialistic objects ORDER (in leaves 5, 8, and 9) and PRICES (in leaves 5 and 8) than the other two post-materialistic objects. In East Germany, judges appeared to support ORDER and

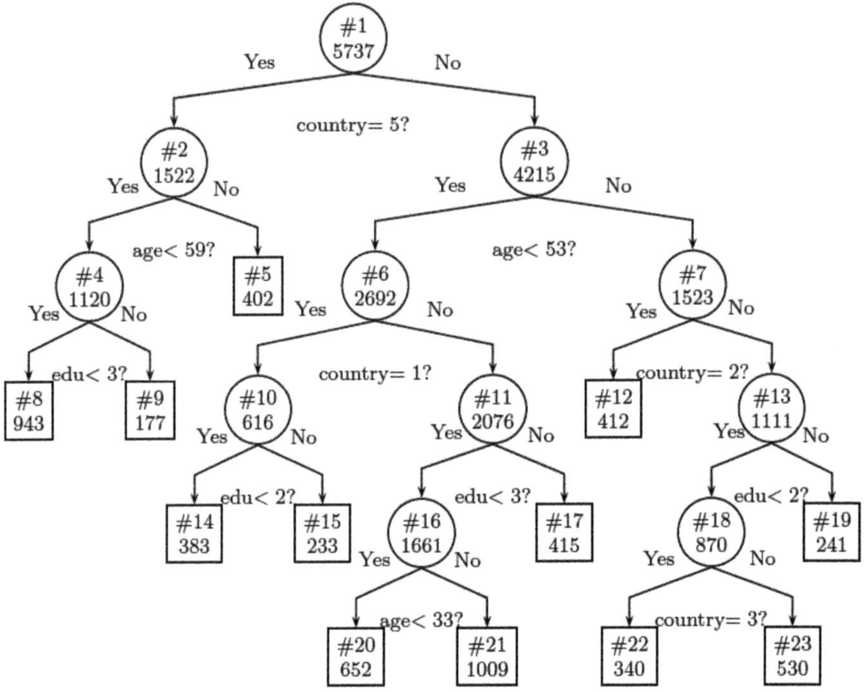


**Fig. 10.3** Fitted tree based on top-2 entropy
Remark: In each node, the node ID and the number of judges are shown. The splitting rule is given under the node. The abbreviation "edu" stands for the variable education

SAY more; particularly those older generations gave higher priority to ORDER (in leaf 12). Respondents of West Germany showed stronger emphasis on SAY. Those better educated West Germans were more post-materialist than the lower educated ones as they preferred SAY and SPEECH, rather than the other two materialist objects (in leaf 15). Mixed value orientations were anchored in British because all the related leaf nodes give us a preference prediction of ORDER $\succ$ SAY or SAY $\succ$ ORDER.

Summarizing, we note that:

(i) Despite some cross-national differences, our findings do not deviate much from Inglehart's theory, which claimed that societies embrace post-materialistic values as they move toward more economic security and affluence. The older European generations experienced economic and social insecurity in their preadult years during World War II. They thus attached more importance to materialistic values compared to younger cohorts. Younger postwar generations developed post-materialist values as they grew up during periods of relative prosperity.
(ii) There is a clear tendency in each country for the higher educated to be the more post-materialistic groups. Duch and Taylor (1993) stated that the

**Table 10.3** Value priority, frequent rankings, and pairwise probabilities in leaf nodes of top-2 entropy tree

| Node($\tau$) | Node Size | Value | Frequent top two ranking | | | Pairwise probabilities | | | | | |
|---|---|---|---|---|---|---|---|---|---|---|---|
| | | | 1st | 2nd | 3rd | $p_w(1,2\|\tau)$ (%) | $p_w(1,3\|\tau)$ (%) | $p_w(1,4\|\tau)$ (%) | $p_w(2,3\|\tau)$ (%) | $p_w(2,4\|\tau)$ (%) | $p_w(3,4\|\tau)$ (%) |
| 5 | 402 | M | 1,3 (40.3%) | 3,1 (24.9%) | 1,2 (7.5%) | 83.1 | 60.8 | 87.3 | 21.0 | 53.7 | 83.1 |
| 8 | 943 | M | 3,1 (25.7%) | 1,3 (22.5%) | 3,2 (13.0%) | 65.3 | 45.9 | 79.3 | 31.0 | 64.8 | 82.1 |
| 9 | 177 | M | 1,3 (17.3%) | 3,1 (13.7%) | 1,2 (12.2%) | 58.8 | 55.9 | 69.2 | 47.7 | 61.9 | 63.8 |
| 12 | 412 | B | 1,3 (27.5%) | 1,2 (22.2%) | 2,1 (18.2%) | 67.5 | 77.7 | 90.0 | 53.9 | 72.0 | 68.3 |
| 14 | 383 | B | 2,1 (17.0%) | 2,3 (12.3%) | 2,4 (12.0%) | 27.4 | 32.1 | 41.0 | 37.6 | 40.1 | 36.0 |
| 15 | 233 | P | 2,4 (25.9%) | 4,2 (15.7%) | 2,3 (11.7%) | 44.1 | 57.8 | 64.8 | 63.6 | 69.5 | 57.3 |
| 17 | 415 | B | 2,4 (15.7%) | 2,1 (14.9%) | 1,2 (14.2%) | 44.5 | 63.1 | 61.0 | 68.3 | 71.1 | 51.0 |
| 19 | 241 | B | 1,3 (22.1%) | 2,1 (14.9%) | 1,2 (13.8%) | 60.8 | 76.3 | 74.5 | 58.3 | 62.2 | 51.9 |
| 20 | 652 | B | 2,1 (20.1%) | 2,3 (19.1%) | 1,2 (13.4%) | 41.6 | 59.7 | 70.6 | 68.9 | 77.3 | 61.5 |
| 21 | 1009 | B | 1,3 (18.0%) | 1,2 (16.1%) | 2,1 (15.8%) | 52.6 | 64.5 | 74.1 | 60.0 | 72.8 | 62.6 |
| 22 | 340 | B | 1,3 (18.1%) | 2,3 (15.9%) | 1,2 (12.3%) | 51.3 | 55.6 | 71.2 | 52.8 | 68.2 | 67.9 |
| 23 | 530 | M | 1,3 (28.2%) | 1,2 (16.2%) | 3,1 (9.6%) | 69.0 | 69.1 | 78.7 | 45.7 | 61.8 | 66.8 |

Remark: The four political goals are labeled as 1=[ORDER], 2 = [SAY], 3 = [PRICES], and 4 = [SPEECH]. In column 3, "Value" shows the value priority group of judges in each leaf node, where B = mixed values, M = materialist, and P = post-materialist. In columns 4–6, "$i$, $j$" implies goal $i \succ$ goal $j$ and the percentage beside indicates the proportion of instances having the corresponding top two ranking in node $\tau$

**Table 10.4** Result of Friedman test and multiple comparison procedures in leaf nodes of top-2 entropy tree

| Node ($\tau$) | Node Size | Mean rank ORDER | SAY | PRICES | SPEECH | Friedman test statistic | Multiple comparison (mean rank difference) † ORDER vs SAY | ORDER vs PRICES | ORDER vs SPEECH | SAY vs PRICES | SAY vs SPEECH | PRICES vs SPEECH | Goal priority[#] |
|---|---|---|---|---|---|---|---|---|---|---|---|---|---|
| 5 | 402 | 1.69 | 3.08 | 1.99 | 3.24 | 285.8** | −1.40 | −0.30 | −1.55 | 1.10 | −0.16 | −1.25 | 1≻3≻2≻4 |
| 8 | 943 | 2.10 | 2.70 | 1.95 | 3.26 | 318.6** | −0.60 | 0.15 | −1.17 | 0.75 | −0.57 | −1.31 | 1≻3≻2≻4 |
| 9 | 177 | 2.16 | 2.49 | 2.40 | 2.95 | 14.0** | −0.33 | – | −0.79 | – | −0.46 | −0.55 | 1≻2≻4; 3≻4 |
| 12 | 412 | 1.65 | 2.42 | 2.63 | 3.30 | 188.6** | −0.77 | −0.98 | −1.66 | −0.22 | −0.89 | −0.67 | 1≻2≻3≻4 |
| 14 | 383 | 2.33 | 2.11 | 2.64 | 2.92 | 35.0** | 0.22 | −0.31 | −0.58 | −0.53 | −0.80 | −0.27 | 2≻1≻3≻4 |
| 15 | 233 | 2.73 | 1.86 | 2.97 | 2.44 | 42.4** | 0.86 | −0.24 | −0.29 | −1.11 | −0.58 | 0.53 | 2≻4≻1≻3 |
| 17 | 415 | 2.31 | 2.05 | 2.80 | 2.83 | 45.4** | 0.26 | −0.49 | −0.52 | −0.75 | −0.78 | – | 2≻1≻3, 4 |
| 19 | 241 | 1.88 | 2.40 | 2.83 | 2.89 | 40.5** | −0.52 | −0.94 | −1.00 | −0.43 | −0.48 | – | 1≻2≻3, 4 |
| 20 | 652 | 2.28 | 1.95 | 2.67 | 3.09 | 127.8** | 0.33 | −0.39 | −0.81 | −0.72 | −1.14 | −0.42 | 2≻1≻3≻4 |
| 21 | 1009 | 2.09 | 2.20 | 2.62 | 3.10 | 165.9** | −0.11 | −0.53 | −1.01 | −0.42 | −0.90 | −0.48 | 1≻2≻3≻4 |
| 22 | 340 | 2.22 | 2.30 | 2.40 | 3.07 | 39.5** | – | −0.19 | −0.85 | – | −0.77 | −0.67 | 1≻3≻4; 2≻4 |
| 23 | 530 | 1.83 | 2.62 | 2.48 | 3.07 | 114.0** | −0.78 | −0.65 | −1.24 | 0.14 | −0.46 | −0.59 | 1≻3≻2≻4 |

Remark: ** The test is significant at 0.01 level

† Only the mean rank differences of those pairwise multiple comparison procedures at $p < 0.05$ were shown

# In the last column, the codes 1–4 represent one of the political goals: 1 = [ORDER], 2 = [SAY], 3 = [PRICES], and 4 = [SPEECH]

post-materialistic objects tap certain fundamental democratic values, such as liberty and rights consciousness. The better educated would have had more opportunity to learn to appreciate such principles, and thus they will prefer post-materialistic objects more.

## 10.2 Statistical Test Approach Based on Intergroup Concordance

In Sect. 10.1, a decision tree for ranking data was constructed in such a way that the leaf nodes are as pure as possible according to a certain impurity measure. In other words, the node splitting in the impurity function approach aims to find the best split such that the two resulting child nodes are as homogeneous as possible. However, this does not guarantee that the rankings of objects between the two child nodes are significantly different.

As mentioned earlier, there is a statistical test approach which can provide an alternative splitting measure based on a test for intergroup concordance on rankings for the selection of the best split during the tree-growing stage. In Sect. 4.2, we have seen several tests for testing for agreement or concordance between two or more groups in ranking a set of objects. Therefore, it is possible to apply those tests in constructing decision tree for ranking data.

### *10.2.1 Building Decision Tree Using Test for Intergroup Concordance*

To construct a decision tree based on the statistical test of intergroup concordance, we follow the same methodology used in the impurity function approach in Sect. 10.1 except for the choice of splitting criterion. For the splitting criterion, the best splitting rule of a node is the one that maximizes the test statistic for testing concordance between two child nodes.

Here, we will make use of two tests of concordance based on Spearman and Kendall statistics described in Sect. 4.2. Note that the combined estimates of the covariance function is used throughout. The notation SP refers to the Spearman statistic and KW to the Kendall statistic.

The partitioning process stops when further splitting does not lead to a statistically significant result. The tree size can then be controlled by setting a threshold significance level on the test procedures. Lower threshold values tend to produce smaller trees.

After the tree is fully constructed, the cost-complexity pruning procedure is executed to avoid the problem of overfitting the ranking data. Most steps are the same as the one used in Sect. 10.1 except that an alternative cost function based on Spearman Footrule distance is considered. For the details of this pruning procedure, see Wan (2011).

### *10.2.2 Analysis of US General Social Survey Data on Job Value Preference*

The general social survey (GSS) has been conducted in US by the National Opinion Research Center (NORC) of the University of Chicago annually since 1972 (except for the years 1979 and 1981) and biennially since 1994 (Davis and Smith 2009). Each year the GSS consisted of a 90-min in-person interview with a full-probability sample of English- or Spanish-speaking persons aged 18 years or above who lived in households. It is a multidimensional social survey that gathers sociodemographic characteristics and replicated core measurements on social and political attitudes and behaviors, plus topics of special interest. Many of the core questions have remained unchanged since 1972 to facilitate time-trend studies as well as replication of earlier findings. In relation to job value, the characteristics were measured by the GSS in an ipsative approach. The respondents were asked to rank in order of preference from (i) "most preferred," (ii) "second most important," to (v) "fifth most important" five aspects about a job:

1. High income (JINC)
2. No danger of being fired (JSEC)
3. Working hours are short, lots of free time (JHOUR)
4. Chances for advancement (JPRO)
5. Work important and gives a feeling of accomplishment (JMEAN)

Job values, as defined by Kalleberg (1977), are what individuals hold as desirable with respect to their work activity and the attitudes are central to the social psychology of work. Under many other names (including work/occupational value/attribute/characteristic), they refer to the importance people place on occupational rewards and play a key role in conditioning a range of work-related outcomes, such as job satisfaction and commitment, work centrality, and occupational choice and stability. Theoretically, the perceived job attributes have been conceptualized into two value dimensions, either *intrinsic* or *extrinsic*. Intrinsic values concern the rewards emanating directly from the work activity and experience itself (e.g., job autonomy, challenge, use of abilities, expression of interest and creativity, workplace cooperation, job useful to society). In contrast, extrinsic values involve the rewards derived from the job but external to the work itself (e.g., job security, pay, fringe benefit, prestige, promotional opportunities, pleasant working environment, good hours, no excessive amount of works). Among the five work values listed in the GSS, the first four attributes (i)–(iv) represent extrinsic factors of the job, while the last value (v) is considered intrinsic.

An ongoing interest among researchers in work value preference has been witnessed over the decades. Previous studies have shown that work values are not externally given, but rather come as a result of socialization processes throughout an individual's life (Johnson 2002; Mortimer and Lorence 1979). They are developed initially as a function of parents' social origin and socioeconomic positions, during schooling and the early years of work. Past research has consisted of comparing and explaining the differences in job value preferences between white and blue collar workers (Weaver 1975), male and female (Lacy et al. 1983), older and younger generations (Loscocco and Kalleberg 1988), as well as blacks and whites (Martin and Tuch 1993). In the following section, we exploit the data from the General Social Survey to revisit the priority of occupational values in the US using the proposed tree model.

This study utilized data from three samples ($N = 3744$) collected in 1973, 1985, and 2006, in order to examine the role of social class origins and socioeconomic characteristics in shaping one's job value orientation. Table 10.5 shows the eight individual attributes (sex, race, birth cohort, highest educational degree attained, family income, marital status, number of children that the respondent ever had, and household size) and three properties of work conditions (working status, employment status, and occupation) that are involved in the model building process. The attribute "year," referring to the year of survey, is also included to address the question of changing work values over time.

Complete rankings of the five job characteristics were obtained from the entire sample. Summarized results of their preferences by years are provided in Table 10.6. The pattern is remarkably consistent over the 30-year period. Meaningful work (JMEAN) was far more important than any other value, taking over 40 % of the top rank order in each year of survey. The next two attributes, high income (JINC) and having opportunities for advancement (JPROMO), were fairly close to each other in importance. Less than 5 % of the respondents regarded short working hours and more leisure time (JHOUR) as the most important value, while 51.3 % placed it as the least important.

Next, the ranking data are analyzed using the decision tree model described earlier. Following the methodology presented in Sect. 10.1.4, we randomly divide the data into 2 parts: (1) the learning set constituting 70 % of the data to grow the initial tree with a threshold significance level of 0.5 and search the best pruned subtree for each of the test statistic and (2) the testing set containing the remaining data to evaluate the tree performance and select the best splitting measure to build the final tree. Note that the entire sample will be included to produce the final model.

Preliminary attempts at learning from the data are not promising because the overall value placed on short working hours in all leaf nodes is found to be the least important to people than meaningful work, high income, and advancement opportunities. Given the limited significance of "JHOUR," we removed it from further analysis and reduced the number of ranked objects to four. The performance of the best pruned subtree for each statistical measure is reported in columns 2 and 3 of Table 10.7. The averaged AUC and the standard error obtained over 50 replications do not differ much among the measures. Indeed, the trees developed have similar structure; the splitting rules at the first, second, and third level are found to be the same. For this reason, we restrict attention to the SP tree model in the coming discussion.

Figure 10.4 depicts the ROC curves of six object pairs that arise from the SP tree. The predictive performance of the classifier is found to be superior on the object pairs "JMEAN vs JSEC" and "JMEAN vs JINC" and inferior on the object pairs

**Table 10.5** Description of US general social survey data of job value preference

| Covariate | Description | Code | Type |
|---|---|---|---|
| Year | Year of survey | 1973, 1985, 2006 | Nominal |
| Sex | Sex | 1 = male, 2 = female | Binary |
| Race | Race | 1 = white, 2 = black, 3 = others | Nominal |
| Cohort | Birth cohort (actual year of birth) | 1 = 1883–1889, 2 = 1890–1899, 3 = 1900–1909, 4 = 1910–1919, 5 = 1920–1929, 6 = 1930–1939, 7 = 1940–1949, 8 = 1950–1959, 9 = 1960–1969, 10 = 1970–1979, 11 = 1980–1989 | Ordinal |
| Educ | Highest educational degree attained | 0 = less than high school, 1 = high school, 2 = associate/junior college, 3 = bachelor's, 4 = graduate | Nominal |
| Finc | Family income | 1 = $1–999, 2 = $1,000–1,999, 3 = $2,000–2,999, 4 = $3,000–3,999, 5 = $4,000–4,999, 6 = $5,000–5,999, 7 = $6,000–6,999, 8 = $7,000–7,999, 9 = $8,000–8,999, 10 = $9,000–9,999, 11 = $10,000–10,999, 12 = $11,000–11,999, 13 = $12,000–14,999, 14 = $15,000–19,999, 15 = $20,000–24,999, 16 = $25,000–29,999, 17 = $30,000–39,999, 18 = $40,000–49,999, 19 = $50,000–74,999, 20 = $75,000–99,999, 21 = >$100,000 | Ordinal |
| Marital | Marital status | 1 = married, 2 = widowed, 3 = divorced, 4 = separated, 5 = never married | Nominal |
| Child | No. of children ever had | 0, 1, 2, 3, 4, 5, 6, 7, 8, or more | Ordinal |
| Hsize | Household size | Value ranges from 1 to 16 | Interval |
| Wkstat | Working status | 1 = working full time, 2 = working part time, 3 = with a job, but not at work because of temporary illness, vacation, and strike, 4 = unemployed, laid off, looking for work, 5 = retired, 6 = in school, 7 = keeping house, 8 = others | Nominal |
| Employ | Employment status | 1 = self-employed, 2 = someone else | Binary |
| Occ | Occupation | 1 = professional, technical, and related workers, 2 = administrative and managerial workers, 3 = clerical and related workers, 4 = sales workers, 5 = service workers, 6 = agriculture, animal husbandry and forestry workers, fishermen, 7 = production and related workers, transport, equipment operators, and laborers | Nominal |

**Table 10.6** Importance of five job values in the US General Social Survey

| Year | JMEAN | JINC | JPRO | JSEC | JHOUR | Sample size |
|---|---|---|---|---|---|---|
| | *Top choice* | | | | | |
| 1973 | 620 (53.7%) | 221 (18.4%) | 223 (19.2%) | 77 (6.4%) | 59 (4.9%) | 1,200 |
| 1985 | 643 (48.8%) | 255 (19.4%) | 287 (21.8%) | 91 (6.9%) | 41 (3.1%) | 1,317 |
| 2006 | 506 (41.2%) | 283 (23.1%) | 244 (19.9%) | 131 (10.7%) | 63 (6.4%) | 1,227 |
| Total | 1769 (47.2%) | 759 (20.3%) | 754 (20.1%) | 299 (8.0%) | 163 (4.4%) | 3,744 |
| | *Last choice* | | | | | |
| 1973 | 90 (7.5%) | 79 (6.6%) | 115 (9.6%) | 359 (29.9%) | 557 (46.4%) | 1,200 |
| 1985 | 67 (5.1%) | 47 (3.6%) | 77 (5.8%) | 351 (26.7%) | 775 (58.8%) | 1,317 |
| 2006 | 85 (6.9%) | 85 (6.9%) | 122 (9.9%) | 345 (28.1%) | 590 (48.1%) | 1,227 |
| Total | 242 (6.5%) | 211 (5.6%) | 314 (8.4%) | 1055 (28.2%) | 1922 (51.3%) | 3,744 |

**Table 10.7** Summary of the best pruned subtrees by two statistical significance measures

| Method | Avg. AUC | SE | AUC | No. of leaves | Depth |
|---|---|---|---|---|---|
| Spearman | 0.65045 | 0.0056 | 0.64689 | 28 | 9 |
| Kendall | 0.64844 | 0.0062 | 0.64229 | 22 | 12 |

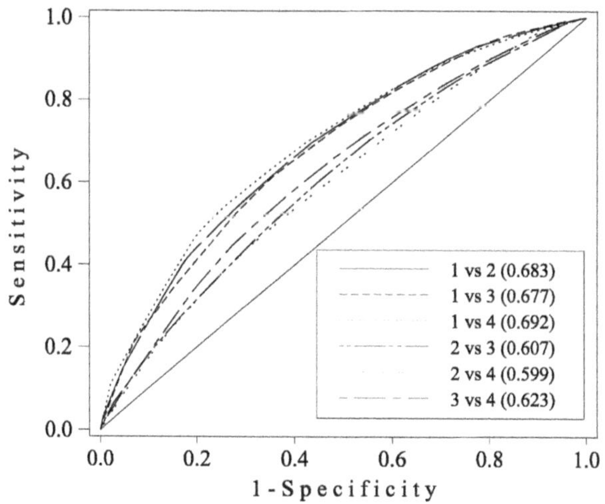


**Fig. 10.4** ROC curves of SP tree
Remark: The four value objects are coded as follows: 1 = [JMEAN], 2 = [JINC], 3 = [JPRO], and 4 = [JSEC]. The 45° diagonal line connecting (0,0) and (1,1) is the ROC curve corresponding to random chance. The areas under the corresponding *dashed* ROC curves appear in brackets

"JINC vs JSEC" and "JINC vs JPRO." As displayed in Figure 10.5, the nine-level SP tree has 28 leaf nodes (in square box). Inside each node, the node ID and the number of judges are shown, whereas the splitting rule is given under the node. Race is

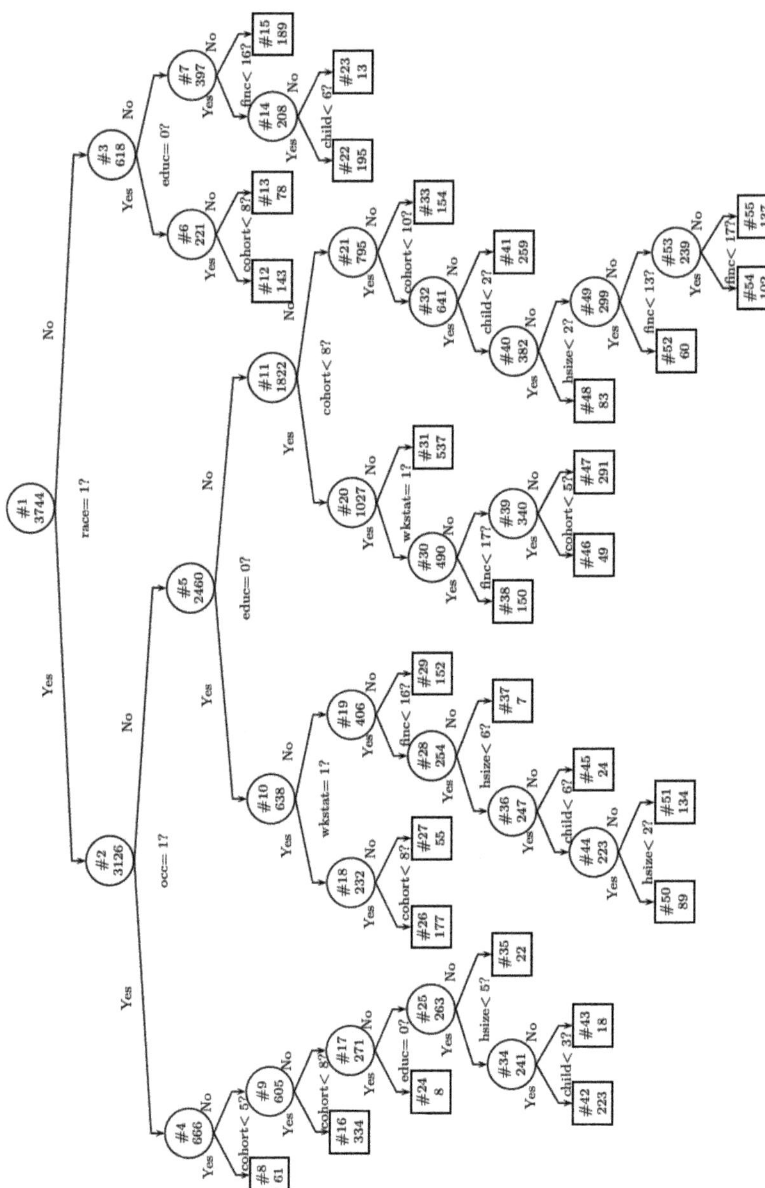


**Fig. 10.5** Fitted SP tree

found to be the most important factor, splitting the entire sample into white (race=1) and black (race=2) Americans. At the second level, white Americans are separated based on their occupation (occ=1 vs occ≠1), while for black Americans, they are divided according to their highest education level attained (educ=0 or not). Other predictor variables appearing in the partitions in lower levels include birth cohort, family income, number of children ever had, working status, and household size. Surprisingly, the demographic variable sex has no contribution in the classification problem, while marital status and employment status are found not to be influential as well. In addition, the factor "year" does not appear in any splitting rule. This suggests that after controlling for the personal and work-related characteristics, the work value orientations have not varied over the past few decades.

Leaf statistics of the SP tree are given in Tables 10.8 and 10.9, which present the mean rank, the three most frequent rankings and the pairwise probability of the six object pairs. For every leaf node, the Friedman test is applied to examine if the rank order differences are significant across four job values and the test results are summarized in column 7 of Table 10.9. It is found that all tests performed are statistically significant at 0.1 level, indicating that at least one value tends to be ranked higher. We conducted multiple comparisons between all pairs of values to determine differences in terms of preferences. The induced ranking in the 28 leaf nodes is provided in the last column of Table 10.9.

It appears that black Americans placed more emphasis on the extrinsic rewards than whites, especially among those less educated (in leaf 12) and with lower family income (in leaf 22). These racial differences may be attributed to their persistent disadvantaged economic status and wage gap in the labor market (Martin and Tuch 1993). On the other hand, resulting from anti-discrimination legislation and improvements in the economic and social conditions in the US since the 1980s, a greater emphasis on intrinsic job value was found among the younger black cohort (in leaf 13).

For white Americans, the occupation effect is important. The result indicates that professional, technical, and related workers attach relatively higher importance to the intrinsic value "JMEAN" than workers in other occupational groups (in leaves 8, 16, 35, and 42). Given their greater skills and credentials, they felt more secure in their ability to seek alternative employment and, thus, showed less favor to the value "JSEC." Nevertheless, professionals with more children have less propensity to take risk in the labor market (in leaf 43).

Consistent with a hierarchy of needs perspective, nonprofessional part-time working white Americans who are less educated, earn less, and have a large household size have higher valuation on extrinsic value of job security and high income when compared to their counterparts (in leaf 7). Meanwhile, the younger cohort in their early stages of working careers tended to desire a job with good prospects for advancement (in leaves 13, 24, and 27).

In summary, the study reveals that significant racial disparities in extrinsic and intrinsic job value preferences existed in US throughout the last three decades. The race effect in conjunction with other variables including education level, family income, age, household size, number of children, and working status explains

**Table 10.8** Frequent rankings and pairwise probabilities in leaf nodes of SP tree

| Node ($\tau$) | Node size | Frequent ranking | | | Pairwise probabilities | | | | | |
|---|---|---|---|---|---|---|---|---|---|---|
| | | 1st | 2nd | 3rd | $p_W(1,2\|\tau)$ (%) | $p_W(1,3\|\tau)$ (%) | $p_W(1,4\|\tau)$ (%) | $p_W(2,3\|\xi)$ (%) | $p_W(2,4\|\tau)$ (%) | $p_W(3,4\|\tau)$ (%) |
| 8 | 61 | 1,3,4,2 (19.7%) | 1,3,2,4 (16.4%) | 1,2,3,4 (14.8%) | 91.8 | 63.9 | 86.9 | 23.0 | 54.1 | 82.0 |
| 12 | 143 | 2,4,3,1 (12.6%) | 2,4,1,3 (8.4%) | 2,3,4,1 (7.7%) | 33.6 | 50.3 | 52.4 | 65.7 | 69.9 | 54.5 |
| 13 | 78 | 3,1,2,4 (9.0%) | 4,3,1,2 (9.0%) | 2,3,1,4 (9.0%) | 37.2 | 23.1 | 50.0 | 43.6 | 64.1 | 67.9 |
| 15 | 189 | 1,2,3,4 (14.8%) | 1,3,2,4 (14.3%) | 2,3,1,4 (9.5%) | 57.7 | 55.6 | 76.7 | 48.7 | 71.4 | 74.1 |
| 16 | 334 | 1,3,2,4 (29.0%) | 1,2,3,4 (24.3%) | 1,3,4,2 (8.4%) | 83.2 | 82.6 | 94.6 | 46.1 | 80.5 | 83.2 |
| 22 | 195 | 2,3,1,4 (9.7%) | 2,3,4,1 (9.7%) | 3,2,1,4 (9.7%) | 37.9 | 31.8 | 60.0 | 52.3 | 79.0 | 74.9 |
| 23 | 13 | 1,3,4,2 (23.1%) | 3,1,4,2 (15.4%) | 3,1,2,4 (15.4%) | 76.9 | 38.5 | 53.8 | 15.4 | 30.8 | 69.2 |
| 24 | 8 | 3,1,2,4 (25.0%) | 3,2,1,4 (12.5%) | 1,3,2,4 (12.5%) | 62.5 | 37.5 | 75.0 | 37.5 | 87.5 | 87.5 |
| 26 | 177 | 1,3,2,4 (10.7%) | 1,2,3,4 (10.2%) | 2,1,3,4 (8.5%) | 54.2 | 59.3 | 70.6 | 58.8 | 69.5 | 63.8 |
| 27 | 55 | 3,2,1,4 (16.4%) | 2,3,1,4 (10.9%) | 2,4,3,1 (7.3%) | 29.1 | 30.9 | 58.2 | 49.1 | 70.9 | 67.3 |
| 29 | 152 | 1,3,2,4 (16.4%) | 3,1,2,4 (9.9%) | 1,2,3,4 (7.9%) | 63.2 | 54.6 | 75.0 | 40.8 | 67.8 | 72.4 |
| 31 | 537 | 1,3,2,4 (26.3%) | 1,2,3,4 (14.2%) | 1,3,4,2 (11.5%) | 78.6 | 69.6 | 87.5 | 34.8 | 70.0 | 81.8 |
| 33 | 154 | 2,1,3,4 (9.7%) | 1,3,4,2 (9.7%) | 3,2,1,4 (9.1%) | 48.1 | 51.3 | 68.2 | 50.6 | 65.6 | 69.5 |
| 35 | 22 | 1,3,4,2 (27.3%) | 1,2,3,4 (22.7%) | 1,3,2,4 (13.6%) | 77.3 | 77.3 | 86.4 | 40.9 | 68.2 | 66.8 |
| 37 | 7 | 4,2,1,3 (28.6%) | 2,4,3,1 (28.6%) | 2,4,1,3 (14.3%) | 14.3 | 57.1 | 14.3 | 85.7 | 42.9 | 14.3 |
| 38 | 150 | 1,3,2,4 (14.7%) | 1,2,3,4 (10.7%) | 1,2,4,3 (8.7%) | 65.3 | 59.3 | 71.3 | 48.7 | 65.3 | 60.7 |

| | | | | | | | | | | |
|---|---|---|---|---|---|---|---|---|---|---|
| 41 | 259 | 1,3,2,4 (17.4%) | 1,2,3,4 (14.3%) | 2,1,3,4 (8.1%) | 57.1 | 66.4 | 76.1 | 54.8 | 76.8 | 71.4 |
| 42 | 223 | 1,3,2,4 (22.0%) | 1,2,3,4 (20.2%) | 1,2,4,3 (12.6%) | 77.6 | 83.0 | 87.4 | 58.3 | 75.8 | 66.8 |
| 43 | 18 | 1,2,4,3 (22.2%) | 1,4,2,3 (22.2%) | 1,2,3,4 (16.7%) | 88.9 | 94.4 | 88.9 | 72.2 | 50.0 | 33.3 |
| 45 | 24 | 1,3,4,2 (20.8%) | 1,3,2,4 (12.5%) | 1,4,3,2 (8.3%) | 70.8 | 54.2 | 70.8 | 16.7 | 37.5 | 66.7 |
| 46 | 49 | 1,3,2,4 (24.5%) | 1,3,4,2 (12.2%) | 1,2,4,3 (12.2%) | 71.4 | 73.5 | 71.4 | 42.9 | 65.3 | 67.3 |
| 47 | 291 | 1,3,2,4 (23.4%) | 1,2,3,4 (18.9%) | 2,1,3,4 (8.3%) | 67.7 | 72.2 | 86.6 | 51.9 | 81.4 | 78.4 |
| 48 | 83 | 1,3,2,4 (18.1%) | 1,2,3,4 (16.9%) | 1,2,4,3 (7.2%) | 66.3 | 72.3 | 79.5 | 53.0 | 72.3 | 65.1 |
| 50 | 89 | 3,4,1,2 (11.2%) | 1,3,4,2 (10.1%) | 1,3,2,4 (10.1%) | 64.0 | 47.2 | 56.2 | 36.0 | 48.3 | 62.9 |
| 51 | 134 | 1,3,2,4 (15.7%) | 3,1,2,4 (9.0%) | 1,4,3,2 (7.5%) | 58.2 | 49.3 | 67.2 | 38.8 | 67.2 | 61.2 |
| 52 | 60 | 1,3,2,4 (18.3%) | 1,2,4,3 (8.3%) | 1,3,4,2 (6.7%) | 68.3 | 68.3 | 85.0 | 46.7 | 75.0 | 73.3 |
| 54 | 102 | 1,3,4,2 (10.8%) | 3,1,2,4 (9.8%) | 3,2,1,4 (9.8%) | 51.0 | 43.1 | 71.6 | 42.2 | 69.6 | 79.4 |
| 55 | 137 | 1,3,2,4 (20.4%) | 1,2,3,4 (13.9%) | 1,3,4,2 (9.5%) | 65.0 | 59.9 | 80.3 | 41.6 | 73.7 | 81.8 |

Remark: The four job values are labeled as 1 = [JMEAN], 2 = [JINC], 3 = [JPRO], and 4 = [JSEC]. In columns 3–5, "$i, j$" implies value $i \succ$ value $j$ and the percentage beside indicates the proportion of instances having the corresponding ranking in node $\tau$

**Table 10.9** Result of Friedman test and multiple comparison procedures in leaf nodes of SP tree

| Node ($\tau$) | Node size | Mean rank | | | | Friedman test statistic $T_F$ | Multiple comparison (mean rank difference) † | | | | | | Job value preference[#] |
|---|---|---|---|---|---|---|---|---|---|---|---|---|---|
| | | JMEAN | JINC | JPRO | JSEC | | JMEAN vs JINC | JMEAN vs JPRO | JMEAN vs JSEC | JINC vs JPRO | JINC vs JSEC | JPRO vs JSEC | |
| 8 | 61 | 1.57 | 3.15 | 2.05 | 3.23 | 73.66** | −1.57 | −0.48 | −1.66 | 1.10 | – | −1.18 | 1≻3≻2, 4 |
| 12 | 143 | 2.64 | 1.98 | 2.62 | 2.77 | 32.24** | 0.66 | – | – | −0.64 | −0.79 | – | 2≻1, 3, 4 |
| 13 | 78 | 2.90 | 2.29 | 1.99 | 2.82 | 26.48** | 0.60 | 0.91 | – | – | −0.53 | −0.83 | 3, 2≻1, 4 |
| 15 | 189 | 2.10 | 2.38 | 2.30 | 3.22 | 83.46** | −0.28 | – | −1.12 | – | −0.85 | −0.92 | 1≻2; 1, 3, 2≻4 |
| 16 | 334 | 1.40 | 2.57 | 2.46 | 3.58 | 481.28** | −1.17 | −1.06 | −2.19 | – | −1.02 | −1.13 | 1≻3, 2≻4 |
| 22 | 195 | 2.70 | 2.07 | 2.09 | 3.14 | 93.91** | 0.64 | 0.61 | −0.44 | – | −1.07 | −1.05 | 2, 3≻1≻4 |
| 23 | 13 | 2.31 | 3.31 | 1.85 | 2.54 | 8.72* | −1.01 | – | – | 1.46 | – | – | 3, 1≻2 |
| 24 | 8 | 2.25 | 2.38 | 1.88 | 3.50 | 7.05* | – | – | −1.25 | – | – | −1.63 | 3, 1≻4 |
| 26 | 177 | 2.16 | 2.26 | 2.54 | 3.04 | 49.64** | – | −0.38 | −0.88 | −0.28 | −0.78 | −0.50 | 1, 2≻3≻4 |
| 27 | 55 | 2.82 | 2.09 | 2.13 | 2.96 | 20.54** | 0.73 | 0.69 | – | – | −0.87 | −0.84 | 2, 3≻1, 4 |
| 29 | 152 | 2.07 | 2.55 | 2.23 | 3.15 | 62.19** | −0.47 | – | −1.08 | 0.32 | −0.61 | −0.92 | 1, 3≻2≻4 |
| 31 | 537 | 1.64 | 2.74 | 2.23 | 3.39 | 535.98** | −1.09 | −0.58 | −1.75 | 0.51 | −0.66 | −1.17 | 1≻3≻2≻4 |
| 33 | 154 | 2.32 | 2.32 | 2.32 | 3.03 | 114.0** | – | – | −0.71 | – | −0.71 | −0.71 | 1, 2, 3≻4 |
| 35 | 22 | 1.59 | 2.68 | 2.27 | 3.45 | 24.05** | −1.09 | −0.68 | −1.86 | – | −0.77 | −1.18 | 1≻3, 2≻4 |
| 37 | 7 | 3.14 | 1.86 | 3.29 | 1.71 | 8.66* | 1.29 | – | 1.43 | −1.43 | – | 1.57 | 4, 2≻1, 3 |
| 38 | 150 | 2.04 | 2.51 | 2.47 | 2.97 | 39.29** | −0.47 | −0.43 | −0.93 | – | −0.46 | −0.50 | 1≻3, 2≻4 |

| 41 | 259 | 2.00 | 2.25 | 2.50 | 3.24 | 133.44** | −0.25 | −0.49 | −1.24 | −0.24 | −0.99 | −0.75 | 1≻2≻3≻4 |
|---|---|---|---|---|---|---|---|---|---|---|---|---|---|
| 42 | 223 | 1.52 | 2.43 | 2.74 | 3.30 | 222.74** | −0.91 | −1.22 | −1.78 | −0.31 | −0.87 | −0.56 | 1≻2≻3≻4 |
| 43 | 18 | 1.28 | 2.67 | 3.33 | 2.72 | 24.47** | −1.39 | −2.06 | −1.44 | −0.67 | – | – | 1≻2≻3; 1≻4 |
| 45 | 24 | 2.04 | 3.17 | 2.04 | 2.75 | 13.35** | −1.13 | – | −0.71 | 1.13 | – | −0.71 | 1, 3≻4, 2 |
| 46 | 49 | 1.84 | 2.63 | 2.49 | 3.04 | 22.05** | −0.80 | −0.65 | −1.20 | – | – | −0.55 | 1≻3≻4; 1≻2 |
| 47 | 291 | 1.74 | 2.34 | 2.46 | 3.46 | 268.89** | −0.61 | −0.72 | −1.73 | – | −1.12 | −1.01 | 1≻2, 3≻4 |
| 48 | 83 | 1.82 | 2.41 | 2.60 | 3.17 | 46.27** | −0.59 | −0.78 | −1.35 | – | −0.76 | −0.57 | 1≻2, 3≻4 |
| 50 | 89 | 2.33 | 2.80 | 2.20 | 2.67 | 12.71** | −0.47 | – | – | 0.60 | – | −0.47 | 3, 1≻2; 3≻4 |
| 51 | 134 | 2.25 | 2.52 | 2.27 | 2.96 | 25.88** | – | – | −0.70 | – | −0.43 | −0.69 | 1, 3, 2≻4 |
| 52 | 60 | 1.78 | 2.47 | 2.42 | 3.33 | 43.78** | −0.68 | −0.63 | −1.55 | – | −0.87 | −0.92 | 1≻3, 2≻4 |
| 54 | 102 | 2.34 | 2.39 | 2.06 | 3.21 | 44.62** | – | 0.28 | −0.86 | 0.33 | −0.81 | −1.15 | 3≻1, 2≻4 |
| 55 | 137 | 1.95 | 2.50 | 2.20 | 3.36 | 92.97** | −0.55 | – | −1.41 | 0.30 | −0.86 | −1.16 | 1, 3≻2≻4 |

Remark: ** The test is significant at 0.01 level; * the test is significant at 0.1 level

† Only the mean rank differences of those pairwise multiple comparison procedures at $p < 0.05$ were shown

# In the last column, the code 1–4 represent each of the political goals: 1 = [JMEAN], 2 = [JINC], 3 = [JPRO], and 4 = [JSEC]

preferences on work values among Americans. Individuals with advantaged academic experiences, more income, larger household size, less children, and full-time work were more concerned with the intrinsic value of jobs. All these interaction effects have a stronger relationship to job attribute priority than gender.

Although the data utilized in this study covers a 30-year span, the cross-sectional nature of the data limits our work to an incomplete picture from individuals' transition in occupational values over time. Of course, the use of panel data will help to address the changing trajectories in a life course perspective. Another limitation of this study stems from the unavailability of time-series data. With only three time points, it is difficult to confirm whether our conclusion of social and work influences on job value preferences is short term or evidence of longer trends in society. Collecting such data will help to place the findings in a larger perspective and hence should be a goal of subsequent research.

# Chapter 11
# Extension of Distance-Based Models for Ranking Data

## 11.1 Weighted Distance-Based Models

Recall from Sect. 8.3 that distance-based models assume that the probability of observing a ranking $\boldsymbol{\pi}$ is inversely proportional to its distance from a modal ranking $\boldsymbol{\pi}_0$. The closer to the modal ranking $\boldsymbol{\pi}_0$, the more likely the ranking $\boldsymbol{\pi}$ is observed. There are different measures of distances between two rankings as previously noted. Recently, Lee and Yu (2012) proposed new distance-based models by using weighted distance measures to allow different weights for different ranks. In this way, the properties of distance can be retained while at the same time enhancing the model flexibility.

Motivated by Shieh (1998), we define the weighted Kendall distance by

$$d_K(\boldsymbol{\pi}, \boldsymbol{\sigma}; \mathbf{w}) = \sum_{i<j} w_{\pi_0(i)} w_{\pi_0(j)} I\{[\pi(i) - \pi(j)][\sigma(i) - \sigma(j)] < 0\}, \tag{11.1}$$

where the weights are functions related to the modal ranking. It is important to note that this weighted distance satisfies all the usual distance properties, in particular, the symmetric property: $d_K(\boldsymbol{\pi}, \boldsymbol{\sigma}; \mathbf{w}) = d_K(\boldsymbol{\sigma}, \boldsymbol{\pi}; \mathbf{w})$.

Other distance measures can be generalized in a similar manner as follows:
Weighted Spearman distance is

$$d_S(\boldsymbol{\pi}, \boldsymbol{\sigma}; \mathbf{w}) = \sum_{i=1}^{t} w_{\pi_0(i)} [\pi(i) - \sigma(i)]^2. \tag{11.2}$$

Square root of weighted Spearman distance is

$$d_{\sqrt{S}}(\boldsymbol{\pi}, \boldsymbol{\sigma}; \mathbf{w}) = \sqrt{d_S(\boldsymbol{\pi}, \boldsymbol{\sigma}; \mathbf{w})} = \sqrt{\sum_{i=1}^{t} w_{\pi_0(i)} [\pi(i) - \sigma(i)]^2}. \tag{11.3}$$



M. Alvo, P.L.H. Yu, *Statistical Methods for Ranking Data*, Frontiers in Probability and the Statistical Sciences, DOI 10.1007/978-1-4939-1471-5_11

Weighted Spearman Footrule is

$$d_F(\boldsymbol{\pi}, \boldsymbol{\sigma}; \mathbf{w}) = \sum_{i=1}^{t} w_{\pi_0(i)} |\pi(i) - \sigma(i)|. \tag{11.4}$$

See Tarsitano (2009) for other examples.

The probability of observing a ranking $\boldsymbol{\pi}$ under the weighted distance-based ranking model is

$$P(\boldsymbol{\pi}|\mathbf{w}, \boldsymbol{\pi}_0) = \frac{e^{-d(\pi,\pi_0;\mathbf{w})}}{C(\mathbf{w})}. \tag{11.5}$$

Generally speaking, a large value of $w_i$ increases our confidence in the ranking of the object ranked $i$ in $\boldsymbol{\pi}_0$. This is because such disagreement will greatly increase the distance and hence the probability of observing it will become very small. If $w_i$ is close to zero, a change in the rank of the object ranked $i$ in $\boldsymbol{\pi}_0$ will not affect the distance much. If $\mathbf{w} = \mathbf{0}$, this is just a uniform model or a random noise (denoted by model $N$ hereafter).

### *11.1.1 Properties of Weighted Distance-Based Models*

As defined in Sect. 8.5, some properties for ranking models are (1) label invariance, (2) reversibility, (3) $L$-decomposability, (4) strong unimodality, and (5) complete consensus.

It is natural to see that property (1) is essential for all statistical models for ranking data, and it is satisfied by the distances in Sect. 8.3. All the proposed weighted distance-based models satisfy property (2) and only the weighted Spearman and Footrule distances satisfy (3), as shown in the following theorem.

**Theorem 11.1.** *The weighted Spearman and Footrule distances are $L$-decomposable.*

*Proof.* Critchlow et al. (1991) showed that a ranking model is $L$-decomposable if there exist functions $f_r$, $r = 1, \ldots t$ such that

$$d(\boldsymbol{\pi}, \mathbf{e}) = \sum_{r=1}^{t} f_r[\pi^{-1}(r)].$$

Since

$$d_S(\boldsymbol{\pi}, \mathbf{e}) = \sum_{r=1}^{t} w_{\pi_0(\pi^{-1}(r))}[r - \pi^{-1}(r)]^2$$

**Table 11.1** Details of Footrule and weighted Footrule models for the "song" data set

| Ordering | Observed frequency | Expected frequency | |
|---|---|---|---|
| | | Footrule | Weighted Footrule |
| 3≻2≻1≻4≻5 | 19 | 28.478 | 26.592 |
| 3≻1≻2≻4≻5 | 10 | 6.459 | 11.384 |
| 1≻3≻2≻4≻5 | 9 | 1.465 | 5.204 |
| 3≻2≻4≻1≻5 | 8 | 6.459 | 8.203 |
| **1≻2≻3≻4≻5** | **7** | **1.465** | **5.557** |
| 2≻3≻1≻4≻5 | 6 | 6.459 | 5.204 |
| 3≻2≻1≻5≻4 | 6 | 6.459 | 2.001 |
| 3≻2≻4≻5≻1 | 5 | 1.465 | 2.001 |
| **2≻1≻3≻4≻5** | **4** | **1.465** | **2.379** |
| 3≻1≻4≻2≻5 | 3 | 1.465 | 1.503 |
| 2≻3≻4≻1≻5 | 2 | 1.465 | 1.605 |
| 3≻4≻2≻1≻5 | 2 | 1.465 | 1.083 |
| 3≻2≻5≻4≻1 | 2 | 1.465 | 1.583 |
| others | 0 | 16.963 | 8.701 |
| Total | 83 | | |
| Log likelihood | | −234.177 | −212.183 |

and

$$d_F(\boldsymbol{\pi}, \mathbf{e}) = \sum_{r=1}^{t} w_{\pi_0(\pi^{-1}(r))} |r - \pi^{-1}(r)|$$

the result follows. □

All weighted distance-based ranking models do not satisfy properties (4) and (5) unless all the weights are equal. However, rankings that violate properties (4) and (5) are commonly seen.

*Example 11.1.* Consider the "song" data set from Critchlow et al. (1991). Ninety-eight students were asked to rank 5 words, (1) score, (2) instrument, (3) solo, (4) benediction, and (5) suit, according to the association with the word "song." However, only 83 rankings are given in Critchlow et al. (1991) and we fit the Footrule model and the weighted Footrule model for comparison. The details are given in Table 11.1.

The modal ranking is $3 \succ 2 \succ 1 \succ 4 \succ 5$. Note that object 1 is less preferred than object 2 in the modal ranking. It is interesting to examine the ranking pair $(\mathbf{1} \succ \mathbf{2} \succ \mathbf{3} \succ \mathbf{4} \succ \mathbf{5}, \mathbf{2} \succ \mathbf{1} \succ \mathbf{3} \succ \mathbf{4} \succ \mathbf{5})$. By the strong unimodality property, we expect $P(\mathbf{1} \succ \mathbf{2} \succ \mathbf{3} \succ \mathbf{4} \succ \mathbf{5}) \leq P(\mathbf{2} \succ \mathbf{1} \succ \mathbf{3} \succ \mathbf{4} \succ \mathbf{5})$. The observed rankings however do not follow the property, and hence this data set cannot be fitted well using (unweighted) distance-based models. The weighted Footrule model gives a better fit as it is more flexible than its unweighted counterpart. Note that in this data set, there are other ranking pairs which cannot satisfy property (4) or (5). It will be seen that the extension using weighted distance provides a greater flexibility resulting in a better fit when the ranking data do not satisfy properties (4) and (5).

## 11.2 Mixtures of Weighted Distance-Based Models

Distance-based models assume a homogeneous population with a single modal ranking $\boldsymbol{\pi}_0$. In the case of heterogeneous data, one can adopt a mixture modeling framework to produce more sophisticated models. The EM algorithm can fit the mixture models in a quick and simple manner. Murphy and Martin (2003) extended the use of mixture models to distance-based models to describe the presence of heterogeneity among the judges. As a result, the limitation of the assumption of homogeneous population in distance-based models can be relaxed. Inspired by these results, Lee and Yu (2012) considered mixtures of weighted distance-based models for ranking data.

If a population contains $G$ subpopulations with probability mass function (pmf) $P_g(x)$ and the proportion of subpopulation $g$ equals $p_g$, the pmf of the mixture model is

$$P(x) = \sum_{g=1}^{G} p_g P_g(x). \tag{11.6}$$

Hence, the probability of observing a ranking $\boldsymbol{\pi}$ under a mixture of $G$-weighted distance-based ranking models is

$$P(\boldsymbol{\pi}) = \sum_{g=1}^{G} p_g P(\boldsymbol{\pi}|\mathbf{w}_g, \boldsymbol{\pi}_{0g}) = \sum_{g=1}^{G} p_g \frac{e^{-d(\boldsymbol{\pi},\boldsymbol{\pi}_{0g};\mathbf{w}_g)}}{C(\mathbf{w}_g)}, \tag{11.7}$$

and the log-likelihood for $n$ observations is

$$\ell = \sum_{k=1}^{n} \log\left(\sum_{g=1}^{G} p_g \frac{e^{-d(\boldsymbol{\pi}_k,\boldsymbol{\pi}_{0g};\mathbf{w}_g)}}{C(\mathbf{w}_g)}\right). \tag{11.8}$$

Estimating the model parameters can be done by applying the EM algorithm The E-step of an EM algorithm computes, for all observations, the probabilities of belonging to every subpopulation, and the M-step maximizes the conditional expected complete-data log-likelihood given the estimates generated in E-step.

To derive the EM algorithm, we define a latent variable $z_k = (z_{1k}, \ldots, z_{Gk})$ as: $z_{gk} = 1$ if observation $k$ belongs to subpopulation $g$, otherwise $z_{gk} = 0$. The complete-data log-likelihood is

$$L_{com} = \sum_{k=1}^{n} \sum_{g=1}^{G} z_{gk}[\log(p_g) - d(\boldsymbol{\pi}_k, \boldsymbol{\pi}_{0g}; \mathbf{w}_g) - \log(C(\mathbf{w}_g))]. \tag{11.9}$$

We first select the initial parameters for $\mathbf{w}_g$, $\boldsymbol{\pi}_{0g}$, and $p_g$. Then we alternatively run the E-step and M-step until the estimates converge.

In the E-step, $\hat{z}_{gk}$, $g = 1, 2, \ldots, G$ are updated for observations $k = 1, 2, \ldots, n$, by

$$\hat{z}_{gk} = \frac{\hat{p}_g P(\hat{\boldsymbol{\pi}}_k | \hat{\mathbf{w}}_g, \hat{\boldsymbol{\pi}}_{0g})}{\sum_{h=1}^{G} \hat{p}_h P(\hat{\boldsymbol{\pi}}_k | \hat{\mathbf{w}}_h, \hat{\boldsymbol{\pi}}_{0h})}. \tag{11.10}$$

In the M-step, model parameters are updated by maximizing complete-data log-likelihood with $z_{gk}$ replaced by $\hat{z}_{gk}$. The MLE of $\hat{\boldsymbol{\pi}}_{0g}$ and $\hat{\mathbf{w}}_g$ are obtained simultaneously. For a given $g = 1, \ldots, G$, $\hat{\boldsymbol{\pi}}_{0g}$ is obtained by an exhaustive search algorithm. Then $\hat{\mathbf{w}}_g$ satisfies the following equation (Murphy and Martin 2003, pp. 648, Equation (5)):

$$\frac{\sum_{k=1}^{n} \hat{z}_{gk} d(\boldsymbol{\pi}_k, \hat{\boldsymbol{\pi}}_{0g}; \hat{\mathbf{w}}_g)}{\sum_{k=1}^{n} \hat{z}_{gk}} = \sum_{j=1}^{t!} P(\boldsymbol{\sigma}_j | \hat{\mathbf{w}}_g, \hat{\boldsymbol{\pi}}_{0g}) d(\boldsymbol{\sigma}_j, \hat{\boldsymbol{\pi}}_{0g}; \hat{\mathbf{w}}_g)$$

is obtained, where $\boldsymbol{\sigma}_1, \cdots, \boldsymbol{\sigma}_{t!}$ are all possible rankings of the $t$ objects. Using the latest weights, $\hat{\boldsymbol{\pi}}_{0g}$ is recomputed. The model fitting procedure stops when $\hat{\boldsymbol{\pi}}_{0g}$ does not change anymore.

Based on our experience, we found that the parameter estimates are not sensitive to the initialization. Therefore, random numbers drawn from the uniform distribution on the interval (0, 1), $1/G$, and the ranking sorted according to the mean rank were used as initial values for $\mathbf{w}$, $p$, and $\boldsymbol{\pi}_0$, respectively. In our experience, it is found that the EM algorithm can converge within 20 iterations.

There will be two major difficulties in fitting weighted distance-based models when $t$ is large. First, the global search algorithm for the maximum likelihood estimate $\hat{\boldsymbol{\pi}}_0$ is not practical because the number of possible choices is too large. Instead, as suggested in Busse et al. (2007), a local search algorithm should be used. They suggested computing the sum of distances $\sum_{k=1}^{n} d(\boldsymbol{\pi}_k, \hat{\boldsymbol{\pi}}_0; \mathbf{w})$ for all $\hat{\boldsymbol{\pi}}_0 \in \Pi$, where $\Pi$ is a set containing all rankings having a Cayley distance of 0/1 to the "initial ranking." A reasonable choice of initial ranking can be constructed using mean rank.

Second, the numerical computation of the proportionality constant $C(\mathbf{w})$ is time-consuming. Lebanon and Lafferty (2002) proposed an MCMC algorithm for fitting (unweighted) distance-based models, and the simulation study in Klementiev et al. (2008) showed that the performance of this estimation technique is acceptable for $t$=10. Similar methods can be extended to the weighted distance-based models.

To determine the number of mixtures, we use the Bayesian information criterion (BIC). BIC equals

$$-2\ell + v\log(n), \tag{11.11}$$

where $\ell$ is the log-likelihood, $n$ is the sample size, and $v$ is the number of model parameters. The model with the smallest BIC is chosen to be the best model (Murphy and Martin 2003).

To assess the goodness of fit of the model, we use the sum of squares Pearson residuals ($\chi^2$) suggested by Marden (1995). It is given by

$$\chi^2 = \sum_i^{t!} r_i^2, \tag{11.12}$$

where

$$r_i = \frac{(O_i - E_i)}{\sqrt{E_i}} \tag{11.13}$$

is the Pearson residual and $O_i$ and $E_i$ are the observed and expected frequencies of ranking $i$, respectively.

However, if the size of some $E_i$ is smaller than 5, the computed chi-square statistic will be biased. We are likely to encounter this problem when the size of the data set is small and $t$ is large. In this case, we suggest using the truncated sum of squares Pearson residuals criterion described in Erosheva et al. (2007). For a specified level of truncation, the truncated sum of squares Pearson residuals is computed by summing up the residuals in which the expected value is greater than the specified level. When using the truncated sum of squares Pearson residuals, the effective number of observations, which is the number of observations included in the truncated sum of squares Pearson residuals, should also be reported.

Simulation studies were conducted to demonstrate the performance of the EM algorithm and also the effectiveness of using BIC in selecting the number of mixtures (Lee and Yu 2012). These studies showed that the algorithm can accurately estimate the model parameters and the BIC is appropriate in the model selection.

### *11.2.1 Analysis of Croon's Political Goals Data*

To illustrate the applicability of the weighted distance-based models, we make use of the ranking data set obtained from Croon (1989). It consists of 2,262 rankings of four political goals for the government collected from a survey conducted in Germany. The four goals were:

(A) Maintain order in nation.
(B) Give people more say in government decisions.
(C) Fight rising prices.
(D) Protect freedom of speech.

**Table 11.2** BIC of mixture models (unweighted distances)

| # Mixture | Distance | | | |
|---|---|---|---|---|
| | Kendall | $\sqrt{\text{Spearman}}$ | Spearman | Footrule |
| $N$ | 14377.52 | 14377.52 | 14377.52 | 14377.52 |
| 1 | 13052.58 | 13001.36 | 12988.06 | 13163.26 |
| $1+N$ | 13014.91 | 13009.09 | 12889.45 | 13162.11 |
| 2 | 12908.05 | 12848.57 | 12851.75 | 12980.63 |
| $2+N$ | 12860.70 | 12856.28 | 12758.02 | 12944.18 |
| 3 | 12846.88 | 12832.44 | 12754.64 | 12902.74 |
| $3+N$ | 12839.56 | **12733.08** | 12770.09 | 12932.52 |
| 4 | 12851.53 | 12847.89 | 12770.09 | 12918.19 |

**Table 11.3** BIC of mixture models (weighted distances)

| # Mixture | Weighted distance | | | |
|---|---|---|---|---|
| | Kendall | $\sqrt{\text{Spearman}}$ | Spearman | Footrule |
| $N$ | 14377.52 | 14377.52 | 14377.52 | 14377.52 |
| 1 | 12974.28 | 13011.22 | 12951.34 | 13174.30 |
| $1+N$ | 12943.44 | 13018.94 | 12863.46 | 13172.96 |
| 2 | 12797.52 | 12774.10 | 12864.90 | 12806.18 |
| $2+N$ | 12688.72 | 12713.96 | 12691.64 | 12697.92 |
| 3 | 12692.20 | 12678.88 | 12671.24 | **12670.82** |
| $3+N$ | 12678.06 | 12688.20 | 12673.36 | 12843.80 |
| 4 | 12730.74 | 12716.28 | 12709.86 | 12701.08 |

The respondents were classified into three value priority groups according to their top two choices. "Materialist" corresponds to individuals who gave priority to (A) and (C) regardless of the ordering, whereas those who chose (B) and (D) were classified as "post-materialist." The last category consisted of respondents giving all the other combinations of rankings and they were classified as holding "mixed" value orientations.

Weighted distance-based models were fitted for four types of weighted distances with mixing components $G = N, 1, \ldots, 3+N$, and 4, where $N$ represents the noise model, i.e., the one with $\mathbf{w} = \mathbf{0}$. The BIC values are listed in Tables 11.2 and 11.3. The underlined BIC values represent the best number of mixtures within each distance type. In each table, the BIC value in bold type represents the best model within that class of mixture model. Finally, we find that the best model is the weighted Footrule with $G = 3$ (Table 11.3). The BIC is 12670.82 which is better than the strict utility (SU) model (i.e., the Luce model in Sect. 8.1.1) (12670.87) and pendergrass-bradley (PB) model (i.e., the MBT model in Sect. 8.2) (12673.07) discussed in Croon (1989). It is undoubtedly better than the best (unweighted) distance-based model (12733.08, $\sqrt{\text{Spearman}}$, Table 11.2). For all types of distances, both unweighted and weighted, the lowest BIC appear when $G = 3$ or $3{+}N$.

**Table 11.4** Parameters of weighted Footrule mixture model and SU mixture model

| Group | Ordering of goals in $\pi_0$ | $p$ | $w_1$ | $w_2$ | $w_3$ | $w_4$ |
|---|---|---|---|---|---|---|
| 1 | $C \succ A \succ B \succ D$ | 0.352 | 2.030 | 1.234 | $\sim 0$ | 0.191 |
| 2 | $A \succ C \succ B \succ D$ | 0.441 | 1.348 | 0.917 | 0.107 | 0.104 |
| 3 | $B \succ D \succ C \succ A$ | 0.208 | 0.314 | $\sim 0$ | 0.151 | 0.552 |

| Group | $p$ | $A$ | $B$ | $C$ | $D$ |
|---|---|---|---|---|---|
| 1 | 0.449 | 0.590 | −1.071 | 1.730 | −1.249 |
| 2 | 0.326 | 1.990 | −0.920 | 0.060 | −1.130 |
| 3 | 0.225 | −0.691 | 0.630 | −0.010 | 0.071 |

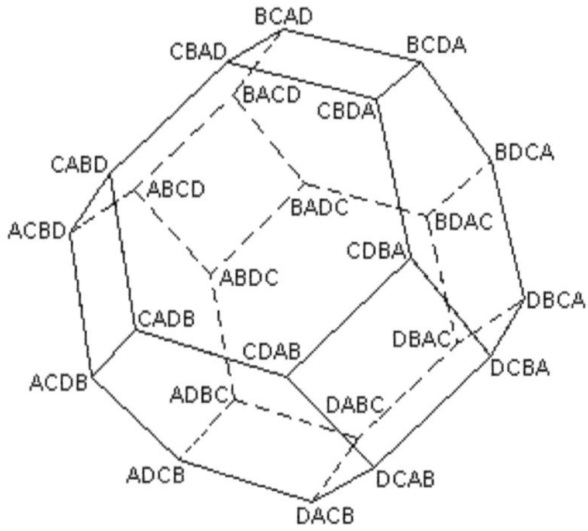


**Fig. 11.1** A truncated octahedron representing rankings of 4 objects

The parameter estimates of the best model, mixtures of three weighted Footrule models, are shown in Table 11.4. The parameter estimates of SU model are provided for comparison. For the SU model, an object with a larger parameter implies that this object is more preferred. The first two groups, which comprised 79 % of respondents, were materialists as they ranked (A) and (C) more important than the other two goals. The third group was post-materialistic as people in this group ranked (B) and (D) more important. Based on our grouping, we may conclude that Inglehart's theory is not appropriate in Germany (c.f. Sect. 10.1.4). We should at least distinguish the two types of materialists, one ranking (A) higher than (C) and the other ranking (C) higher than (A). This conclusion is similar to the findings in Croon (1989) and Moors and Vermunt (2007).

The mixture solution obtained here is slightly different from the SU mixture solution of Croon (1989). This can be evidenced by visualizing the data via a truncated octahedron (Thompson 1993a). An illustration of the truncated octahedron is shown in Fig. 11.1. The 24 rankings are placed on the vertices in a way that the

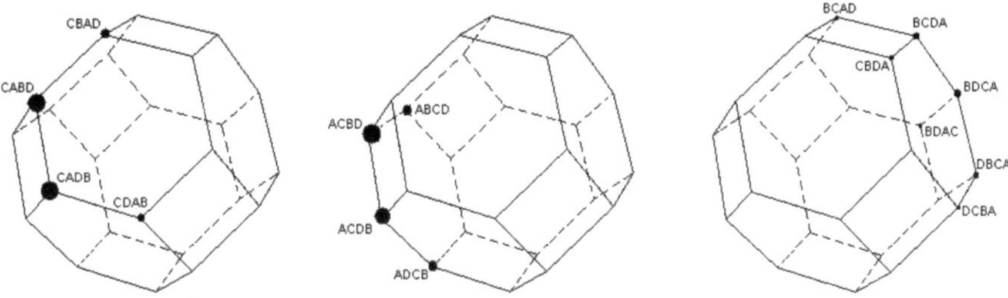


**Fig. 11.2** Representation of Footrule mixture model

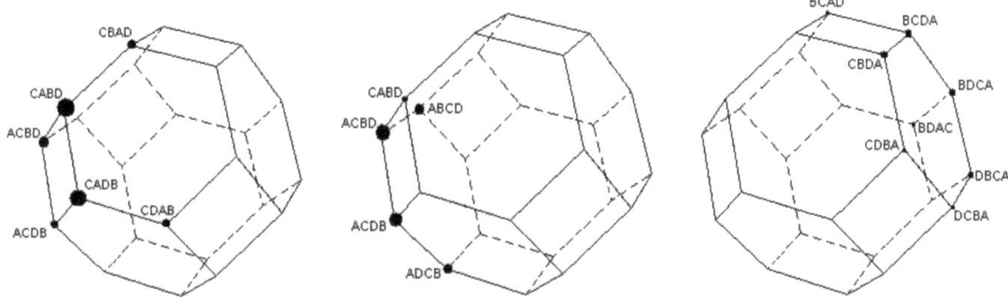


**Fig. 11.3** Representation of SU mixture model

edges represent a Kendall distance of 1. Among all six hexagon surfaces, there are four surfaces where all vertices have the same top choice. For example, the hexagon surface facing the readers represents the six rankings with top choice $C$.

Figures 11.2 and 11.3 show the predicted distributions of the Footrule and SU mixture models, respectively. The three truncated octahedrons represent mixtures having central rankings $C \succ A \succ B \succ D$, $A \succ C \succ B \succ D$, and $B \succ D \succ C \succ A$, respectively. Rankings with frequency greater than 5 % of the total mixture size are plotted. It can be seen that the three mixtures produced using the weighted Footrule distance are more separated since the difference between groups 1 and 2 is clearer than in the SU mixture model. For groups 1 and 2, weights $w_3$ and $w_4$ are close to zero while $w_1$ and $w_2$ are much larger, indicating that observations from groups 1 and 2 are mainly $C \succ A \succ ? \succ ?$ and $A \succ C \succ ? \succ ?$, respectively. As compared with that in groups 1 and 2, the weights in group 3 are close to zero, implying that people belonging to this group were less certain about their preferences than people in the other groups. The weight of object $A$ is the largest in group 3, meaning that $A$ has a relatively high probability to be ranked the last. This can be seen from Table 11.5 as well.

Although both models suggest three-mixture solution and their $\chi^2$ values are very close, the constituent of the three mixtures are quite different. Comparing Fig. 11.2 with Fig. 11.3, we see that the weighted Footrule mixtures are much more pure. A detailed frequency table is provided (Table 11.5). In the SU mixtures the estimated proportions of groups 1 and 2 are 0.449 and 0.326, respectively. Our model has a higher estimated proportion of group 2 (0.441). This difference is mainly due to

**Table 11.5** Observed and expected frequencies of weighted Footrule and SU mixture models

| | | Expected frequency for weighted Footrule | | | | Expected frequency for SU model | | | |
|---|---|---|---|---|---|---|---|---|---|
| Ordering | Obs. freq. | 1 | 2 | 3 | Total | 1 | 2 | 3 | Total |
| ABCD | 137 | 1.519 | 107.782 | 8.492 | 117.792 | 11.802 | 101.686 | 13.134 | 126.622 |
| ABDC | 29 | 0.165 | 38.832 | 7.300 | 46.296 | 0.600 | 30.954 | 14.246 | 45.800 |
| ACBD | 309 | 11.566 | 300.104 | 5.334 | 317.005 | 111.084 | 195.125 | 9.180 | 315.388 |
| ACDB | 255 | 9.556 | 242.962 | 3.898 | 256.417 | 92.907 | 158.243 | 5.246 | 256.396 |
| ADBC | 52 | 0.136 | 38.954 | 5.334 | 44.425 | 0.595 | 29.357 | 10.247 | 40.199 |
| ADCB | 93 | 1.037 | 87.535 | 4.535 | 93.106 | 9.778 | 78.211 | 5.399 | 93.389 |
| BACD | 48 | 5.218 | 25.143 | 20.176 | 50.538 | 9.571 | 20.837 | 20.187 | 50.595 |
| BADC | 23 | 0.566 | 9.059 | 17.344 | 26.969 | 0.487 | 6.343 | 21.896 | 28.726 |
| BCAD | 61 | 11.566 | 16.332 | 30.115 | 58.013 | 27.121 | 3.777 | 26.448 | 57.346 |
| BCDA | 55 | 2.782 | 3.822 | 52.288 | 58.892 | 4.310 | 0.167 | 56.638 | 61.115 |
| BDAC | 33 | 0.136 | 2.120 | 30.115 | 32.371 | 0.387 | 1.048 | 30.299 | 31.735 |
| BDCA | 59 | 0.302 | 1.377 | 60.825 | 62.504 | 1.211 | 0.152 | 59.820 | 61.183 |
| CABD | 330 | 302.565 | 31.155 | 7.962 | 341.682 | 286.092 | 43.282 | 10.478 | 339.853 |
| CADB | 294 | 249.985 | 25.223 | 5.819 | 281.026 | 239.280 | 35.102 | 5.988 | 280.369 |
| CBAD | 117 | 88.071 | 7.268 | 18.918 | 114.257 | 86.135 | 4.088 | 19.641 | 109.864 |
| CBDA | 69 | 21.181 | 1.701 | 32.848 | 55.729 | 13.688 | 0.181 | 42.061 | 55.929 |
| CDAB | 70 | 60.121 | 5.902 | 10.103 | 76.126 | 70.159 | 3.283 | 7.427 | 80.869 |
| CDBA | 34 | 17.500 | 1.706 | 24.004 | 43.210 | 13.330 | 0.179 | 27.833 | 41.342 |
| DABC | 21 | 0.387 | 9.116 | 9.262 | 18.765 | 0.479 | 5.964 | 11.988 | 18.431 |
| DACB | 30 | 2.943 | 20.484 | 7.874 | 31.301 | 7.872 | 15.890 | 6.316 | 30.078 |
| DBAC | 29 | 0.113 | 2.127 | 22.007 | 24.246 | 0.385 | 1.039 | 23.060 | 24.484 |
| DBCA | 52 | 0.249 | 1.381 | 44.449 | 46.080 | 1.202 | 0.151 | 45.529 | 46.882 |
| DCAB | 35 | 6.524 | 13.305 | 11.752 | 31.581 | 21.931 | 3.007 | 7.612 | 32.550 |
| DCBA | 27 | 1.899 | 3.846 | 27.923 | 33.668 | 4.167 | 0.164 | 28.525 | 32.856 |
| $\chi^2$ | | | | | 22.811 | | | | 23.073 |

the difference in assigning rankings $A \succ C \succ B \succ D$ and $A \succ C \succ D \succ B$ to group 2. At first glance, these two rankings should be assigned to group 2. Referring to Table 11.5, our mixture model assigns approximately 96 % of these two rankings to group 2, while for the SU mixture model the percentage drops to 63 %. The grouping of weighted Footrule mixture model appears in fact more reasonable.

## 11.3 Distance-Based Tree Models

For distance-based models, the inability to incorporate covariates still remains a major inadequacy. Lee and Yu (2010) provide a plausible solution by combining a decision tree and a distance-based model so as to develop a more flexible distance-based model which can allow the presence of covariates, hence addressing the problem of homogeneous population. Our proposed methodology for constructing distance-based tree models will be explained below.

### *11.3.1 Building Distance-Based Tree Models*

Suppose we have $n$ observations $\{(\pi_k, X_k), k = 1, \ldots, n\}$, where $X_k$ is a collection of $m$-dimensional covariates. Here, the $X_k$'s can be categorical, interval, or ordinal variables. In the following, we will illustrate how we can construct a distance-based tree model using these observations, with the aim of building a model for predicting ranking $\pi$ based on $X$.

The tree-growing stage is similar to the one used in Chap. 10 except for the choice of splitting criterion. In selecting the splitting rule for each internal node, Lee and Yu (2010) chose the one that minimizes the weighted sum of the mean deviance of two child nodes formed, i.e., $n_L D_L + n_R D_R$. For the weighted distance-based model, the deviance of a particular node $\tau$ with size $n_\tau$ is

$$D_\tau = \frac{2}{n_\tau} \sum_{k=1}^{n_\tau} d(\pi_k, \hat{\pi}_0; \hat{\mathbf{w}}) + 2\log C(\hat{\mathbf{w}}) \tag{11.14}$$

which equals $-\frac{2}{n_\tau} \times$ log-likelihood (see (11.5)). An exhaustive search algorithm is used to determine the best splitting rule that gives the smallest weighted sum of mean deviance of the two child nodes.

Lee and Yu (2010) commented that the above mean deviance criterion often results in a split which produces two unbalanced nodes, where one of them is small but pure. To avoid this problem, they suggested to stop splitting when the size of a child node is smaller than one-tenth of the size of its parent node. Besides, in order to avoid overfitting, a node with sample size smaller than one-tenth of the total sample will not be further split and will automatically become a leaf node.

**Table 11.6** Subject covariates for 1999 EVP data

| Covariate | Description/code | Type |
|---|---|---|
| Country | Group 1 = 1, group 2 = 2, group 3 = 3, group 4 = 4 | Nominal |
| Gender | Male = 1, female = 2 | Binary |
| Year of birth | Value ranges from 1909–1981 | Interval |
| Marital status | Married = 1, windowed = 2, divorced = 3, separated = 4, never married = 5 | Nominal |
| Employment status | Ordinal value ranges from 1 to 8 | Ordinal |
| Household income | Ordinal value ranges from 1 to 10 | Ordinal |
| Age of education completion | Interval value ranges from 7 to 50 | Interval |

Note that in the growing stage, we tend to build an overly large tree, hoping not to miss any important features of the tree. As a result, many redundant nodes will be created and interpretation will become difficult. Pruning is necessary to remove these redundant nodes. The pruning procedure of our model makes use of the minimal cost-complexity measure with ten-fold cross-validation to obtain the final tree, where the cost function used is the total deviance, i.e., the sum of deviances (11.14) for all leaf nodes. See Lee and Yu (2010) for details of the pruning procedure.

### *11.3.2 Analysis of 1999 European Value Priority Data*

Consider the ranking data set obtained from the European Values Studies which is a continuing, annual program of cross-national collaboration on surveys covering topics important for social science research. Here, we will examine the survey which was conducted in 1999 in 32 countries in Europe (Vermunt 2004).

The respondents' covariates used in tree building are summarized in Table 11.6. Countries were categorized into four groups as suggested by Vermunt (2004). Table 11.7 shows the four groups of countries. After removing records containing missing in any one of the covariates, we end up with a ranking data of 1,911 respondents. We use 75 % of the data for model building (1,433 observations) and the remaining data (478 observations) for testing model performance.

The survey mainly focused on value orientations, attitudes, beliefs, and knowledge concerning nature and environmental issues and included the so-called Inglehart Index (Inglehart 1977), a collection of four indicators of materialism/post-materialism as well. Respondents were asked to pick the most important and the

**Table 11.7** The four groups of countries

| Group 1 | Group 2 | Group 3 | Group 4 |
|---|---|---|---|
| Italy | Croatia | Luxembourg | Lithuania |
| Sweden | Belgium | Slovenia | Latvia |
| Denmark | Greece | Czechnia | Poland |
| Austria | France | Iceland | Estonia |
| Netherlands | Spain | Finland | Belarus |
| | Northern Ireland | West Germany | Slovakia |
| | Ireland | Portugal | Hungary |
| | | Romania | Ukraine |
| | | Malta | Russia |
| | | East Germany | |
| | | Bulgaria | |

second most important political goals for their government from the following four alternatives (or objects):

(A) Maintain order in nation.
(B) Give people more say in government decisions.
(C) Fight rising prices.
(D) Protect freedom of speech.

In Sect. 11.2, we have studied similar political goals data for Germany (Croon data) using the mixture of weighted distance-based models. However, the Croon data set contains complete rankings with no covariates whereas the 1999 EVP data set contains top 2 rankings with seven covariates. In Sect. 10.1.4, we have studied 1993 EVP data for five countries only, but here the 1999 EVP data set contains data on 32 European countries.

Note that only top two objects are ranked in the 1999 EVP data. We need to modify the likelihood in order to fit the distance-based models to this data set. Since we have no preference information about the non-ranked objects in a top 2 ranking, it is natural to assume that all possible rankings that are compatible with the top 2 ranking are equally likely to be observed. Therefore, the probability of observing a top 2 ranking $\boldsymbol{\pi}^*$ is the sum of the probabilities of all possible complete ranking $\boldsymbol{\pi}$ that are compatible with $\boldsymbol{\pi}^*$.

Four classes of distance-based models are considered. Table 11.8 shows the comparison of different distance-based models and their corresponding tree models and their weighted distance versions. In terms of cross-validation deviance, tree-based extensions of distance-based models with different types of distances are all better than the original models. Similarly, weighted distance-based tree models perform better than $\phi$-component tree models. Obviously, the best among them is the weighted Kendall tree model, as the deviance on testing data (2.2562) is the smallest.

Figure 11.4 displays the weighted Kendall tree. Table 11.9 provides the parameter estimates in the leaf nodes of weighted Kendall tree model. The corresponding

**Table 11.8** Log-likelihood, AIC, and deviance for different models

| Model | Size | Log L | AIC | DEV-CV | DEV-testing |
|---|---|---|---|---|---|
| *(Unweighted) distance-based models* | | | | | |
| Kendall | 1 | −3387 | 6776 | 2.4913 | 2.8536 |
| $\phi$-Component | 1 | −3296 | 6596 | 2.3015 | 2.5231 |
| $\sqrt{\text{Spearman}}$ | 1 | −3365 | 6732 | 2.4920 | 2.8532 |
| Spearman | 1 | −3430 | 6862 | 2.4634 | 2.6810 |
| Footrule | 1 | −3407 | 6816 | 2.4920 | 2.5281 |
| *Weighted distance-based models* | | | | | |
| Kendall | 1 | −3252 | 6512 | 2.2722 | 2.4705 |
| $\sqrt{\text{Spearman}}$ | 1 | −3495 | 6994 | 2.8402 | 2.9930 |
| Spearman | 1 | −3732 | 7468 | 2.6001 | 2.6362 |
| Footrule | 1 | −3762 | 7528 | 2.6267 | 2.6663 |
| *(Unweighted) Distance-based tree models* | | | | | |
| Kendall | 5 | −3460 | 6938 | 2.4578 | 2.3755 |
| $\phi$-Component | 3 | −3271 | 6558 | 2.2854 | 2.3871 |
| $\sqrt{\text{Spearman}}$ | 4 | −3451 | 6916 | 2.4599 | 2.6167 |
| Spearman | 12 | −3545 | 7136 | 2.4389 | 2.4521 |
| Footrule | 5 | −3467 | 6952 | 2.4452 | 2.3906 |
| *Weighted distance-based tree models* | | | | | |
| Kendall | 5 | −3074 | 6196 | 2.1766 | **2.2562** |
| $\sqrt{\text{Spearman}}$ | 2 | −3494 | 7006 | 2.3761 | 2.5377 |
| Spearman | 4 | −3454 | 6946 | 2.4033 | 2.5332 |
| Footrule | 4 | −3495 | 7028 | 2.4222 | 2.5041 |

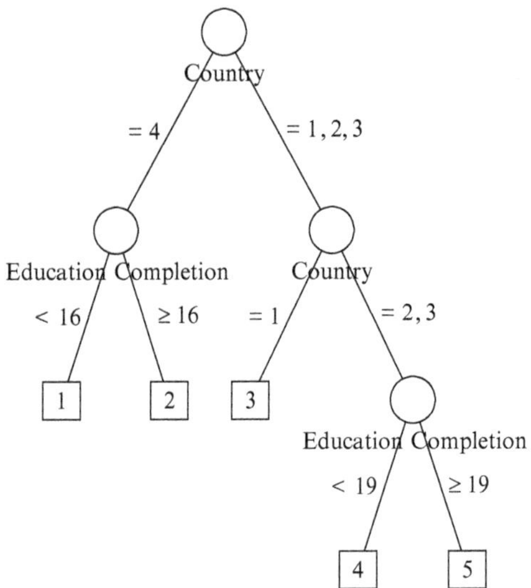


**Fig. 11.4** The fitted weighted Kendall tree model

**Table 11.9** Parameter estimates of the fitted weighted Kendall model and weighted Kendall tree model

| Model | Node | Size | Ordering of goals in $\pi_0$ | $w_1$ | $w_2$ | $w_3$ | $w_4$ |
|---|---|---|---|---|---|---|---|
| Weighted Kendall | | 1433 | $A \succ B \succ C \succ D$ | 0.94 | 0.36 | 0.50 | 0.67 |
| Weighted Kendall tree | 1 | 114 | $A \succ C \succ B \succ D$ | 0.69 | 1.24 | 0.10 | 2.24 |
| | 2 | 487 | $A \succ B \succ C \succ D$ | 0.79 | 0.24 | 0.84 | 1.58 |
| | 3 | 177 | $A \succ C \succ D \succ B$ | 0.64 | 0.55 | 0.26 | 1.40 |
| | 4 | 383 | $A \succ B \succ C \succ D$ | 1.33 | 0.20 | 0.52 | 0.65 |
| | 5 | 272 | $A \succ B \succ D \succ C$ | 0.59 | 1.07 | 0.27 | 1.00 |

weighted Kendall model is shown for comparison. A model fitness was carried out, the results of which appear in Table 11.10. It is evident that there is good agreement between observed and expected frequencies and that hence the weighted Kendall tree model provides a good fit to the data.

In general, the two covariates, country and education level, were important predictors of how Inglehart's objects are ranked. Country was a more important predictor than age of education completion. People in group 1 countries and those people in group 4 countries (mainly former USSR countries) who ended their education before the age of 16 tended to prefer (A) and (C), and they could be classified as materialist type. However, people in group 4 countries believed that the protection of freedom of speech (D) was least important, but people in group 1 countries considered that giving people more say in government decisions (B) was least important. Finally, people in other country groups tended to prefer (A) and (B), and they were of mixed type.

This case study clearly points out that tree-based extensions of distance-based models successfully improve the model fitness, thereby widening their applicability. In particular, our weighted distance-based tree models outperform the unweighted distance-based ranking models.

## Chapter Notes

Distance-based tree model is a kind of model-based decision tree where a statistical model is built in each leaf node of the tree. Models other than distance-based models could also be considered. Lee and Yu (2013) have developed rank-ordered logit (ROL) tree model. Unlike distance-based tree model, this model can handle both linear and nonlinear effects via the ROL regression and the tree, respectively.

The scope of analyzing ranking data has been expanded in the recent decade and new areas of ranking research are established. Examples include the recommendation system (Sun and Lebanon 2012), rank aggregation (Hall and Schimek 2012), institutional ranking (Hall and Miller 2010), gene ranking (Alvo et al. 2010), etc.

**Table 11.10** Sum of squares Pearson residuals of root node and final nodes for different models, evaluated using testing data

| – | *AB* | *AC* | *AD* | *BA* | *BC* | *BD* | *CA* | *CB* | *CD* | *DA* | *DB* | *DC* | $\chi^2$ |
|---|---|---|---|---|---|---|---|---|---|---|---|---|---|
| Observed | 76 | 97 | 41 | 57 | 38 | 34 | 51 | 23 | 11 | 23 | 22 | 5 | |
| Expected ($\phi$-component) | 85.53 | 79.39 | 73.72 | 40.16 | 43.26 | 40.16 | 21.87 | 25.37 | 21.86 | 11.91 | 13.82 | 12.83 | |
| Residual | −1.03 | 1.98 | −3.81 | 2.66 | −0.80 | −0.97 | 6.23 | −0.47 | −2.32 | 3.21 | 2.20 | −2.19 | 92.53 |
| Expected (Kendall tree) | 83.29 | 48.96 | 53.83 | 56.00 | 30.42 | 31.57 | 33.15 | 21.33 | 16.43 | 36.69 | 24.06 | 18.01 | |
| Residual | −0.80 | 6.87 | −1.75 | 0.13 | 1.38 | 0.43 | 3.10 | 0.36 | −1.33 | −2.26 | −0.42 | −3.07 | 79.16 |
| Expected (weighted kendall) | 94.97 | 80.99 | 52.37 | 63.64 | 30.78 | 18.76 | 46.60 | 24.85 | 15.66 | 24.84 | 14.05 | 10.51 | |
| Residual | −1.95 | 1.78 | −1.57 | −0.83 | 1.30 | 3.52 | 0.64 | −0.37 | −1.18 | −0.37 | 2.12 | −1.70 | 33.65 |
| Expected (weighted kendall tree) | 88.99 | 90.52 | 44.44 | 65.03 | 32.54 | 19.86 | 44.71 | 23.72 | 14.09 | 27.49 | 17.66 | 8.29 | |
| Residual | −1.37 | 0.68 | −0.52 | −1.00 | 0.96 | 3.17 | 0.94 | −0.14 | −0.82 | −0.86 | 1.03 | −1.14 | 19.29 |

# Appendix A: Ranking Data Sets

## A.1 Goldberg Data

In the Goldberg (1976) data, 143 graduates were asked to rank ten occupations according to the degree of social prestige. These ten occupations are (i) faculty member in an academic institution (Fac), (ii) mechanical engineer (ME), (iii) operation researcher (OR), (iv) technician (Tech), (v) section supervisor in a factory (Sup), (vi) owner of a company employing more than 100 workers (Own), (vii) factory foreman (For), (viii) industrial engineer (IE), (ix) manager of a production department employing more than 100 workers (Mgr), and (x) applied scientist (Sci). The data are given in Cohen and Mallows (1980) and have been analyzed by many researchers. Fligner and Verducci (1988) and Marden (1992) summarized the findings of these analyses.

Feigin and Cohen (1978) analyzed the Goldberg data and found three outliers due to the fact that the corresponding graduates wrongly presented rankings in reverse order. After reversing these three rankings, the average ranks received by the ten occupations are 8.57, 4.90, 6.29, 1.90, 4.34, 8.13, 1.47, 6.27, 5.29, 7.85, with the convention that higher rank means more prestige. Then the preference of graduates is in the order Fac > Own > Sci > OR > IE > Mgr > ME > Sup > Tech > For.



M. Alvo, P.L.H. Yu, *Statistical Methods for Ranking Data*, Frontiers in Probability and the Statistical Sciences, DOI 10.1007/978-1-4939-1471-5

## A.2 Leisure Time Data

**Table A.1** Sutton's leisure time data

| Ranks on | | | Number of white females | Number of black females |
|---|---|---|---|---|
| A: males | B: females | C: both sexes | | |
| 1 | 2 | 3 | 0 | 1 |
| 1 | 3 | 2 | 0 | 1 |
| 2 | 1 | 3 | 1 | 0 |
| 2 | 3 | 1 | 0 | 5 |
| 3 | 1 | 2 | 7 | 0 |
| 3 | 2 | 1 | 6 | 6 |

The data appear in Hollander and Sethuraman (1978).

## A.3 Data on the Big Four EPL Teams

**Table A.2** Data on ranking of Big Four EPL teams

| | Ranks on | | | |
|---|---|---|---|---|
| Season ended in | Arsenal | Chelsea | Liverpool | Manchester United |
| 1993 | 3 | 4 | 2 | 1 |
| 1994 | 2 | 4 | 3 | 1 |
| 1995 | 4 | 3 | 2 | 1 |
| 1996 | 3 | 4 | 2 | 1 |
| 1997 | 2 | 4 | 3 | 1 |
| 1998 | 1 | 4 | 3 | 2 |
| 1999 | 2 | 3 | 4 | 1 |
| 2000 | 2 | 4 | 3 | 1 |
| 2001 | 2 | 4 | 3 | 1 |
| 2002 | 1 | 4 | 2 | 3 |
| 2003 | 2 | 3 | 4 | 1 |
| 2004 | 1 | 2 | 4 | 3 |
| 2005 | 2 | 1 | 4 | 3 |
| 2006 | 4 | 1 | 3 | 2 |
| 2007 | 4 | 2 | 3 | 1 |
| 2008 | 3 | 2 | 4 | 1 |
| 2009 | 4 | 3 | 2 | 1 |
| 2010 | 3 | 1 | 4 | 2 |
| 2011 | 3 | 2 | 4 | 1 |
| 2012 | 2 | 3 | 4 | 1 |
| 2013 | 3 | 2 | 4 | 1 |

The data are obtained from Wikipedia.

## A.4 MovieLens Data

MovieLens data set (http:\www.grouplens.org) consists of one million online ratings of 4,000 movies by 6,000 raters on a scale of 1–5 (Resnick et al. 1994). The one that we studied in this book is a subset of the original data. By ordering the ratings of movies given by each rater, it ends up with a ranking data containing 72,979 possibly incomplete and tied rankings of 55 movies made by 5,625 raters.

## A.5 Language and Arithmetic Scores Data

**Table A.3** Language and arithmetic scores

| Student | 1 | 2 | 3 | 4 | 5 | 6 | 7 | 8 | 9 |
|---|---|---|---|---|---|---|---|---|---|
| Language | 50 | 23 | 28 | 34 | 14 | 54 | 46 | 52 | 53 |
| Arithmetic | 38 | 28 | 14 | 26 | 18 | 40 | 23 | 30 | 27 |
| Language ranks | 6 | 2 | 3 | 4 | 1 | 9 | 5 | 7 | 8 |
| Arithmetic ranks | 8 | 6 | 1 | 4 | 2 | 9 | 3 | 7 | 5 |

The data appear in Lehmann (1975).

**Table A.4** Incomplete language and arithmetic scores

| Student | 3 | 5 | 7 | 4 | 9 | 6 | 6 | 1 | 9 |
|---|---|---|---|---|---|---|---|---|---|
| Arithmetic (2) | 14 | 18 | 23 | 26 | 27 | 30 | 40 | – | – |
| Language (1) | 28 | 14 | 46 | – | 53 | – | 54 | 50 | – |
| Arithmetic ranks | 1 | 2 | 3 | 4 | 5 | 6 | 7 | – | – |
| Language ranks | 2 | 1 | 3 | – | 5 | – | 6 | 4 | – |

The data appear in Lehmann (1975).

## A.6 Public Opinion Survey Data

**Table A.5** Data from the public opinion survey

| Education level | Response | | | | | | |
|---|---|---|---|---|---|---|---|
| | 1 | 2 | 3 | 4 | 5 | Missing | Subtotal |
| Primary or below | 2 | 35 | 23 | 7 | 3 | 33 | 103 |
| Secondary | 2 | 72 | 129 | 37 | 6 | 53 | 299 |
| Matriculated | 0 | 9 | 9 | 7 | 0 | 3 | 28 |
| Tertiary, nondegree | 1 | 9 | 6 | 6 | 0 | 5 | 27 |
| Tertiary, degree | 0 | 22 | 28 | 7 | 6 | 6 | 69 |
| Missing | 0 | 2 | 3 | 0 | 0 | 1 | 6 |
| Subtotal | 5 | 149 | 198 | 64 | 15 | 101 | 532 |

## A.7 Wind Direction Data

Wind directions were recorded at 6 a.m. and at 12 noon on each day at a weather station for 21 consecutive days.

**Table A.6** Wind direction in degree

| | | | | | | | | | | | |
|---|---|---|---|---|---|---|---|---|---|---|---|
| 6 a.m. | 356 | 97 | 211 | 262 | 343 | 292 | 157 | 302 | 324 | 85 | 324 |
| Noon | 119 | 162 | 221 | 259 | 270 | 29 | 97 | 292 | 40 | 313 | 94 |
| 6 a.m. | 85 | 324 | 340 | 157 | 238 | 254 | 146 | 232 | 122 | 329 | |
| Noon | 45 | 47 | 108 | 221 | 248 | 270 | 45 | 23 | 270 | 119 | |

The data appear in Johnson and Wehrly (1977).

In another example, wind direction and ozone concentration were measured.

**Table A.7** Wind direction and ozone concentration

| | | | | | | | | | | |
|---|---|---|---|---|---|---|---|---|---|---|
| Wind direction | 327 | 91 | 88 | 305 | 344 | 270 | 67 | 21 | 281 | |
| Ozone concentration | 28.0 | 85.2 | 80.5 | 4.7 | 45.9 | 12.7 | 72.5 | 56.6 | 31.5 | |
| Wind direction | 8 | 204 | 86 | 333 | 18 | 57 | 6 | 11 | 27 | 84 |
| Ozone concentration | 112.0 | 20.0 | 72.5 | 16.0 | 45.9 | 32.6 | 56.6 | 52.6 | 91.8 | 55.2 |

The data appear in Johnson and Wehrly (1977).

## A.8 Lymph Heart Pressure Data

**Table A.8** Lymph heart pressure in mm Hg taken over a 24 h period at 6 h intervals on eight toads during dehydration

| | | Time | | | |
|---|---|---|---|---|---|
| Toad ID | Block | 6 h | 12 h | 18 h | 24 h |
| 21 | 1 | 11.865 | 9.832 | 7.567 | 10.168 |
| 22 | 2 | 5.601 | 4.892 | 4.032 | 3.126 |
| 23 | 3 | | 14.415 | 14.185 | 7.800 |
| 24 | 4 | 13.267 | | | 9.953 |
| 25 | 5 | 8.006 | 7.793 | | 7.582 |
| 27 | 6 | 17.692 | 16.644 | 15.327 | 11.573 |
| 28 | 7 | 9.027 | 7.973 | 11.855 | 6.820 |
| 29 | 8 | 9.789 | 7.967 | 7.758 | 7.849 |

The data appear in Alvo and Cabilio (1995b).

## A.9 Mortality Data for South Africa 2000–2008

**Table A.9** Mortality statistics for South Africa 2000–2008

| Year | Number of deaths | Population size |
|---|---|---|
| 2000 | 416,155 | 43,789,115 |
| 2001 | 454,882 | 43,997,828 |
| 2002 | 502,050 | 44,187,637 |
| 2003 | 556,779 | 44,344,136 |
| 2004 | 576,709 | 42,718,530 |
| 2005 | 598,131 | 42,768,678 |
| 2006 | 612,778 | 43,647,658 |
| 2007 | 603,094 | 43,586,097 |
| 2008 | 592,073 | 43,421,021 |

The data appear in Alvo and Berthelot (2012).

## A.10 *E. coli* Data for Six Beaches in Hong Kong

**Table A.10** Annual geometric mean *E. coli* level (per 100 ml) in the Sai Kung District. Beaches: Clear Water Bay First (1), Clear Water Bay Second (2), Hap Mun Bay (3), Kiu Tsui (4), Silverstrand (5), Trio (6), and number of good beaches (7)

| Year | (1) | (2) | (3) | (4) | (5) | (6) | (7) |
|---|---|---|---|---|---|---|---|
| 1986 | 102 | 69 | 9 | 18 | 255 | 49 | 2 |
| 1987 | 133 | 52 | 6 | 9 | 62 | 32 | 2 |
| 1988 | 39 | 35 | 4 | 3 | 129 | 35 | 2 |
| 1989 | 80 | 38 | 3 | 5 | 192 | 23 | 3 |
| 1990 | 51 | 42 | 4 | 5 | 89 | 31 | 2 |
| 1991 | 30 | 14 | 2 | 4 | 106 | 14 | 4 |
| 1992 | 52 | 42 | 2 | 5 | 94 | 32 | 2 |
| 1993 | 31 | 16 | 3 | 4 | 56 | 20 | 4 |
| 1994 | 30 | 35 | 3 | 3 | 72 | 14 | 3 |
| 1995 | 55 | 39 | 6 | 3 | 226 | 16 | 3 |
| 1996 | 34 | 43 | 5 | 5 | 126 | 29 | 2 |
| 1997 | 62 | 66 | 3 | 5 | 148 | 30 | 2 |
| 1998 | 41 | 44 | 2 | 4 | 99 | 21 | 3 |
| 1999 | 11 | 12 | 2 | 4 | 32 | 17 | 5 |
| 2000 | 16 | 26 | 2 | 5 | 61 | 10 | 4 |
| 2001 | 28 | 22 | 1 | 5 | 100 | 12 | 4 |
| 2002 | 28 | 14 | 2 | 4 | 133 | 6 | 4 |
| 2003 | 17 | 21 | 4 | 5 | 97 | 10 | 5 |
| 2004 | 9 | 10 | 3 | 17 | 74 | 2 | 5 |
| 2005 | 16 | 19 | 4 | 14 | 67 | 6 | 5 |
| 2006 | 20 | 13 | 4 | 11 | 30 | 5 | 5 |
| 2007 | 14 | 9 | 3 | 6 | 33 | 2 | 5 |
| 2008 | 11 | 19 | 5 | 12 | 35 | 12 | 5 |
| 2009 | 15 | 27 | 3 | 19 | 31 | 5 | 4 |

The data appear in Alvo and Berthelot (2012).

## A.11 Umbrella Alternative Data

Example discussed (Alvo 2008) on the Wechsler Adult intelligence scale scores on males by age groups

| Age group | | | | |
|---|---|---|---|---|
| 16–19 | 20–34 | 35–54 | 55–69 | >70 |
| 8.62 | 9.85 | 9.98 | 9.12 | 4.80 |
| 9.94 | 10.43 | 10.69 | 9.89 | 9.18 |
| 10.06 | 11.31 | 11.40 | 10.57 | 9.27 |

## A.12 APA Election Data

In 1980, the American Psychological Association (APA) conducted an election in which five candidates ($A, B, C, D$, and $E$) were running for president and voters were asked to rank order all of the candidates. Candidates $A$ and $B$ are research psychologists, $C$ is a community psychologist, and $D$ and $E$ are clinical psychologists. Among those voters, 5,738 gave complete rankings. These complete rankings are considered here (Diaconis 1988). Note that lower rank implies more favorable. Then the average ranks received by candidates $A, B, C, D$, and $E$ are 2.84, 3.16, 2.92, 3.09, and 2.99, respectively. This means that voters generally prefer candidate $A$ the most, candidate $C$ the second, etc.

## A.13 Job Selection Data

In 1997, a mainland marketing research firm conducted a survey on people's attitude toward career and living style in three major cities in Mainland China—Beijing, Shanghai, and Guangzhou. Five hundred responses from each city were obtained. A question regarding the behavior, conditions, and criteria for job selection of the 500 respondents in Guangzhou will be discussed here. In the survey, respondents were asked to rank the three most important criteria on choosing a job among 13 criteria:

(1) Favorable company reputation
(2) Large company scale
(3) More promotion opportunities
(4) More training opportunities
(5) Comfortable working environment
(6) High income
(7) Stable working hours

(8) Fringe benefits
(9) Well matched with employees' profession or talent
(10) Short distance between working place and home
(11) Challenging
(12) Corporate structure of the company
(13) Low working pressure

## A.14 1993 European Value Priority Data

The ranking data set was obtained from the International Social Service Programme (ISSP) in 1993 (Jowell et al. 1993), which is a continuing, annual program of cross-national collaboration on surveys covering a wide spectrum of topics for social science research. The survey was conducted using standardized questionnaire in 1993 at 20 countries around the world, such as Great Britain, Australia, the USA, Bulgaria, the Philippines, Israel, and Spain. It mainly focused on value orientations, attitudes, beliefs, and knowledge concerning nature and environmental issues and included the so-called Inglehart Index, a collection of four indicators of materialism/post-materialism as well. Respondents were asked to pick the most important (rank "1") and the second most important (rank "2") goals for their government from the following four alternatives:

1. Maintain order in nation (ORDER).
2. Give people more to say in Government decisions (SAY).
3. Fight rising prices (PRICES).
4. Protect freedom of speech (SPEECH).

After removing those invalid responses, the survey gave a ranked data set of 5,737 observations with top choice and top two rankings. In addition, the data provide some judge-specific characteristics and they are applied in tree partitioning. The candidate splitting variables are summarized in Table A.11.

Respondents can be classified into value priority groups on the basis of their top two choices among the four goals. "Materialist" corresponds to an individual who

**Table A.11** Description of European ranking data of political values

| Covariate | Description/code | Type |
|---|---|---|
| Country | West Germany=1, East Germany=2, Great Britain=3, Italy=4, Poland=5 | Nominal |
| Gender | Male=1, female=2 | Binary |
| Education | 0–10 years=1, 11–13 years=2, 14 or more years=3 | Ordinal |
| Age | Value ranges from 15 to 91 | Interval |
| Religion | Catholic and Greek Catholic=1, Protestant=2, others=3, none=4 | Nominal |

gives priority to ORDER and PRICES regardless of the ordering, whereas those who choose SAY and SPEECH will be termed "post-materialist." The last category comprises of judges giving all the other combinations of rankings, and they will be classified as holding "mixed" value orientations.

## A.15 US General Social Survey Data

The General Social Survey (GSS) has been conducted in the USA by the National Opinion Research Center (NORC) of the University of Chicago annually since 1972 (except for the years 1979 and 1981) and biennially since 1994 (Davis and Smith 2009). Each year the GSS consisted of a 90-min in-person interview with a full-probability sample of English- or Spanish-speaking persons aged 18 years or above who lived in households. It is a multidimensional social survey that gathers sociodemographic characteristics and replicated core measurements on social and political attitudes and behaviors, plus topics of special interest. Many of the core questions have remained unchanged since 1972 to facilitate time-trend studies as well as replication of earlier findings. In relation to job value, the characteristics were measured by the GSS in an ipsative approach. The respondents were asked to rank in order of preference from (i) "most preferred," (ii) "second most important," to (v) "fifth most important" five aspects about a job:

1. High income (JINC)
2. No danger of being fired (JSEC)
3. Working hours are short, lots of free time (JHOUR)
4. Chances for advancement (JPRO)
5. Work important and gives a feeling of accomplishment (JMEAN)

Job values, as defined by Kalleberg (1977), are what individuals hold as desirable with respect to their work activity and the attitudes are central to the social psychology of work. Under many other names (including work/occupational value/attribute/characteristic), they refer to the importance people place on occupational rewards and play a key role in conditioning a range of work-related outcomes, such as job satisfaction and commitment, work centrality, and occupational choice and stability. Theoretically, the perceived job attributes have been conceptualized into two value dimensions, either *intrinsic* or *extrinsic*. Intrinsic values concern the rewards emanating directly from the work activity and experience itself (e.g., job autonomy, challenge, use of abilities, expression of interest and creativity, workplace cooperation, job useful to society). In contrast, extrinsic values involve the rewards derived from the job but external to the work itself (e.g., job security, pay, fringe benefit, prestige, promotional opportunities, pleasant working environment, good hours, no excessive amount of works). Among the five work values listed in the GSS, the first four attributes (i)–(iv) represent extrinsic factors of the job, while the last value (v) is considered intrinsic.

## A.16 Song Data

Consider the "song" data set from Critchlow et al. (1991). Ninety-eight students were asked to rank five words, (1) score, (2) instrument, (3) solo, (4) benediction, and (5) suit, according to the association with the word "song." Critchlow et al. (1991) reported that the average ranks for words (1)–(5) are 2.72, 2.27, 1.60, 3.71, and 4.69, respectively. However, the available data given in Critchlow et al. (1991) is in grouped format and the ranking of 15 students is unknown and hence discarded, resulting in 83 rankings, as shown below.

**Table A.12** "Song" data set

| Ordering | Observed frequency |
|---|---|
| $3 \succ 2 \succ 1 \succ 4 \succ 5$ | 19 |
| $3 \succ 1 \succ 2 \succ 4 \succ 5$ | 10 |
| $1 \succ 3 \succ 2 \succ 4 \succ 5$ | 9 |
| $3 \succ 2 \succ 4 \succ 1 \succ 5$ | 8 |
| $1 \succ 2 \succ 3 \succ 4 \succ 5$ | 7 |
| $2 \succ 3 \succ 1 \succ 4 \succ 5$ | 6 |
| $3 \succ 2 \succ 1 \succ 5 \succ 4$ | 6 |
| $3 \succ 2 \succ 4 \succ 5 \succ 1$ | 5 |
| $2 \succ 1 \succ 3 \succ 4 \succ 5$ | 4 |
| $3 \succ 1 \succ 4 \succ 2 \succ 5$ | 3 |
| $2 \succ 3 \succ 4 \succ 1 \succ 5$ | 2 |
| $3 \succ 4 \succ 2 \succ 1 \succ 5$ | 2 |
| $3 \succ 2 \succ 5 \succ 4 \succ 1$ | 2 |
| Others | 0 |
| Total | 83 |

## A.17 Croon's Political Goals Data

This ranking data set is obtained from Croon (1989) which consists of 2,262 rankings of four political goals for the government collected from a survey conducted in Germany. The four goals were:

(A) Maintain order in nation.
(B) Give people more say in Government decisions.
(C) Fight rising prices.
(D) Protect freedom of speech.

## A.18 1999 European Value Priority Data

Similar to the 1993 European Value Priority data, this data set is collected in the survey which was conducted in 1999 in 32 countries in Europe (Vermunt 2004). The respondents' covariates used in tree building are summarized in Table A.14. Countries were categorized into four groups as suggested by Vermunt (2004). Table A.15 shows the four groups of countries. After removing records containing missing in any one of the covariates, we end up with a ranking data of 1,911 respondents.

**Table A.13** Croon's political goals data

| Ordering | Frequency | Ordering | Frequency |
|---|---|---|---|
| A≻B≻C≻D | 137 | C≻A≻B≻D | 330 |
| A≻B≻D≻C | 29 | C≻A≻D≻B | 294 |
| A≻C≻B≻D | 309 | C≻B≻A≻D | 117 |
| A≻C≻D≻B | 255 | C≻B≻D≻A | 69 |
| A≻D≻B≻C | 52 | C≻D≻A≻B | 70 |
| A≻D≻C≻B | 93 | C≻D≻B≻A | 34 |
| B≻A≻C≻D | 48 | D≻A≻B≻C | 21 |
| B≻A≻D≻C | 23 | D≻A≻C≻B | 30 |
| B≻C≻A≻D | 61 | D≻B≻A≻C | 29 |
| B≻C≻D≻A | 55 | D≻B≻C≻A | 52 |
| B≻D≻A≻C | 33 | D≻C≻A≻B | 35 |
| B≻D≻C≻A | 59 | D≻C≻B≻A | 27 |

**Table A.14** Subject covariates for 1999 EVP data

| Covariate | Description/code | Type |
|---|---|---|
| Country | Group 1=1, group 2=2, group 3=3, group 4=4 | Nominal |
| Gender | Male=1, female=2 | Binary |
| Year of birth | Value ranges from 1909 to 1981 | Interval |
| Marital status | Married=1, windowed=2, divorced=3, separated=4, never married=5 | Nominal |
| Employment status | Ordinal value ranges from 1 to 8 | Ordinal |
| Household income | Ordinal value ranges from 1 to 10 | Ordinal |
| Age of education completion | Interval value ranges from 7 to 50 | Interval |

**Table A.15** The four groups of countries

| Group 1 | Group 2 | Group 3 | Group 4 |
|---|---|---|---|
| Italy | Croatia | Luxembourg | Lithuania |
| Sweden | Belgium | Slovenia | Latvia |
| Denmark | Greece | Czechnia | Poland |
| Austria | France | Iceland | Estonia |
| Netherlands | Spain | Finland | Belarus |
| | Northern Ireland | West Germany | Slovakia |
| | Ireland | Portugal | Hungary |
| | | Romania | Ukraine |
| | | Malta | Russia |
| | | East Germany | |
| | | Bulgaria | |

# Appendix B: Limit Theorems

## B.1 Hoeffding's Combinatorial Central Limit Theorem

Let $(Y_{n1}, ..., Y_{nn})$ be a random vector which takes the n! permutations of $(1, ..., n)$ with equal probabilities $1/n!$. Set

$$S_n = \sum_{i=1}^{n} a_n(i)\, b_n(Y_{ni})$$

and

$$d_n(i,j) = c_n(i,j) - \frac{1}{n}\sum_{g=1}^{n} c_n(g,j) - \frac{1}{n}\sum_{h=1}^{n} c_n(i,h) + \frac{1}{n^2}\sum_{g=1}^{n}\sum_{h=1}^{n} c_n(g,h).$$

**Theorem B.1 (Hoeffding 1951).** *The distribution of $S_n$ is asymptotically normal with mean*

$$ES_n = \frac{1}{n}\sum\sum c_n(i,j)$$

*and variance*

$$VarS_n = \frac{1}{n-1}\sum\sum d_n^2(i,j)$$

*provided*

$$lim_{n\to\infty} n \frac{max\,(a_n(i) - \bar{a}_n)^2}{\sum (a_n(i) - \bar{a}_n)^2} \frac{max\,(b_n(i) - \bar{b}_n)^2}{\sum (b_n(i) - \bar{b}_n)^2} = 0.$$



M. Alvo, P.L.H. Yu, *Statistical Methods for Ranking Data*, Frontiers in Probability and the Statistical Sciences, DOI 10.1007/978-1-4939-1471-5

## B.2 Multivariate Central Limit Theorem

A random m-vector Y is said to have a multivariate normal distribution with mean vector $\mu$ and variance-covariance matrix $\Sigma$ if its density is given by

$$f(y) = (2\pi)^{-m/2} |\Sigma|^{-1/2} exp - \frac{1}{2} (y - \mu)' \Sigma^{-1} (y - \mu)$$

written $Y \sim N_m(\mu, \Sigma)$.

The maximum likelihood estimators of $\mu, \Sigma$ based on a random sample of size n are given respectively by

$$\bar{Y}_n = \frac{1}{n} \sum Y_i$$

$$\hat{\Sigma} = \frac{1}{n} \sum \left(Y_i - \bar{Y}_n\right) \left(Y_i - \bar{Y}_n\right)'.$$

If $Y \sim N_m(\mu, \Sigma)$ and $Z = AY + b$ for some $q \times m$ matrix of constants $A$ of rank q$\leq m$ and b is a constant q-vector, then

$$Z \sim N_q\left(A\mu + b, A\Sigma A'\right).$$

**Theorem B.2.** *Let $Y_1, \ldots, Y_n$ be a random sample from some m-variate distribution with mean and variance-covariance matrix $\mu, \Sigma$, respectively. Then, as n$\rightarrow \infty$, the asymptotic distribution of $\sqrt{n}\left(\bar{Y}_n - \mu\right)$ is multivariate normal with mean 0 and variance-covariance matrix $\Sigma$.*

**Corollary B.1.** *Let $T$ be an r$\times$m matrix. Then $\sqrt{n}\left(T\bar{Y}_n - T\mu\right)$ is multivariate normal with mean 0 and variance-covariance matrix $T\Sigma T'$.*

## B.3 Quadratic Forms

If $Y \sim N_m(\mu, \Sigma)$ and $A$ is a symmetric matrix of rank r, then

$$Y'AY \sim \chi_r^2(\delta),$$

$\delta = \mu' A \mu$ if and only if $A\Sigma$ is idempotent or if $A\Sigma A = A$.

## B.4 Asymptotic Efficiency

### *B.4.1 One-Sided Tests*

We follow closely the development in Hájek and Sidak (1967, p. 267). Knowledge of the asymptotic power of a test forms the basis for a comparison of two tests. Let $F$ be a cumulative distribution function and let $f$ be its density. Define the Fisher information function

$$I(f) = \int_{-\infty}^{\infty} \left( \frac{f'(x)}{f(x)} \right)^2 f(x)\, dx$$

and set

$$\varphi(u, f) = \frac{f'\left(F^{-1}(u)\right)}{f\left(F^{-1}(u)\right)}, 0 < u < 1.$$

Suppose that the asymptotically most powerful test of $H_0$ against $H_1$is based on a test statistic $Z_0$ and that we wish to compare another test statistic $Z_1$ to it. Suppose moreover that asymptotically we have the following characteristics:

Under $H_0$

$$Z_0 \approx N\left(0, \sigma_0^2\right)$$
$$Z_1 \approx N\left(0, \sigma_1^2\right)$$

whereas under $H_1$

$$Z_0 \approx N\left(\mu_0, \sigma_0^2\right)$$
$$Z_1 \approx N\left(\mu_1, \sigma_1^2\right)$$

with $\mu_1 \geq 0$. Then the asymptotic efficiency of the test based on $Z_1$ is defined to be

$$e = \left( \frac{\mu_1 \sigma_0}{\mu_0 \sigma_1} \right)^2.$$

Suppose now that the likelihood function of the data under the alternative is given by the product

$$\prod f_0\left(x_{ii} - d_i\right) \tag{B.1}$$

and that we are interested in the linear rank statistic

$$Z_1 = \sum \left(c_i - \overline{c}\right) a_N\left(R_i\right).$$

Assume:

1. $max\left(d_i - \overline{d}\right)^2 \to 0$
2. $I\left(f_0\right) max\left(d_i - \overline{d}\right)^2 \to b^2 < \infty$
3. $\frac{\sum(c_i-\overline{c})^2}{max(c_i-\overline{c})^2} \to \infty$
4. $\frac{\int \varphi(u)\varphi(u,f_0)du}{\sqrt{\int \varphi^2(u,f_0)du \int(\varphi(u)-\overline{\varphi})^2 du}} = Q_1$
5. $\frac{\sum(c_i-\overline{c})\left(d_i-\overline{d}\right)}{\sqrt{\sum(c_i-\overline{c})^2 \sum\left(d_i-\overline{d}\right)^2}} \to Q_2$

Then it can be shown under these assumptions and (B.1) that

$$e = (Q_1 Q_2)^2 .$$

# Appendix C: Review on Decision Trees

## C.1 Introduction

A decision tree model is a rule for predicting the class of an object based on its covariates. It is widely used in data mining because it is easy to interpret and it can handle both categorical and interval measurement. The decision tree is constructed by recursively partitioning the data into different nodes.

Among the numerous tree building strategies, the classification and regression trees (CART) procedure (Breiman et al. 1984) is the most popular one. Trees for categorical target are named classification trees, and trees for continuous target are named regression trees.

## C.2 CART Algorithm

Suppose we have a learning sample of size $N$ with measurements $(Y_i, \boldsymbol{X}_i)$, $i = 1, ..., N$, where $Y$ is our target variable and $\boldsymbol{X}$ is the vector of $Q$ predictors $X^q, q = 1, ..., Q$. $\boldsymbol{X}$ and $Y$ can be interval, ordinal, or categorical variables. The goal is to predict $Y$ based on $\boldsymbol{X}$ via tree-structured classification. CART consists of two stages: growing and pruning. We will discuss them in the following.

### *C.2.1 Growing Stage of Decision Tree*

CART is a decision tree that is constructed by recursively partitioning the $N$ learning sample into different subsets, beginning with the root node that contains the whole learning sample. Each subset is represented by a node in the tree. In a binary tree

 
M. Alvo, P.L.H. Yu, *Statistical Methods for Ranking Data*, Frontiers in Probability and the Statistical Sciences, DOI 10.1007/978-1-4939-1471-5

structure, all internal nodes have two *child/descendant* nodes whereas the nodes with no descendants are called *terminal/leaf* nodes.

At each partition process, a splitting rule $s(\tau)$, comprised of a splitting variable $X^q$ and a split point, is used to split a group of $N(\tau)$ cases in node $\tau$ to the left node $N_L(\tau)$ and the right node $N_R(\tau)$. The decision tree identifies the best split by exhaustive search. The number of possible splits of a categorical predictor $X^q$ of $I$ categories is $2^{I-1} - 1$. For an interval $X^q$ with $F$ distinct values or an ordinal predictor with $F$ ordered categories, $F - 1$ possible splits will be produced on $X^q$.

The key step of tree growing is to choose a split among all possible splits at each node so that the resulting child nodes are the "purest." To measure the purity of a node $\tau$, Breiman et al. (1984) proposed a measure called impurity function $i(\tau)$. Let $p(j \mid \tau), j \in 1, ..., t$ be the conditional probability of having class $j$ in the learning sample in node $\tau$, $\sum_{j=1}^{t} p(j \mid t) = 1$. Impurity function should satisfy the following three properties:

1. It is minimum when the node is pure ($p(j \mid \tau) = 1$ for one $j \in \{1, ..., t\}$).
2. It is maximum when the node is the most impure ($p(1 \mid \tau) = ... = p(t \mid \tau) = \frac{1}{t}$).
3. Renaming of objects does not change the node impurity.

It can be shown that if the impurity function is concave, properties 1 and 2 will be satisfied. Property 3 is required because labeling of classes is arbitrary. CART includes various impurity criteria for classification trees, namely the Gini criterion, twoing criterion and entropy:

$$\text{Gini:} \quad i(\tau) = 1 - \sum_{j=1}^{t} p(j \mid \tau)^2 \tag{C.1}$$

$$\text{Twoing:} \quad i(\tau) = \frac{p_L \cdot p_R}{4} \left[ \sum_{j=1}^{t} |p(j \mid \tau_L) - p(j \mid \tau_R)| \right]^2 \tag{C.2}$$

$$\text{Entropy:} \quad i(\tau) = -\sum_{j=1}^{t} p(j \mid \tau) \log_2 p(j \mid \tau) \tag{C.3}$$

where $p_L = N_L(\tau)/N(\tau)$ and $p_R = N_R(\tau)/N(\tau)$ are the proportion of data cases in $\tau$ to the left child node $\tau_L$ and to the right child node $\tau_R$, respectively. Modification of existing measures of node homogeneity is essential for building decision tree model for ranking data and the two more popular impurity measures—Gini and entropy are adopted.

Based on the impurity measure for a node, a splitting criterion $\Delta i(s, \tau)$ can be defined as the reduction in impurity resulting from the split $s$ of node $\tau$:

$$\Delta i(s, \tau) = i(\tau) - p_L i(\tau_L) - p_R i(\tau_R). \tag{C.4}$$

The best split is chosen to maximize a splitting criterion. The concavity property of $i(\tau)$ assures that further splitting does not increase the impurity, so we can continue growing a tree until every node is pure, and some may contain only one observation. This would lead to a very large tree that would over-fit the data. To eliminate nodes that are overspecialized, pruning is required so that the best pruned subtree can be obtained.

### C.2.2 Pruning Stage of Decision Tree

In order to shape the tree into a reasonable size for easy understanding and interpretation, two common approaches of tree pruning are introduced, namely *pre-pruning* and *post-pruning*. Pre-pruning terminates the tree construction earlier and this can be done by imposing some *stop splitting criteria* such as:

1. The number of observations in one or more resulting child node(s) is less than the minimum number of observations for a parent node.
2. The maximum tree depth has been reached.
3. The reduction in impurity after splitting is less than a certain threshold (for impurity based splitting measure).
4. The purity of the node reached a certain bound (for impurity-based splitting measure).
5. The test value of the statistical test is less than a certain threshold (for splitting measure based on statistical test).

Thus, the splitting node will turn to a leaf node once any of the predefined stop splitting rules are met.

On the other hand, post-pruning emphasized removing subtrees after a full tree has been created. CART uses a sophisticated method called minimal cost-complexity pruning to do the task. Before proceeding to the algorithmic framework, some notations are first defined. Let $\tilde{\Upsilon}$ be the set of leaf nodes of tree $\Upsilon$ and the number of leaf nodes, denoted by $|\tilde{\Upsilon}|$, be defined as the complexity of $\Upsilon$. Define $C(\tau)$ to be the cost induced by node $\tau$. An obvious candidate of $C(\tau)$ is the misclassification rate; there are also other choices for the cost function. In a class probability tree, Breiman et al. (1984) considered pruning with the mean square error, which corresponds to take $C(\tau)$ as the Gini diversity index. For entropy tree, it is natural to take $C(\tau)$ as deviance. Chou (1991) developed a class of divergences in the form of expected loss function and it was shown that Gini, entropy, and misclassification rate can be written in the proposed form. In Chaps. 10 and 11, we specify the cost functions $C(\tau)$ such that they coincide with impurity functions for ranking data.

For any tree $\Upsilon$, the cost-complexity function $C_\theta(\Upsilon)$ is formulated as a linear combination of the cost of $\Upsilon$, $C(\Upsilon)$, and its complexity, $\theta|\tilde{\Upsilon}|$:

$$C_\theta(\Upsilon) = \sum_{\tau \in \tilde{\Upsilon}} (C(\tau) + \theta) = C(\Upsilon) + \theta|\tilde{\Upsilon}|.$$

The complexity parameter $\theta$ measures how much additional accuracy a split must add to the entire tree to warrant the addition of one more leaf node. Now consider $\Upsilon_{\tau'}$ as the subtree with root $\tau'$, $\Upsilon_{\tau'}$ will be retained as long as

$$C_\theta(\Upsilon_{\tau'}) < C_\theta(\tau'),$$

the branch $\Upsilon_{\tau'}$ contributes less complexity cost to tree $\Upsilon$ than node $\tau'$. This occurs for small $\theta$.

When $\theta$ increases to a certain value, the equality of the two cost-complexities is achieved. At this point, the subtree $\Upsilon_{\tau'}$ will be removed since it no longer helps improving the classification. The strength of the link from node $\tau$, $g(\tau)$, is therefore defined as

$$g(\tau) = \frac{C(\tau) - C(\Upsilon_\tau)}{|\tilde{\Upsilon}_\tau| - 1}.$$

The $V$-fold cross-validation cost-complexity pruning algorithm works as follows: The full learning data set $L$ is divided randomly into $V$ equal-size subsets $L_1, L_2, ..., L_V$ and the $v$th learning sample is denoted to be $L^v = L - L_v$. Using the full learning data set $L$, an overly large tree $\Upsilon^0$ is built. The function $g(\tau)$ are calculated for all internal nodes in $\Upsilon^0$ and the node with the minimum value $g(\tau^1)$ is located. A pruned tree $\Upsilon^1$ is created by turning the weakest-linked internal node $\tau^1$ into a leaf node. This process is repeated until $\Upsilon^0$ is pruned up to the root $\Upsilon^\tau$.

Denote by $\theta_i$ the value of $g(\tau)$ at the $i$th stage. A sequence of nested trees $\Upsilon^0 \supseteq \Upsilon^1 \supseteq \Upsilon^2 \supseteq ... \supseteq \Upsilon^\tau$ is generated, such that each pruned tree $\Upsilon^i$ is optimal for $\theta \in [\theta_i, \theta_{i+1})$. Here, the word "nested" means that each subsequent tree in the sequence is obtained from its predecessor by cutting one or more subtrees, and thus the accuracy of the sequence of progressively smaller pruned trees decreases monotonically.

Next, for $v = 1, ..., V$, the $v$th auxiliary maximal tree $\Upsilon_v^0$ is constructed based on $L^v$ and the nested sequence of pruned subtrees of $\Upsilon_v^0$

$$\Upsilon_v^0 \supseteq \Upsilon_v^1 \supseteq \Upsilon_v^2 \supseteq ... \supseteq \Upsilon_v^\tau$$

is generated. The cross-validation estimate of the cost $C^{CV}(\Upsilon^i)$ is then evaluated as

$$C^{CV}(\Upsilon^i) = \frac{1}{V} \sum_{v=1}^{V} C(\Upsilon_v(\sqrt{\theta_i\,\theta_{i+1}}))$$

where $\Upsilon_v(\theta)$ is equal to the pruned subtree $\Upsilon_v^i$ in the $i$th stage such that $\theta_i \leq \theta \leq \theta_{i+1}$. Note that the cost of the pruned subtree $\Upsilon_v(\sqrt{\theta_i\,\theta_{i+1}})$ is estimated by the independent subset $L_v$. Finally, the right-sized tree $\Upsilon^*$ is selected from $\{\Upsilon^0, \Upsilon^1, ..., \Upsilon^\tau\}$ such that

$$C^{CV}(\Upsilon^*) = \min_i C^{CV}(\Upsilon^i).$$

Nonetheless, the position of the minimum $C^{CV}(\Upsilon^*)$ is unstable. The uncertainties in the estimates $C^{CV}(\Upsilon^i)$ can be gauged by estimating their standard errors. As a result, the 1-standard error (1-SE) rule

$$C^{CV}(\Upsilon^{**}) \leq C^{CV}(\Upsilon^*) + SE(C^{CV}(\Upsilon^*))$$

is adopted to choose the simplest subtree $\Upsilon^{**}$ as the final tree model.

### *C.2.3 Class Assignment of Leaf Nodes of Decision Tree*

Each leaf node of the final selected tree $\Upsilon^*$ carries with it a class label $j^* \in \{1, ..., t\}$, which represents the predicted class for target $Y$, of the samples which fall within this node. The class label is usually determined by the plurality rule

$$p\,(j^* \,|\, \tau) = \max_j\,[p\,(j \,|\, \tau)]$$

so that the misclassification rate of the tree is minimized. The decision tree classifies a new observation by first passing it down the tree from the root node till it ends up in a leaf node and then the new observation will be assigned to the class according to the above plurality rule.

## C.3 Other Decision Tree Models

Apart from CART, there are many tree building strategies proposed in the literature, for example, C4.5 (Quinlan 1992), FACT (Loh and Vanichsetakul 1988), and QUEST (Loh and Shih 1997).

CART is capable of handling categorical and continuous data only. There are many extensions proposed in the literature which enable CART to handle other kinds of data, for example, longitudinal data, (Segal 1992), ordinal data (Piccarreta 2008; Kim and Baets 2003), and time-series data (Meek et al. 2002). CART was originally designed for univariate data and it was later extended to handle multivariate data as well, for example, Siciliano and Mola (2000) and Zhang (1998).

# Bibliography

Adkins, L., & Fligner, M. (1998). A non-iterative procedure for maximum likelihood estimation of the parameters of Mallows' model based on partial rankings. *Communications in Statistics: Theory and Methods, 27*(9), 2199–2220.

Allison, P. D., & Christakis, N. A. (1994). Logit models for sets of ranked items. *Sociological Methodology, 24*, 199–228.

Alvo, M. (1998). On non-parametric measures of correlation for directional data. *Environmetrics, 9*, 645–656.

Alvo, M. (2008). Nonparametric tests of hypotheses for umbrella alternatives. *Canadian Journal of Statistics, 36*, 143–156.

Alvo, M., & Berthelot, M.-P. (2012). Nonparametric tests of trend for proportions. *International Journal of Statistics and Probability, 1*, 92–104.

Alvo, M., & Cabilio, P. (1984). A comparison of approximations to the distribution of average Kendall tau. *Communications in Statistics: Theory and Methods, 13*, 3191–3216.

Alvo, M., & Cabilio, P. (1985). Average rank correlation statistics in the presence of ties. *Communications in Statistics: Theory and Methods, 14*, 2095–2108.

Alvo, M., & Cabilio, P. (1992). *Correlation methods for incomplete rankings*. (Technical Report 200) Laboratory for Research in Statistics and Probability: Carleton University and University of Ottawa.

Alvo, M., & Cabilio, P. (1993). *Tables of critical values of rank tests for trend when the data is incomplete*. (Technical Report 230) Laboratory for Research in Statistics and Probability: Carleton University and University of Ottawa.

Alvo, M., & Cabilio, P. (1994). Rank test of trend when data are incomplete. *Environmetrics, 5*, 21–27.

Alvo, M., & Cabilio, P. (1995a). Rank correlation methods for missing data. *Canadian Journal of Statistics, 23*, 345–358.

Alvo, M., & Cabilio, P. (1995b). Testing ordered alternatives in the presence of incomplete data. *Journal of the American Statistical Association, 90*, 1015–1024.

Alvo, M., & Cabilio, P. (1996). Analysis of incomplete blocks for rankings. *Statistics and Probability Letters, 29*, 177–184.

Alvo, M., & Cabilio, P. (1998). Applications of Hamming distance to the analysis of block data. In B. Szyszkowicz (Ed.), *Asymptotic methods in probability and statistics: A volume in honour of Miklós Csörgõ* (pp. 787–799). Amsterdam: Elsevier Science.

Alvo, M., & Cabilio, P. (2000). Calculation of hypergeometric probabilities using Chebyshev polynomials. *The American Statistician, 54*, 141–144.

 
M. Alvo, P.L.H. Yu, *Statistical Methods for Ranking Data*, Frontiers in Probability and the Statistical Sciences, DOI 10.1007/978-1-4939-1471-5

Alvo, M., & Cabilio, P. (2005). General scores statistics on ranks in the analysis of unbalanced designs. *The Canadian Journal of Statistics, 33*, 115–129.

Alvo, M., Cabilio, P., & Feigin, P. (1982). Asymptotic theory for measures of concordance with special reference to Kendall's tau. *The Annals of Statistics, 10*, 1269–1276.

Alvo, M., & Ertas, K. (1992). Graphical methods for ranking data. *Canadian Journal of Statistics, 20*(4), 469–482.

Alvo, M., Liu, Z., Williams, A., & Yauk, C. (2010). Testing for mean and correlation changes in microarray experiments: An application for pathway analysis. *BMC Bioinformatics, 11*(60), 1–10.

Alvo, M., & Pan, J. (1997). A general theory of hypothesis testing based on rankings. *Journal of Statistical Planning and Inference, 61*, 219–248.

Alvo, M., & Park, J. (2002). Multivariate non-parametric tests of trend when the data are incomplete. *Statistics and Probability Letters, 57*, 281–290.

Alvo, M., & Smrz, P. (2005). An arc model for ranking data. *Journal of Statistical Research, 39*, 43–54.

Anderson, R. (1959). Use of contingency tables in the analysis of consumer preference studies. *Biometrics, 15*, 582–590.

Arbuckle, J., & Nugent, J. H. (1973). A general procedure for parameter estimation for the law of comparative judgement. *British Journal of Mathematical and Statistical Psychology, 26*, 240–260.

Armitage, P. (1955). Tests for linear trends in proportions and frequencies. *Biometrics, 11*, 375–386.

Baba, Y. (1986). Graphical analysis of rank data. *Behaviormetrika, 19*, 1–15.

Barnes, S. H., & Kaase, M. (1979). *Political action: Mass participation in five western countries.* London: Sage.

Beckett, L. A. (1993). Maximum likelihood estimation in Mallows' model using partially ranked data. In M. A. Fligner & J. S. Verducci (Eds.), *Probability models and statistical analyses for ranking data* (pp. 92–108). New York: Springer.

Beggs, S., Cardell, S., & Hausman, J. (1981). Assessing the potential demand for electric cars. *Journal of Econometrics, 16*, 1–19.

Benard, A., & van Elteren, P. H. (1953). A generalization of the method of m rankings. *Indagationes Mathematicae, 15*, 358–369.

Benter, W. (1994). Computer-based horse race handicapping and wagering systems: A report. In W. T. Ziemba, V. S. Lo, & D. B. Haush (Eds.), *Efficiency of racetrack betting markets* (pp. 183–198). San Diego: Academic.

Biernacki, C., & Jacques, J. (2013). A generative model for rank data based on insertion sort algorithm. *Computational Statistics and Data Analysis, 58*, 162–176.

Bockenholt, U. (1993). Applications of Thurstonian models to ranking data. In M. Fligner & J. Verducci (Eds.), *Probability models and statistical analyses for ranking data.* New York: Springer.

Bockenholt, U. (2001). Mixed-effects analysis of rank-ordered data. *Psychometrika, 66*(1), 45–62.

Borg, I., & Groenen, P. J. F. (2005). *Modern multidimensional scaling: Theory and applications* (2nd ed.). New York: Springer.

Box, G., & Cox, D. (1964). An analysis of transformations. *Journal of the American Statistical Association, 26*, 211–252.

Brady, H. E. (1989). Factor and Ideal Point Analysis for Interpersonally Incomparable Data, Psychometrika, *54*(2), 181–202.

Bradley, A. P. (1997). The use of the area under the ROC curve in the evaluation of machine learning algorithms. *Pattern Recognition, 30*, 1145–1159.

Bradley, R. A., & Terry, M. (1952). Rank analysis of incomplete block designs: I. The method of paired comparisons. *Biometrika, 39*(3/4), 324–345.

Breckling, J. (1989). *The analysis of directional time series.* Lecture Notes in Statistics (Vol. 61). Berlin: Springer.

Breiman, L., Friedman, J. H., Olshen, R. A., & Stone, C. J. (1984). *Classification and regression trees*. Belmont, CA: Wadsworth.

Brook, D., & Upton, G. (1974). Biases in local government elections due to position on the ballot paper. *Applied Statistics, 23*, 414–419.

Brunden, M., & Mohberg, N. (1976). The Bernard-van Elteren statistic and nonparametric computation. *Communications in Statistics: Simulation and Computation, 4*, 155–162.

Bu, J., Cabilio, P., & Zhang, Y. (2009). Tests of concordance between groups of incomplete rankings. *International Journal of Statistical Sciences, 9*, 97–112.

Bunch, D. (1991). Estimability in the multinomial probit model. *Transportation Research Part B: Methodological, 25*(1), 1–12.

Busse, L. M., Orbanz, P., & Buhmann, J. M. (2007). Cluster analysis of heterogeneous rank data. In *Proceedings of the 24th International Conference on Machine Learning*, ACM New York, NY, USA (pp. 113–120).

Cabilio, P., & Tilley, J. (1999). Power calculations for tests of trend with missing observations. *Environmetrics, 10*, 803–816.

Carroll, J. D. (1972). Individual differences and multidimensional scaling. In R. N. Shepard, R. A. Kimball, & S. B. Nerlove (Eds.), *Multidimensional scaling: Theory and applications in the behavioral sciences, Volume I: Theory*. New York: Seminar Press.

Cayley, A. (1849). A note on the theory of permutations. *Philosophical Magazine, 34*, 527–529.

Chapman, R., & Staelin, R. (1982). Exploiting rank ordered choice set data within the stochastic utility model. *Journal of Marketing Research, 19*, 288–301.

Chintagunta, P. K. (1992). Estimating a multinomial probit model of brand choice using the method of simulated moments. *Marketing Science, 11*(4), 386–407.

Chou, P. A. (1991). Optimal partitioning for classification and regression trees. *IEEE Transactions on Pattern Analysis and Machine Intelligence, 13*, 340–354.

Cochran, W. G. (1954). Some methods of strengthening the common chi-square tests. *Biometrics, 10*, 417–451.

Cohen, A., & Mallows, C. (1980). *Analysis of ranking data*. (Technical memorandum). Murray Hill, NJ: AT&T Bell Laboratories.

Conover, W. J. (1999). *Practical nonparametric statistics* (3rd ed.). New York: Wiley.

Coombs, C. (1950). Psychological scaling without a unit of measurement. *Psychological Review, 57*, 147–158.

Cox, T., & Cox, M. (2001). *Multidimensional scaling* (2nd ed.). Boca Raton: Chapman and Hall.

Craig, B. M., Busschbach, J. J. V., & Salomon, J. A. (2009). Modeling ranking, time trade-off, and visual analog scale values for eq-5d health states: A review and comparison of methods. *Medical Care, 47*(6), 634–641.

Critchlow, D. (1985). *Metric methods for analyzing partially ranked data*. New York: Springer.

Critchlow, D., & Verducci, J. (1992). Detecting a trend in paired rankings. *Applied Statistics, 41*, 17–29.

Critchlow, D. E., Fligner, M. A., Verducci, J. S. (1991). Probability models on rankings. *Journal of Mathematical Psychology, 35*, 294–318.

Croon, M. A. (1989). Latent class models for the analysis of rankings. In G. D. Soete, H. Feger, & K. C. Klauer (Eds.), *New developments in psychological choice modeling* (pp. 99–121). North-Holland: Elsevier Science.

Daniels, H. (1950). Rank correlation and population models. *Journal of the Royal Statistical Society Series B, 12*, 171–181.

Dansie, B. R. (1985). Parameter estimability in the multinomial probit model. *Transportation Research Part B: Methodological, 19*(6), 526–528.

David, H. A. (1988). *The method of paired comparisons*. New York: Oxford University Press.

Davis, J. A., & Smith, T. W. (2009). *General social surveys, 1972–2008 [machine-readable data file]*. National Data Program for the Social Sciences, No. 18.

de Leeuw, J., & Mair, P. (2009). Multidimensional scaling using majorization: SMACOF in R. *Journal of Statistical Software, 31*(3), 1–30.

Decarlo, L. T., & Luthar, S. S. (2000). Analysis and class validation of a measure of parental values perceived by early adolescents: An application of a latent class models for rankings. *Educational and Psychological Measurement, 60*(4), 578–591.

Devroye, L. (1986). *Non-uniform random variate generation*. New York: Springer.

Diaconis, P. (1988). *Group representations in probability and statistics*. Hayward: Institute of Mathematical Statistics.

Diaconis, P. (1989). A generalization of spectral analysis with application to ranked data. *Annals of Statistics, 17*, 949–979.

Diaconis, P., & Graham, R. (1977). Spearman's footrule as a measure of disarray. *Journal of the Royal Statistical Society Series B, 39*, 262–268.

Dittrich, R., Katzenbeisser, W., & Reisinger, H. (2000). The analysis of rank ordered preference data based on Bradley-Terry type models. *OR Spektrum, 22*, 117–134.

Doignon, J.-P., Pekec, A., & Regenwetter, M. (2004). The repeated insertion model for rankings: Missing link between two subset choice models. *Psychometrika, 69*(1), 33–54.

Downs, T. (1973). *Rotational angular correlations*. New York: Wiley.

Duch, R. M., & Taylor, M. A. (1993). Postmaterialism and the economic condition. *American Journal of Political Science, 37*, 747–778.

Duncan, O. D., & Brody, C. (1982). Analyzing n rankings of three items. In R. M. Hauser, D. Mechanic, A. O. Haller, & T. S. Hauser (Eds.), *Social structure and behavior* (pp. 269–310). New York: Academic. .

Erosheva, E. A., Fienberg, S. E., & Joutard, C. (2007). Describing disability through individual-level mixture models for multivariate binary data. *The Annals of Applied Statistics, 1*(2), 502–537.

Feigin, P. D. (1993). Modelling and analysing paired ranking data. In M. A. Fligner & J. S. Verducci (Eds.), *Probability models and statistical analyses for ranking data* (pp. 75–91). New York: Springer.

Feigin, P. D., & Alvo, M. (1986). Intergroup diversity and concordance for ranking data: An approach via metrics for permutations. *The Annals of Statistics, 14*, 691–707.

Feigin, P. D., & Cohen, A. (1978). On a model for concordance between judges. *Journal of the Royal Statistical Society Series B, 40*, 203–213.

Fligner, M. A., & Verducci, J. S. (1986). Distance based ranking models. *Journal of the Royal Statistical Society Series B, 48*(3), 359–369.

Fligner, M. A., & Verducci, J. S. (1988). Multi-stage ranking models. *Journal of the American Statistical Association, 83*, 892–901.

Fok, D., Paap, R., & van Dijk, B. (2012). A rank-ordered logit model with unobserved heterogeneity in ranking capabilities. *Journal of Applied Econometrics, 27*, 831–846.

Friedman, M. (1937). The use of ranks to avoid the assumption of normality implicit in the analysis of variance. *Journal of the American Statistical Association, 32*, 675–701.

Gao, X., & Alvo, M. (2005a). A nonparametric test for interaction in two-way layouts. *The Canadian Journal of Statistics, 33*, 1–15.

Gao, X., & Alvo, M. (2005b). A unified nonparametric approach for unbalanced factorial designs. *Journal of the American Statistical Association, 100*, 926–941.

Gao, X., & Alvo, M. (2008). Nonparametric multiple comparison procedures for unbalanced two-way layouts. *Journal of Statistical Planning and Inference, 138*, 3674–3686.

Gao, X., Alvo, M., Chen, J., & Li, G. (2008). Nonparametric multiple comparison procedures for unbalanced one-way factorial designs. *Journal of Statistical Planning and Inference, 138*, 2574–2591.

Genz, A. (1992). Numerical computation of multivariate normal probabilities. *Journal of Computational and Graphical Statistics, 1*, 141–149.

Geweke, J. (1991). Efficient simulation from the multivariate normal and student-t distributions subject to linear constraints. In *Computer Science and Statistics: Proceedings of the 23rd Symposium on the Interface* (pp. 571–578). Alexandria: American Statistical Association.

Gibbons, J. D., & Chakraborti, S. (2011). *Nonparametric statistical inference* (5th ed.). New York: Chapman Hall.

Goldberg, A. I. (1975). The relevance of cosmopolitan local orientations to professional values and behavior. *Sociology of Work and Occupation, 3*, 331–356.

Gormley, I. C., & Murphy, T. B. (2008). Exploring voting blocs within the Irish electorate: A mixture modeling approach. *Journal of the American Statistical Association, 103*, 1014–1027.

Hajek, J. (1968). Asymptotic normality of simple linear rank statistics under alternatives. *The Annals of Mathematical Statistics, 39*, 325–346.

Hájek, J., & Sidak, Z. (1967). *Theory of rank tests*. New York: Academic.

Hajivassiliou, V. (1993). Simulation estimation methods for limited dependent variable models. In G. S. Maddala, C. R. Rao, & H. D. Vinod (Eds.), *Handbook of statistics: Econometrics* (Vol. 11, pp. 519–543). Amsterdam: North-Holland.

Hall, P., & Miller, H. (2010). Modeling the variability of rankings. *The Annals of Statistics, 38*, 2652–2677.

Hall, P., & Schimek, M. G. (2012). Moderate-deviation-based inference for random degeneration in paired rank lists. *Journal of the American Statistical Association, 107*, 661–672.

Han, S. T., & Huh, M. H. (1995). Biplot of ranked data. *Journal of the Korean Statistical Society, 24*(2), 439–451.

Hand, D. J., & Till, R. J. (2001). A simple generalisation of the area under the ROC curve for multiple class classification problems. *Machine Learning, 45*, 171–186.

Hausman, J., & Ruud, P. A. (1987). Specifying and testing econometric models for rank-ordered data. *Journal of Econometrics, 34*, 83–104.

Henery, R. J. (1981). Permutation probabilities as models for horse races. *Journal of the Royal Statistical Society Series B, 43*, 86–91.

Henery, R. J. (1983). Permutation probabilities for gamma random variables. *Applied Probability, 20*, 822–834.

Higgins, J. J. (2004). *An introduction to modern nonparametric statistics*. Pacific Grove: Brooks Cole-Thomson.

Hirst, D., & Naes, T. (1994). A graphical technique for assessing differences among a set of rankings. *Journal of Chemometrics, 8*, 81–93.

Hoeffding, W. (1951). A combinatorial central limit theorem. *Annals of Mathematical Statistics, 22*, 558–566.

Holland, D., & Wessells, C. R. (1998). Predicting consumer preferences for fresh salmon: The influence of safety inspection and production method attributes. *Agricultural and Resource Economics Review, 27*, 1–14.

Hollander, M., & Sethuraman, J. (1978). Testing for agreement between two groups of judges. *Biometrika, 65*(2), 403–411.

Imai, K., & van Dyk, D. A. (2005). MNP: R package for fitting the multinomial probit model. *Journal of Statistical Software, 14*(3), 32.

Iman, R. L., & Davenport, J. M. (1980). Approximations of the critical region of the friedman statistic. *Communications in Statistics - Theory and Methods, 9*, 571–595.

Inglehart, R. (1977). *The silent revolution: Changing values and political styles among western publics*. Princeton: Princeton University Press.

Jensen, D., & Solomon, H. (1972). A gaussian approximation to the distribution of a definite quadratic form. *Journal of the American Statistical Association, 67*, 898–902.

Jin, W. R., Riley, R. M., Wolfinger, R. D., White, K. P., Passador-Gundel, G., & Gibson, G. (2001). The contribution of sex, genotype and age to transcriptional variance in *Drosophila melanogaster*. *Nature Genetics, 29*, 389–395.

Joe, H. (2001). Multivariate extreme value distributions and coverage of ranking probabilities. *Journal of Mathematical Psychology, 45*, 180–188.

John, J., & Williams, E. (1995). *Cyclic designs*. New York: Chapman Hall.

Johnson, M. K. (2002). Social origins, adolescent experiences, and work value trajectories during the transition to adulthood. *Social Forces, 80*, 1307–1340.

Johnson, R., & Wehrly, T. (1977). Measures and models for angular correlation and angular-linear correlation. *Journal of the Royal Statistical Society Series B, 39*, 222–229.

Johnson, T. R., & Kuhn, K. M. (2013). Bayesian Thurstonian models for ranking data using JAGS. *Behavior R, 45*(3), 857–872.

Jonckheere, A. (1954). A test of significance for the relation between m rankings and k ranked categories. *The British Journal of Statistical Psychology, 7*, 93–100.

Jowell, R., Brook, L., & Dowds, L. (1993). *International social attributes: The 10th BSA Report.* Aldershot: Dartmouth Publishing.

Jupp, P., & Mardia, K. (1989). A unified view of the theory of directional statistics, 1975–1988. *International Statistical Review, 57*, 261–294.

Kalleberg, A. L. (1977). Work values and job rewards: A theory of job satisfaction. *American Sociological Association, 42*, 124–143.

Kamishima, T., & Akaho, S. (2006). Efficient clustering for orders. In *Proceedings of the 2nd International Workshop on Mining Complex Data*, Hong Kong, China (pp. 274–278).

Kannemann, K. (1976). An incidence test for k related samples. *Biometrische Zeitschrift, 18*, 3–11.

Keane, M. P. (1994). A computationally practical simulation estimator for panel data. *Econometrica, 62*, 95–116.

Kendall, M., & Gibbons, J. (1990). *Rank correlation methods*. London: Edward Arnold.

Kidwell, P., Lebanon, G., & Cleveland, W. S. (2008). Visualizing incomplete and partially ranked data. *IEEE Transactions on Visualization and Computer Graphics, 14*(6), 1356–1363.

Kim, C. V., & Baets, B. D. (2003). Growing decision trees in an ordinal setting. *International Journal of Intelligent System, 18*, 733–750.

Klementiev, A., Roth, D., & Small, K. (2008). Unsupervised rank aggregation with distance-based models. In *Proceedings of the 25th International Conference on Machine Learning*, ACM New York, NY, USA, (pp. 472–479).

Koop, G., & Poirier, D. J. (1994). Rank-ordered logit models: An empirical analysis of ontario voter preferences. *Journal of Applied Econometrics, 9*(4), 69–388.

Krabbe, P. F. M., Salomon, J. A., & Murray, C. J. L. (2007). Quantification of health states with rank-based nonmetric multidimensional scaling. *Medical Decision Making, 27*, 395–405.

Lacy, W. B., Bokemeier, J. L., & Shepard, J. M. (1983). Job attribute preferences and work commitment of men and women in the United States. *Journal of the American Statistical Association, 36*, 315–329.

Lawley, D. N., & Maxwell, A. E. (1971). *Factor analysis as a statistical method* (2nd ed.). London: Butterworth.

Lebanon, G., & Lafferty, J. (2002). Cranking: Combining rankings using conditional probability models on permutations. In *Proceedings of the 19th International Conference on Machine Learning*, ACM New York, NY, USA (pp. 363–370).

Lee, H., & Yu, P. (2013). *Rank-ordered logit tree regression.* (Technical report). The University of Hong Kong.

Lee, P. H., & Yu, P. L. H. (2010). Distance-based tree models for ranking data. *Computational Statistics and Data Analysis, 54*, 1672–1682.

Lee, P. H., & Yu, P. L. H. (2012). Mixtures of weighted distance-based models for ranking data with applications in political studies. *Computational Statistics and Data Analysis, 56*, 2486–2500.

Lehmann, E. (1975). *Nonparametrics: Statistical methods based on ranks.* New York: McGraw-Hill.

Leung, H. L. (2003). *Wandering ideal point models for single or multi-attribute ranking data: A Bayesian approach.* (Master's thesis). The University of Hong Kong.

Loh, W. Y., & Shih, Y. S. (1997). Split selection methods for classification trees. *Statistica Sinica, 7*, 815–840.

Loh, W. Y., & Vanichsetakul, N. (1988). Tree-structured classification via generalized discriminant analysis. *Journal of the American Statistical Association, 83*, 715–728.

Loscocco, K. A., & Kalleberg, A. L. (1988). Age and the meaning of work in the United States and Japan. *Social Forces, 67*, 337–356.

Luce, R. D. (1959). *Individual choice behavior*. New York: Wiley.

Mallows, C. L. (1957). Non-null ranking models. I. *Biometrika, 44*, 114–130.

Marden, J. I. (1992). Use of nested orthogonal contrasts in analyzing rank data. *Journal of the American Statistical Association, 87*, 307–318.

Marden, J. I. (1995). *Analyzing and modeling rank data*. New York: Chapman Hall.

Mardia, K. (1975). Statistics of directional data. *Journal of the Royal Statistical Society Series B, 37*, 349–393.

Mardia, K. (1976). Linear-circular correlation coefficients and rhythnometry. *Biometrika, 63*, 403–405.

Marley, A. A. J. (1968). Some probabilistic models of simple choice and ranking. *Journal of Mathematical Psychology, 5*, 311–332.

Martin, J. K., & Tuch, S. A. (1993). Black-white differences in the value of job rewards revisited. *Social Science Quarterly, 74*, 884–901.

Maydeu-Olivares, A., & Bockenholt, U. (2005). Structural equation modeling of paired-comparison and ranking data. *Psychological Methods, 10*(3), 285–304.

McCabe, C., Brazier, J., Gilks, P., Tsuchiya, A., Roberts, J., O'Hagan, A., & Stevens, K. (2006). Use rank data to estimate health state utility models. *Journal of Health Economics, 25*, 418–431.

McCullagh, P. (1993a). Models on spheres and models for permutations. In M. A. Fligner & J. S. Verducci (Eds.), *Probability models and statistical analyses for ranking data* (pp. 278–283). New York: Springer.

McCullagh, P. (1993b). Permutations and regression models. In M. Fligner & J. Verducci (Eds.), *Probability models and statistical analyses for ranking data* (pp. 196–215). New York: Springer.

McCulloch, R. E. & Rossi, P.E. (1994). An exact likelihood analysis of the multinomial probit model. *Journal of Econometrics, 64*, 207–240.

McFadden, D. (1974). Conditional logit analysis of qualitative choice behavior. In P. Zarembka (Ed.), *Frontiers in econometrics* (pp. 105–142). New York: Academic.

McFadden, D. (1978). Modeling the choice of residential location. In F. Snickars & J. Weibull (Eds.), *Spatial interaction theory and planning models* (pp. 75–96). North Holland: Amsterdam.

McFadden, D., & Train, K. (2000). Mixed MNL models for discrete response. *Journal of Applied Econometrics, 15*, 447–470.

Meek, C., Chickering, D. M., & Heckerman, D. (2002). Autoregressive tree models for time-series analysis. In *Proceedings of the Second International SIAM Conference on Data Mining*, Arlington, VA, USA (pp. 229–244).

Meng, X. L., & Wong, W. H. (1996). Simulating ratios of normalizing constants via a simple identity: A theoretical exploration. *Statistica Sinica, 6*, 831–860.

Moors, G., & Vermunt, J. (2007). Heterogeneity in post-materialists value priorities. Evidence from a latent class discrete choice approach. *European Sociological Review, 23*, 631–648.

Mortimer, J. T., & Lorence, J. (1979). Work experience and occupational value socialization: A longitudinal study. *The American Journal of Sociology, 84*, 1361–1385.

Mosteller, F. (1951). Remarks on the method of paired comparisons. I. The least squares solution assuming equal standard deviations and equal correlations. *Psychometrika, 16*, 3–9.

Murphy, T. B., & Martin, D. (2003). Mixtures of distance-based models for ranking data. *Computational Statistics and Data Analysis, 41*, 645–655.

Nombekela, S. W., Murphy, M. R., Gonyou, H. W., & Marden, J. I. (1993). Dietary preferences in early lactation cows as affected by primary tastes and some common feed flavors. *Journal of Diary Science, 77*, 2393–2399.

Page, E. (1963). Ordered hypotheses for multiple treatments: A significance test for linear ranks. *Journal of the American Statistical Association, 58*, 216–230.

Pendergrass, R. N., & Bradley, R. A. (1960). Ranking in triple comparisons. In O. Olkin, S. G. Ghurye, W. Hoeffding, W. G. Madow, & H. B. Mann (Eds.), *Contributions to probability and statistics* (pp. 331–351). Stanford: Stanford University Press.

Piccarreta, R. (2008). Classification trees for ordinal variables. *Computational Statistics, 23*, 407–427.

Plumb, A. A. O., Grieve, F. M., & Khan, S. H. (2009). Survey of hospital clinicians' preferences regarding the format of radiology reports. *Clinical Radiology, 64*, 386–394.

Poon, W. Y., & Xu, L. (2009). On the modelling and estimation of attribute rankings with ties in the thurstonian framework. *British Journal of Mathematical and Statistical Psychology, 62*, 507–527.

Quade, D. (1972). *Average internal rank correlation*. (Technical report) Amsterdam: Mathematical Centre.

Quinlan, J. R. (1992). *C4.5 programs for machine learning*. San Mateo, CA: Morgan Kaufmann.

Ratcliffe, J., Brazaier, J., Tsuchiya, A., Symonds, T., & Brown, M. (2006). Estimation of a preference based single index from the sexual quality of life questionnaire (SQOL) using ordinal data. *Discussion Paper Series, Health Economics and Decision Science, The University of Sheffield, 06*, 6.

Ratcliffe, J., Brazaier, J., Tsuchiya, A., Symonds, T., & Brown, M. (2009). Using DCE and ranking data to estimate cardinal values for health states for deriving a preference-based single index from the sexual quality of life questionnaire. *Health Economics, 18*, 1261–1276.

Regenwetter, M., Ho, M. H. R., & Tsetlin, I. (2007). Sophisticated approval voting, ignorance priors, and plurality heuristics: A behavioral social choice analysis in a Thurstonian framework. *Psychological Review, 114*(4), 994–1014.

Resnick, P., Iacovou, N., Suchak, M., Bergstrom, P., & Riedl, J. (1994). Grouplens: An open architecture for collaborative filtering of netnews. In *Proceedings of the 1994 ACM Conference on Computer Supported Cooperative Work (CSCW)* (pp. 175–186). New York, NY: ACM.

Riketta, M., & Vonjahr, D. (1999). Multidimensional scaling of ranking data for different age groups. *Experimental Psychology, 46*(4), 305–311.

Salomon, J. A. (2003). Reconsidering the use of rankings in the valuation of health states: A model for estimating cardinal values from ordinal data. *Population Health Metrics, 1*, 1–12.

Savage, I. R. (1956). Contributions to the theory of rank order statistics: The two-sample case. *Annals of Mathematical Statistics, 27*, 590–615.

Savage, I. R. (1957). Contributions to the theory of rank order statistics: The "trend" case. *Annals of Mathematical Statistics, 28*, 968–977.

Schulman, R. S. (1979). A geometric model of rank correlation. *The American Statistician, 33*(2), 77–80.

Segal, M. R. (1992). Tree-structured methods for longitudinal data. *Journal of the American Statistical Association, 87*, 407–418.

Sen, P. (1968). Asymptotically efficient tests by the method of n rankings. *Journal of the Royal Statistical Society, Series B, 30*, 312–317.

Shach, S. (1979). A generalization to the friedman test with certain optimality properties. *The Annals of Statistics, 7*, 537–550.

Shi, J. Q., & Lee, S. Y. (1997a). A Bayesian estimation of factor scores in confirmatory factor model with polytomous, censored or truncated data. *Psychometika, 62*, 29–50.

Shi, J. Q., & Lee, S. Y. (1997b). Estimation of factor scores with polytomous data by the EM algorithm. *British Journal of Mathematical and Statistical Psychology, 50*, 215–226.

Shieh, G. S. (1998). A weighted Kendall's tau statistic. *Statistics and Probability Letters, 39*, 17–24.

Siciliano, R., & Mola, F. (2000). Multivariate data analysis and modeling through classification and regression trees. *Computational Statistics and Data Analysis, 32*, 285–301.

Skrondal, A., & Rabe-Hesketh, S. (2003). Multilevel logistic regression for polytomous data and rankings. *Psychometrika, 68*(2), 267–287.

Smith, B. B. (1950). Discussion of Professor Ross's paper. *Journal of the Royal Statistical Society Series B, 12*, 53–56.

Stern, H. (1990a). A continuum of paired comparisons models. *Biometrika, 77*, 265–273.

Stern, H. (1990b). Models for distributions on permutations. *Journal of the American Statistical Association, 85*, 558–564.

Stern, H. (1993). Probability models on rankings and the electoral process. In M. A. Fligner & J. S. Verducci (Eds.), *Probability models and statistical analyses for ranking data* (pp. 173–195). New York: Springer.

Sun, M., & Lebanon, G. (2012). Estimating probabilities in recommendation systems. *Applied Statistics, 61*(3), 471–492.

Tallis, G., & Dansie, B. (1983). An alternative approach to the analysis of permutations. *Applied Statistics, 32*, 110–114.

Tanner, M. A. (1997). *Tools for statistical inference: Methods for the exploration of posterior distributions and likelihood functions* (3rd ed.). New York: Springer.

Tarsitano, A. (2009). *Comparing the effectiveness of rank correlation statistics*. (Working Papers 200906). Università della Calabria, Dipartimento di Economia e Statistica.

Thompson, G. L. (1993a). Generalized permutation polytopes and exploratory graphical methods for ranked data. *The Annals of Statistics, 21*, 1401–1430.

Thompson, G. L. (1993b). Graphical techniques for ranked data. In M. A. Fligner & J. S. Verducci (Eds.), *Probability models and statistical analyses for ranking data* (pp. 294–298). New York: Springer.

Thurstone, L. L. (1927). A law of comparative judgement. *Psychological Reviews, 34*, 273–286.

Timm, N. H. (1975). *Multivariate analysis with applications in education and psychology*. Monterey: Wadsworth Publishing Company, Inc.

Train, K. (2003). *Discrete choice methods with simulation*. Cambridge: Cambridge University Press.

Tversky, A. (1972). Elimination by aspects: A theory of choice. *Psychological Review, 79*(4), 281–299.

Vermunt, J. K. (2004). Multilevel latent class models. *Sociological Methodology, 33*, 213–239.

Vigneau, E., Courcoux, P., & Semenou, M. (1999). Analysis of ranked preference data using latent class models. *Food Quality and Preference, 10*, 201–207.

Wan (2011). *On some topics in modeling and mining ranking data*. (Ph.D. thesis). The University of Hong Kong.

Weaver, C. N. (1975). Job preferences of white collar and blue collar workers. *The Academy of Management Journal, 18*, 167–175.

Wormleighton, R. (1959). Some tests of permutation symmetry. *Annals of Mathematical Statistics, 30*, 1005–1017.

Xu, L. (2000). A multistage ranking model. *Psychometrika, 65*(2), 217–231.

Yai, T., Iwakura, S., & Morichi, S. (1997). Multinomial probit with structured covariance for route choice behavior. *Transportation Research Part B: Methodological, 31*(3), 195–207.

Ye, J., & McCullagh, P. (1993). Matched pairs and ranked data. In M. A. Fligner & J. S. Verducci (Eds.), *Probability models and statistical analyses for ranking data* (pp. 299–306). New York: Springer.

Yellot, J. (1977). The relationship between Luce's choice axiom, Thurstone's theory of comparative judgment and the double exponential distribution. *Journal of Mathematical Psychology, 15*, 109–144.

Yu, P. L. H. (2000). Bayesian analysis of order-statistics models for ranking data. *Psychometrika, 65*(3), 281–299.

Yu, P. L. H., & Chan, L. K. Y. (2001). Bayesian analysis of wandering vector models for displaying ranking data. *Statistica Sinica, 11*, 445–461.

Yu, P. L. H., Lam, K. F., & Alvo, M. (2002). Nonparametric rank test for independence in opinion surveys. *Austrian Journal of Statistics, 31*, 279–290.

Yu, P. L. H., Lam, K. F., & Lo, S. M. (2005). Factor analysis for ranked data with application to a job selection attitude survey. *Journal of the Royal Statistical Society Series A, 168*(3), 583–597.

Yu, P. L. H., Lee, P. H., & Wan, W. M. (2013). Factor analysis for paired ranked data with application on parent-child value orientation preference data. *Computational Statistics, 28*, 1915–1945.

Yu, P. L. H., Wan, W. M., & Lee, P. H. (2010). Preference learning. In J. Furnkranz & E. Hullermeier (Eds.), *Decision tree modelling for ranking data* (pp. 83–106). New York: Springer.

Zhang, H. P. (1998). Classification trees for multiple binary responses. *Journal of the American Statistical Association, 93*, 180–193.

# Index





M. Alvo, P.L.H. Yu, *Statistical Methods for Ranking Data*, Frontiers in Probability and the Statistical Sciences, DOI 10.1007/978-1-4939-1471-5

GPSR Compliance
The European Union's (EU) General Product Safety Regulation (GPSR) is a set of rules that requires consumer products to be safe and our obligations to ensure this.

If you have any concerns about our products, you can contact us on

ProductSafety@springernature.com

In case Publisher is established outside the EU, the EU authorized representative is:

Springer Nature Customer Service Center GmbH
Europaplatz 3
69115 Heidelberg, Germany

www.ingramcontent.com/pod-product-compliance
Ingram Content Group UK Ltd.
Pitfield, Milton Keynes, MK11 3LW, UK
UKHW021833270726
14058UKWH00001B/131

* 9 7 8 1 4 9 3 9 4 7 8 1 2 *